J.-L. Verdegay (Ed.)

Fuzzy Sets Based Heuristics for Optimization

Springer

*Berlin
Heidelberg
New York
Hong Kong
London
Milano
Paris
Tokyo*

Studies in Fuzziness and Soft Computing, Volume 126

http://www.springer.de/cgi-bin/search_book.pl?series=2941

Editor-in-chief
Prof. Janusz Kacprzyk
Systems Research Institute
Polish Academy of Sciences
ul. Newelska 6
01-447 Warsaw
Poland
E-mail: kacprzyk@ibspan.waw.pl

Further volumes of this series
can be found on our homepage

Vol. 106. P. Matsakis and L.M. Sztandera (Eds.)
Applying Soft Computing in Defining Spatial Relations", 2002
ISBN 3-7908-1504-7

Vol. 107. V. Dimitrov and B. Hodge
Social Fuzziology, 2002
ISBN 3-7908-1506-3

Vol. 108. L.M. Sztandera and C. Pastore (Eds.)
Soft Computing in Textile Sciences, 2003
ISBN 3-7908-1512-8

Vol. 109. R.J. Duro, J. Santos and M. Graña (Eds.)
Biologically Inspired Robot Behavior Engineering, 2003
ISBN 3-7908-1513-6

Vol. 110. E. Fink l. 112. Y. Jin
Advanced Fuzzy Systems Design and Applications, 2003
ISBN 3-7908-1523-3

Vol. 111. P.S. Szczepaniak, J. Segovia, J. Kacprzyk and L.A. Zadeh (Eds.)
Intelligent Exploration of the Web, 2003
ISBN 3-7908-1529-2

Vol. 112. Y. Jin
Advanced Fuzzy Systems Design and Applications, 2003
ISBN 3-7908-1537-3

Vol. 113. A. Abraham, L.C. Jain and J. Kacprzyk (Eds.)
Recent Advances in Intelligent Paradigms and Applications", 2003
ISBN 3-7908-1538-1

Vol. 114. M. Fitting and E. Orowska (Eds.)
Beyond Two: Theory and Applications of Multiple Valued Logic, 2003
ISBN 3-7908-1541-1

Vol. 115. J.J. Buckley
Fuzzy Probabilities, 2003
ISBN 3-7908-1542-X

Vol. 116. C. Zhou, D. Maravall and D. Ruan (Eds.)
Autonomous Robotic Systems, 2003
ISBN 3-7908-1546-2

Vol 117. O. Castillo, P. Melin
Soft Computing and Fractal Theory for Intelligent Manufacturing, 2003
ISBN 3-7908-1547-0

Vol. 118. M. Wygralak
Cardinalities of Fuzzy Sets, 2003
ISBN 3-540-00337-1

Vol. 119. Karmeshu (Ed.)
Entropy Measures, Maximum Entropy Principle and Emerging Applications, 2003
ISBN 3-540-00242-1

Vol. 120. H.M. Cartwright, L.M. Sztandera (Eds.)
Soft Computing Approaches in Chemistry, 2003
ISBN 3-540-00245-6

Vol. 121. J. Lee (Ed.)
Software Engineering with Computational Intelligence, 2003
ISBN 3-540-00472-6

Vol. 122. M. Nachtegael, D. Van der Weken, D. Van de Ville and E.E. Kerre (Eds.)
Fuzzy Filters for Image Processing, 2003
ISBN 3-540-00465-3

Vol. 123. V. Torra (Ed.)
Information Fusion in Data Mining, 2003
ISBN 3-540-00676-1

Vol. 125. M. Inuiguchi, S. Hirano and S. Tsumoto (Eds.)
Rough Set Theory and Granular Computing, 2003
ISBN 3-540-00574-9

José-Luis Verdegay (Ed.)

Fuzzy Sets Based Heuristics for Optimization

 Springer

Prof. Dr. José-Luis Verdegay
University of Granada
Dept. Computer Science and A.I.
18071 Granada
Spain
e-mail: verdegay@ugr.es

ISSN 1434-9922
ISBN 3-540-00551-X Springer-Verlag Berlin Heidelberg New York

Library of Congress Cataloging-in-Publication-Data
Fuzzy sets based heuristics for optimization / José-Luis Verdegay (ed.).
p. cm. -- (Studies in fuzziness and soft computing ; 126)
Includes bibliographical references.
ISBN 3-540-00551-X (alk. paper)
1. Heuristic programming. 2. Fuzzy sets. 3. Computer algorithms.
4. Mathematical optimization. I. Verdegay, José-Luis. II. Series
T57.84.F89 2003
511.3'22--dc21

This work is subject to copyright. All rights are reserved, whether the whole or part of the material is concerned, specifically the rights of translation, reprinting, reuse of illustrations, recitations, broadcasting, reproduction on microfilm or in any other way, and storage in data banks. Duplication of this publication or parts thereof is permitted only under the provisions of the German copyright Law of September 9, 1965, in its current version, and permission for use must always be obtained from Springer-Verlag. Violations are liable for prosecution under the German Copyright Law.

Springer-Verlag Berlin Heidelberg New York
a member of BertelsmannSpringer Science+Business Media GmbH
http://www.springer.de

© Springer-Verlag Berlin Heidelberg 2003
Printed in Germany

The use of general descriptive names, registered names trademarks, etc. in this publication does not imply, even in the absence of a specific statement, that such names are exempt from the relevant protective laws and regulations and therefore free for general use.

Typesetting: data delivered by editor
Cover design: E. Kirchner, Springer-Verlag, Heidelberg
Printed on acid free paper   62/3020/M - 5 4 3 2 1 0

Heuristics and Soft Computing

Preface

It is assumed generally that in the first level, the principals constituent of Soft Computing are the Approximate Reasoning and the Functional Approximation/Randomized Search. Then in a second level Probabilistic Models, Fuzzy Sets and Systems, Evolutionary Algorithms and Neural Networks appear. On the one hand, it is evident that since the famous "Fuzzy Boom" of the 90s, Fuzzy Sets and Systems have settled permanently in all the areas of Research, Development and Innovation. Their applications can be found in all the fields of our daily life: health, the banking sector, home, ... and they are a subject of study in different educational levels. On the other hand, there is no doubt that thanks to the technological potential that we have nowadays, we are witness to discoveries that were unpredictable just only a decade ago. Computers in particular, do efficiently tasks that seemed to be very laborious, when not impossible, just a short time ago, allowing us to approach problems of great complexity, both in comprehension as well as in dimension, in a great variety of fields (Bioinformatics, Data Mining, Knowledge Engineering,...). In those fields the Evolutionary Algorithms are showing themselves as very valuable methods to find solutions for specific problems.

However Evolutionary Algorithms are just one more class of heuristics, or of Metaheuristics, as well as they are Taboo Search, Simulated Annealing, Variable Neighbourhood Search, Scatter Search, GRASP or Reactive Search, to cite just a few. All these techniques usually give solutions that are not the very best, but they have quality enough to satisfy the decision maker. When this idea that it is better to satisfy than to optimize takes place, it makes sense the famous phrase from Professor L.A. Zadeh: "...in contrast to traditional hard computing, soft computing exploits the tolerance for imprecision, uncertainty, and partial truth to achieve tractability, robustness, low solution-cost, and better rapport with reality". Consequently, between the constituents of Soft Computing, instead of the Evolutionary Algorithms, that might represent only a part of the methods of search and improvement that are used, it must appear Heuristic Algorithms or Metaheuristics.

Usually there is a lot of controversy and debate about the difference between the concept of Metaheuristic and the one of Heuristic. It is not our intention here to enter in this debate but we are interested in making a slight thought about both concepts. In this respect, the European Chapter on Metaheuristics EU/ME, http://tew.ruca.ua.ac.be/eume/welcome.htm, sends those who are interested to the pages of the National Institute of Standards and Technology (NIST), http://www.nist.gov, where it can be found the following definitions. One defines heuristic as "an algorithm that usually, but not always, works or that gives nearly the right answer". On its turn, metaheuristic is defined as: "a state space search that looks for good solutions by changing a current set of solutions". And so, it is obvious that both concepts are defined through vague characteristics and for that they can be perfectly modelled and developed from methodologies based on the fuzzy sets and systems. From this point of view it makes perfect sense to speak of heuristics or metaheuristics based on fuzzy sets.

However, in spite of the huge success achieved by the Fuzzy Sets and Systems, of the important progress produced by the heuristics and the metaheuristics in a practical way, and of the close relationship between both methodologies as we have just made evident, if we do not consider the superarea of the Genetic Algorithms, or more generally of the Evolutionary Algorithms, not much work has been conducted in the past in combining "the best of these two big worlds", thus, in the development of Fuzzy Sets-based Heuristics. But moreover, what was done so far, due to the great variety of methods and fields of application, it is very scattered in the specialized literature.

Just for this reason the Publishing house Springer Verlag considered appropriate making a book like this one, that apart from giving credibility to the considered topic, it could serve as a catalyst of the theoretic and applied research. However, the field to explore was so large that we were forced to mark its boundaries, concentrating our efforts in those aspects mainly related with algorithms for optimization, what consequently lead us to the subject of this volume: Fuzzy Sets based Heuristics for Optimization. Therefore, the aim of this volume is triple: first, one wants to show how fuzzy sets and systems can help to provide robust and adaptive heuristic optimization algorithms in a variety of situations. To present the state of the art in the concerned area is the second objective, and the last but of course not least is to give a view, as broad as possible, on the real practical applications that fuzzy sets based heuristic algorithms have. To this end, this book is organized in four sections, collecting twenty two contributions kindly prepared by recognized researchers from all over the world.

The book starts with a section devoted to Soft Computing and Metaheuristics, which collects six contributions. In the first one, A. Blanco, D. Pelta and J.L. Verdegay present FANS, a fuzzy adaptive neighborhood search algorithm. Then, J. Cheng, H. Kise, G. Steiner and P. Stephenson discuss branch-and-bound algorithms using fuzzy heuristics, and M. Demirhan and L. Özdamar show FAPA, a fuzzy adaptive partitioning algorithm for global optimization. Fourth contribution, by N. Krasnogor and D. Pelta proposes the integration of the so called fuzzy adaptive neighbourhood search algorithms with multimeme algorithms. P. Lucic and D. Teodorovic present a bee system for solving vehicle routing problems with uncertain demand at nodes, and closing this first part, J.A. Moreno and J. M. Moreno introduce the so-called Fuzzy Constructive Heuristics.

Second section focuses on Heuristics and Mathematical Programming. First, J.M. Cadenas, M.C. Garrido and F. Jimenez present two approaches for solving optimization problems with heuristics. Then, C. Carlsson and R. Fuller consider optimization problems addressed with linguistic variables. The next contribution, by C. Mohan and S.K. Verma proposes interactive algorithms using fuzzy concepts for solving real life optimization problems. S. Lertworasirikul, S.-Ch. Fang, J.A. Joines and H.L.W. Nuttle develop a credibility approach for fuzzy data envelopment analysis, and finally R. Ribeiro and L. R. Varela show examples of fuzzy optimization methods using Simulated Annealing.

Practical Heuristic Algorithms is the title of the third part of the book. M. Gen and A. Syarif open this section approaching multi-stage supply chain network by hybrid genetic algorithms. Then a fuzzy evolutionary approach for multiobjective combinatorial optimization, applied to scheduling problems, is proposed by I. Kacem, S. Hammadi and P. Borne. The following contribution, by F. Jimenez, A. F. Gomez-Skarmeta, G. Sanchez and J.M Cadenas, focuses on multi-objective evolutionary fuzzy modeling, and next the well known fuzzy satisficing methods in mathematical programming are shown by M. Sakawa, K. Kato, S. Ushiro and M. Inaoka in the particular case of nonlinear 0-1 problems. Last contribution in this part, by Y.S. Yun and Mitsuo Gen is devoted to show an adaptive hybrid genetic algorithm with fuzzy logic controller.

A section collecting applications to real world problems ends the volume. The following six contributions are presented: Finding Satisfactory Near-Optimal Solutions in Location Problems, by M.C. Canós, C. Ivorra and V. Liern. Route Choice Making under Uncertainty: A Fuzzy Logic based Approach, by Vincent Henn. A new technique to electrical distribution system load flow based on fuzzy sets, by C. A. F. Murari, M. A. Pereira and M.M.P. Lima. FAPA for Global Optimization: Implementation in Environmental Site Characterization, by L. Özdamar and M. Demirhan. Capacitated Vehicle Routing Problem with Fuzzy Demands, by B. Werners and M. Drawe, and finally an Adaptive Intelligent Control System for Slag Foaming by E. L. Wilson and C. L. Karr.

As a whole these twenty two contributions give an overview on the state of the art on the topic dealt with, and show that heuristic algorithms and metaheuristics benefit from hybridizing them with Fuzzy Sets and Systems methodologies. This gives a clear indication that a closer cooperation between both sides would be highly beneficial.

In the preparation of this volume have participated several institutions and different people. Without their collaboration and help this project would not have been carried out. Among the first ones my gratefulness goes to the Spanish Ministry of Science and Technology, which finances the project "Application of Fuzzy Methodologies to the Development of Metaheuristics-based Decision Support Systems" (HeuriFuzzy), TIC2002-04242-CO3-02, in which this volume is included, and to the Department of Computer Science and Artificial Intelligence of the University of Granada for providing me all the material resources that I needed. Among the second ones, I must mention the Editor in chief of this series, Prof. Janusz Kacprzyk, who from the beginning welcomed the topic of this book with great enthusiasm, and to Katherine Wentzel and Thomas Ditzinger as well, from Springers office, who helped me continuous and efficaciously from the beginning. I also want to express my sincere gratitude to all the members of the Working Group on Models of Decision and Optimization (MODO) for their constant support, encouragement and generous help.

In the end, to emphasize it, I want to put on record clear and explicit that this book has been published thanks to the effort and patience of my collaborator but nevertheless my friend, Dr. David Pelta. David was in charge of the final preparation of the manuscript according to the Publishing House instructions and he was the person who spoke with the authors to coordinate the size of their contributions, getting finally a volume of great level. For all that, I feel deeply obliged to him.

José Luis Verdegay
Granada, December 2002

Contents

Part 1. Soft Computing and Metaheuristics

Fuzzy Adaptive Neighborhood Search: Examples of Application 1
Armando Blanco, David Pelta, José L. Verdegay
1 Introduction 1
2 Description of *FANS* 2
3 On the Usefulness of Fuzzy Valuations 6
4 Examples of Application 9
5 Conclusions 19
References 19

**Branch-and-bound algorithms using fuzzy heuristics
for solving large-scale flow shop scheduling problems** 21
Jinliang Cheng, Hiroshi Kise, George Steiner, Paul Stephenson
1 Introduction 21
2 Lower bounds 23
3 Dominance rules and fuzzy approximation 27
4 Branch-and-bound algorithm 29
5 Computational experiments 30
6 Concluding remarks 33
References 34

**A Fuzzy Adaptive Partition Algorithm (FAPA)
for Global Optimization** 37
Melek Basak Demirhan, Linet Özdamar
1 Introduction 37
2 Fuzzy Adaptive Partitioning Approach 38
3 Computational Results 41
4 Conclusion 44
References 45

**Fuzzy Memes in Multimeme Algorithms:
a Fuzzy-Evolutionary Hybrid** 49
Natalio Krasnogor, David A. Pelta
1 Introduction 49
2 The Protein Structure Prediction Problem 50
3 Memetic Algorithms 52
4 Fuzzy Memes for Multimeme Algorithms 53
5 Experiments setup and results 57
6 Conclusions 63
References 64

Vehicle Routing Problem With Uncertain Demand at Nodes: The Bee System and Fuzzy Logic Approach 67
Panta Lučić, Dušan Teodorović
1   Introduction 67
2   Statement of the problem 68
3   A Proposed solution of the problem 68
4   Results obtained using the "intelligent" vehicle routing system 77
5   Conclusion 79
References 80

Fuzzy Constructive Heuristics 83
José A. Moreno Pérez, J. Marcos Moreno Vega
1   Introduction 83
2   Fuzzy Constructive Methods 84
3   Fuzzy Stopping Rules 85
4   The non-guillotine rectangular two-dimensional cutting problem ... 86
5   Computational experiments 91
References 95

Part 2. Heuristics and Mathematical Programming

Heuristics for Optimization: two approaches for problem resolution 97
J.M. Cadenas, M.C. Garrido, F. Jimenez
1   Introduction 97
2   Fuzzy rule learning in FMP 99
3   Fuzzy rule learning in MFGN framework 106
4   Analysis of results 110
5   Conclusions 110
References 111

Optimization with linguistic variables 113
Christer Carlsson, Robert Fullér
1   Introduction 113
2   Optimization with linguistic variables 116
3   Examples 117
4   Extensions 118
5   Summary 120
References 120

Interactive Algorithms Using Fuzzy Concepts for Solving Mathematical Models of Real Life Optimization Problems 123
C. Mohan, S.K. Verma
1   Introduction 123
2   Non-Symmetric Treatment of Linear and a Class of Non-Linear Multiobjective Fuzzy Programming Problems 126

3 PL-Pareto Optimal Solution	129
4 Calculation of Fuzzy Aspiration Levels	131
5 The Proposed IMMOFP Interactive Algorithm	132
6 Illustrative Examples	133
7 Concluding Observations	136
References	137

Fuzzy Data Envelopment Analysis: A Credibility Approach ... 141
Saowanee Lertworasirikul, Shu-Cherng Fang, Jeffrey A. Joines, Henry L. W. Nuttle

1 Introduction	141
2 Fuzzy DEA Model	143
3 Possibility, Necessity, and Credibility Measures	145
4 CP-DEA model	154
5 Numerical Examples	155
6 Concluding Remarks	156
References	157

Fuzzy Optimization using Simulated Annealing:
An Example Set 159
Rita Almeida Ribeiro, Leonilde Rocha Varela

1 Introduction	159
2 Basics on the approach used for formulating and solving the example set	160
3 Set of linear examples tested and discussion of the results	164
4 Set of non-linear examples tested and discussion of the results	174
5 Conclusions	179
References	179

Part 3. Practical Heuristic Algorithms

Multi-stage Supply Chain Network
by Hybrid Genetic Algorithms 181
Mitsuo Gen, Admi Syarif

1 Introduction	181
2 Mathematical Model	184
3 Design of the Algorithm	185
4 Overall Procedure	192
5 Numerical Examples	193
6 Conclusion	194
References	195

Fuzzy evolutionary approach for multiobjective combinatorial optimization: application to scheduling problems 197
Imed Kacem, Slim Hammadi, Pierre Borne

1 Introduction	197

2	Multiobjective Optimization: The State Of The Art	197
3	Fuzzy Evolutionary Approach	200
4	Application: Case Of Flexible Job-shop Scheduling Problem (FJSP)	207
5	Discussions And Conclusions	216
References		217

Fuzzy Sets based Heuristics for Optimization: Multi-objective Evolutionary Fuzzy Modeling 221
Fernando Jiménez, Antonio F. Gómez Skarmeta, Gracia Sánchez, José M. Cadenas

1	Introducction	221
2	Fuzzy model identification	223
3	A technique to improve transparency and compactness of the fuzzy rule sets	224
4	Training of the RBF neural networks	225
5	Criteria for fuzzy modeling	225
6	Multi-objective neuro-evolutionary algorithm	226
7	Experiments and results	230
8	Conclusions and future research	232
References		233

An Interactive Fuzzy Satisficing Method for Multiobjective Operation Planning in District Heating and Cooling Plants through Genetic Algorithms for Nonlinear 0-1 Programming . 235
Masatoshi Sakawa, Kosuke Kato, Satoshi Ushiro, Mare Inaoka

1	Introduction	235
2	Operational Planning of a DHC Plant	236
3	An Interactive Fuzzy Satisficing Method	241
4	Genetic Algorithms for Nonlinear 0-1 Programming	242
5	Numerical Experiments	247
6	Conclusion	248
References		249

Adaptive Hybrid Genetic Algorithm with Fuzzy Logic Controller 251
YoungSu Yun, Mitsuo Gen

1	Introduction	251
2	Adaptive Genetic Operators (AGOs)	253
3	Proposed Hybrid Concepts and Logics	256
4	Proposed Algorithms for Experimental Comparison	258
5	Numerical Example	260
6	Conclusion	262
References		262

Part 4. Applications to Real World Problems

**Finding Satisfactory Near-Optimal Solutions
in Location Problems** 265
María J. Canós, Carlos Ivorra, Vicente Liern
1 Introduction.. 265
2 The fuzzy p-median problem 266
3 Calculating the satisfaction level 268
4 An interchange heuristic procedure 271
5 Computational results 272
6 Conclusions 275
References 275

**Route Choice Making Under Uncertainty:
a Fuzzy Logic Based Approach** 277
Vincent Henn
1 Introduction.. 277
2 Modeling Traffic Assignment 278
3 Fuzzy Sets as a Basis for Representing Imperfections 280
4 Fuzzy Sets Based Heuristics for Traffic Assignment 282
5 Analysis of the Model 287
6 Conclusions and Perspectives 290
References 291

**A New Technique to Electrical Distribution
System Load Flow Based on Fuzzy Sets** 293
Carlos A. F. Murari, Marcelo A. Pereira, Marcelo M. P. Lima
1 Introduction.. 293
2 Fuzzy Number 294
3 Fuzzy Numbers' Operators 295
4 Fuzzy Load Flow 298
5 Tests 299
6 Conclusions 302
References 305

**A Fuzzy Adaptive Partitioning Algorithm (FAPA)
for Global Optimization: Implementation in
Environmental Site Characterization** 307
Linet Özdamar, Melek Basak Demirhan
1 Introduction.. 307
2 Environmental Site Characterization: Problem Definition.......... 308
3 Implementation of Fuzzy Adaptive Partitioning Algorithm (FAPA)
   in Site Characterization...................................... 309
4 Numerical Results on Hypothetical Sites 313
5 Conclusion 314
References 315

Capacitated Vehicle Routing Problem with Fuzzy Demand ... 317
Brigitte Werners, Michael Drawe
1 Introduction 317
2 A Fuzzy Multi-Criteria Modeling Approach 318
3 A Fuzzy Multi-Criteria Savings Heuristic 326
4 Conclusion 333
References 334

An Adaptive, Intelligent Control System for Slag Foaming 337
Eric L. Wilson, Charles L. Karr
1 Introduction 337
2 The Problem Environment: Slag Foaming in an Electric Arc Steel Furnace 340
3 An Architecture for Achieving Intelligent Adaptive Control 342
4 Results 347
5 Summary 350
References 350

Fuzzy Adaptive Neighborhood Search: Examples of Application

Armando Blanco, David Pelta, and José L. Verdegay

Depto. de Ciencias de la Computación
e Inteligencia Artificial
E.T.S. Ingeniería Informática
Universidad de Granada, Spain

Abstract. Since a couple of years, we are involved in the development of a fuzzy sets-based heuristic called *FANS*, by Fuzzy Adaptive Neighborhood Search, whose main motivation was to show the benefits of merging basic ideas from the Fuzzy Set and Systems area with simple optimization methods.

A very remarkable feature of the method is its ability to capture the qualitative behavior of other neighborhood search methods, what is achieved through appropriated handling of the membership function defining the so called "fuzzy valuation".

Here, we propose to review the main concepts of *FANS* and to describe four applications of the algorithm to problems of quite different domains.

1 Introduction

Combinatorial optimization is one of the most interesting and active fields of discrete mathematics, and is even its core research today. Real world combinatorial optimization applications are being developed to improve the quality of life of human beings and Fuzzy Sets and Systems are employed with great success in the conception, design, construction and utilisation of a wide range of products and systems whose functioning is directly based on the ways human beings reason. Because of the upmost relevance of these two fields, there is an increasing interest at the interface between them, that is, Fuzzy Sets-based solution methodologies for combinatorial optimization problems are being widely exploited [24].

In this work we will show how Fuzzy Sets and Systems helped to design and obtain a novel (fuzzy set-based) heuristic method oriented to deal with difficult and well defined problems.

The method is called *FANS*, by Fuzzy Adaptive Neighborhood Search, and its basic ideas were previously presented in [5,24,20]. *FANS* is termed Fuzzy, because solutions are also evaluated by means of fuzzy valuations and Adaptive because its behavior is adapted as a function of the search state. Fuzzy valuations are represented in *FANS* by fuzzy sets, and are used as a mechanism to represent abstract concepts like "Acceptability" or "Similarity". Within *FANS*, the fuzzy valuation is also used to define a semantic neighborhood and to obtain, for example, the degree of "Acceptability" of a neighborhood solution. Such degrees are used to guide the search.

FANS enables decision maker to start searching for solutions from simple schemes, increasing its complexity as his/her knowledge of the problem increases. This tool, suitable to be embedded on a Decision Support System, is basically a local search optimization heuristic which, differently from other heuristic algorithms, takes advantage of using fuzzy valuations to analyze and control the selection of new solutions.

Being a neighborhood search method, *FANS* is simple to understand, easy to implement and could be tailored to specific applications at low cost.

A very remarkable feature of the method is its ability to capture the qualitative behavior of other neighborhood search methods, what is achieved through appropriated handling of the membership function defining the fuzzy valuation. So, we may refer to *FANS* as a local search framework [5].

Here, we propose to review the main concepts of *FANS* and to describe four applications of the algorithm to problems of quite different domains. In order to do this, the chapter is organized as follows: in Section 2, the characteristics of *FANS* are described. Then, in Section 3 we show how *FANS* is able to obtain the same qualitative behavior of other local search methods by means of adequated definitions of the fuzzy valuation component. Section 4 is divided in four parts, each one devoted to show how *FANS* could be applied to several optimization problems: knapsack problems with single and multiple restrictions, minimization of real functions and an application to the protein structure prediction problem. The description of each application is a small review based on published results. The corresponding references are provided for those interested in the details.

2 Description of *FANS*

Two elements make *FANS* different from other simple local search schemes: the first one is about how solutions are evaluated: within *FANS* a fuzzy valuation is used together with the objective function to obtain a "semantic evaluation" of the solution. In this way we consider the neighborhood as a fuzzy set, being the solutions the elements of the set and the fuzzy valuation, its membership function.

The second difference appears when the search is trapped in a local optimum. Now, the operator used to construct solutions is changed, so a different set of solutions will be explored. This mechanism is tried several times (as much as the number of operators available) until some criterion is met. Then a classical restart mechanism could be applied. Both novel elements will be described in detail below.

We will use the following conventions for *FANS* presentation: $s_i \in \mathcal{S}$ is a solution belonging to the search space; $\mathcal{O}_i \in \mathcal{M}$ is a Modification or Move operator from the space of operators; $\mathcal{P}$ stands for the space of parameters (tuples of values) and $\mathcal{F}$ stands for the space of fuzzy sets with elements μ_i. FANS is fully defined with the following 7-tuple:

$$FANS(\mathcal{NS}, \mathcal{O}, \mathcal{OS}, \mu(), Pars, (cond, action)) \quad (1)$$

where $\mathcal{NS}$ is the Neighborhood Scheduler, $\mathcal{O}$ is an operator used to construct solutions, $\mathcal{OS}$ is the operator scheduler and $\mu()$ is a fuzzy valuation. These four elements are viewed as separate components. Finally, $Pars$ is a set of parameters and the pair $(cond, action)$ is a rule IF cond THEN action used to detect and perform certain action when the search converged.
Below, we will briefly describe the main characteristics of each element.

2.1 The Modification Operator

Given a reference solution $s \in \mathcal{S}$, this operator $\mathcal{O}_i \in \mathcal{M}$, $\mathcal{O} : \mathcal{S} \times \mathcal{P} \to \mathcal{S}$, will generate new solutions s_i from s. The operator has some tunable parameters $t \in \mathcal{P}$ to control its operation; in this sense we may use $\mathcal{O}^t$ to reference the operator with a set of parameters t.

2.2 The Fuzzy Valuation

The fuzzy valuation is represented in FANS by a fuzzy set $\mu() \in \mathcal{F}$, with $\mu : \mathcal{R} \to [0, 1]$. It is also called fuzzy concept or fuzzy property, so we can talk about solutions verifying the property in some degree. A similar idea was used to obtain fuzzy termination criteria for exact algorithms [25].

Besides the objective function, FANS evaluates the solutions with the fuzzy valuation: i.e. we evaluate the membership degree of the solution to a fuzzy set. For example, having the fuzzy set of "good" solutions, we will consider the goodness of the solution of interest.

Thus, given two solutions $a, b \in \mathcal{S}$ we could think about how Similar a and b are, or how Close they are, or also how Different b is from a. Similar, Close, Different will be fuzzy sets represented by appropriated membership functions $\mu()$.

2.3 The Operator Scheduler

This component encapsulates the strategy defined for the adaptation of the operator $\mathcal{O}$ and will be executed when certain conditions are met.

As response the tunable parameters of the operator will be adapted and then, a modified version of it will be returned: $\mathcal{OS}\ (\mathcal{O}^{ti}) \Rightarrow \mathcal{O}^{tj}$.

Instead of modifying a set of parameters, another option is to have a family of operators $\{\mathcal{O}_1, \mathcal{O}_2, \ldots, \mathcal{O}_k\}$ available. Under this situation the operator scheduler will define their order of application. Each operator implies a different neighborhood, so we obtain a structure which resembles that of variable neighborhood search (VNS) [9]. However, we do not have an explicit local search method and we do not make use or need a distance measure between solutions as the one needed in VNS.

2.4 The Neighborhood Scheduler

This component is responsible for the generation and selection of a new solution from the neighborhood. It may be viewed as a function:

$$\mathcal{NS} : \mathcal{S} \times \mathcal{F} \times \mathcal{M} \times \mathcal{P} \Rightarrow \mathcal{S}$$

FANS uses two types of neighborhood: the *operational* and the *semantic* neighborhood of s. Given the current operator $\mathcal{O}$ and the current solution s, the operational neighborhood is:

$$\mathcal{N}(s) = \{\hat{s}_i | \ \hat{s}_i = \mathcal{O}_i(s)\} \quad (2)$$

where $\mathcal{O}_i(s)$ stands for the i-th application of $\mathcal{O}$ over s. Now, using the fuzzy valuation $\mu()$, the *semantic neighborhood* of s is defined as:

$$\hat{\mathcal{N}}(s) = \{\hat{s} \in \mathcal{N}(s) | \ \mu(\hat{s}) \geq \lambda\} \quad (3)$$

Here, the neighborhood $\hat{\mathcal{N}}(s)$ represents the λ-cut of the fuzzy set of solutions represented by $\mu()$.

The scheduler operation is quite simple. First, a *generator* is executed to obtain solutions from the semantical neighborhood sampling the search space with several applications of the operator $\mathcal{O}$. After that, a *selector* procedure has to decide which one is returned, taking as basis the degrees of membership of the obtained solutions, their cost or a combination of both values. If we are using a fuzzy valuation of "Similarity" with respect to the current solution, we may consider selection rules like:

- *Best:* Return the most similar solution available,
- *Worst:* Return the less similar solution,
- *First:* Return the first solution found with enough similarity.

Of course, we can also use the cost of those "Similar" solutions to obtain selection rules like:

- *MaxMax:* From those similar solutions, return the one with higher cost,
- *MaxMin:* From those similar solutions, return the one with lower cost.

2.5 Dealing with local optima

Because *FANS* is a memory-less (by now) heuristic, it has to provide some mechanism to deal with the problem of being trapped in local optima. *FANS* provides two escaping mechanism which are described below.

The first mechanism implemented is based on the fact that the modification operator determines the landscape or search space [12]. When the neighborhood scheduler fails to obtain a "good enough" solution, the modification operator is changed. The semantic neighborhood definition is kept fix, but now a new operational neighborhood is defined. The local optimality of the solution does not hold anymore, so the search can progress to other regions of the search space.

We can imagine that each operator defines a pattern of connectivity between the current solution and each other of the corresponding operational neighborhood. In this way, when the search becomes trapped and the operator is changed, the pattern of connectivity is modified making possible to escape from that local optimum.

The other mechanism is captured in the pair $(cond, action)$ where $cond$, called $TrappedSituation : \mathcal{P} \to [True, False]$, is used to determine when there is enough evidence of being definitely trapped in a local optimum. When $cond1$ holds, then the $action = doEscape()$ procedure will be executed. The user must decide the action to take: restart the algorithm from a new initial solution, perform a special modification of the current solution and resume the search, or anything else.

2.6 Description of the Algorithm

The scheme of *FANS* is shown in Figure 1. The execution of the algorithm finishes when some external condition holds, typically when the number of cost function evaluations reached some limit.

Each iteration begins with a call to the neighborhood scheduler NS with parameters S_{cur} (the current solution), $\mu()$ (the fuzzy valuation), and $\mathcal{O}$ (the modification operator). Two results are possible: an "acceptable" neighborhood solution S_{new} was found or not.

In the first case S_{new} is taken as the current solution and $\mu()$ parameters are adapted. In this way, we are varying our fuzzy valuation as a function of the context or, in other terms, as a function of the state of the search. If NS failed to return an acceptable solution (no solution was good enough in the neighborhood induced by the operator) the escaping mechanism is applied: the operator scheduler OS is executed, returning a modified version of $\mathcal{O}$. The next time NS is executed, it will have a modified operator to search for solutions.

The $TrappedSituation()$ condition will hold (for example) when Top calls to OS were done without improvements in the current solution. In this case, the $doEscape()$ procedure is executed, the cost of the current solution is reevaluated and $\mu()$ is adapted.

The reader must note that what varies at each iteration are the parameters used in the $\mathcal{NS}$ call. The algorithm starts with $\mathcal{NS}\ (s_0, \mathcal{O}^{t_0}, \mu_0)$. If $\mathcal{NS}$ could retrieve an acceptable neighborhood solution, the next iteration the call will be $\mathcal{NS}\ (s_1, \mathcal{O}^{t_0}, \mu_1)$, the current solution is changed and the fuzzy valuation is adapted. If $\mathcal{NS}$ failed to retrieve an acceptable neighborhood solution (at certain iteration l), the operator scheduler will be executed returning a modified version of the operator, so the call will be $\mathcal{NS}\ (s_l, \mathcal{O}^{t_1}, \mu_l)$.

```
Procedure FANS:
Begin
  InitVariables();
  While ( not-end ) Do
    /* The neighborhood scheduler NS is called */
    S_new = NS(O, μ, S_cur);
    If (S_new is good enough in terms of μ()) Then
       S_cur := S_new;
       adaptFuzzyValuation(μ(), S_cur);
    Else
       /* The operator failed to return a good solution */
       /* It will be changed or modified */
       /* calling the op. scheduler */
       O := OpSchedul(O);
    endIf
    If (TrappedSituation()) Then
       doEscape();
    endIf
  endDo
End.
```

Fig. 1. Scheme of *FANS*

Two "problematical" situations may arise: first, several calls were performed with $\mathcal{NS}(s_j, \mathcal{O}^{t_i}, \mu_j)$ with $i \in [1, 2, \ldots, k]$, which means that k different ways were tried to obtain an acceptable solution from s_j without success. Second, the search is moving across acceptable or good enough solutions but no improvements in the best solution ever found were obtained within the last m calls.

When any of both situations occurs, the *doEscape* procedure is applied leading to the one of the following calls: $\mathcal{NS}\ (\hat{s}_0, \mathcal{O}^{t_j}, \hat{\mu}_0)$ or $\mathcal{NS}\ (\hat{s}_0, \mathcal{O}^{t_0}, \hat{\mu}_0)$, where $\hat{s}_0$ is a new solution randomly generated or anyway obtained, and $\hat{\mu}_0$ is the fuzzy valuation reevaluated as a function of $\hat{s}_0$. The operator may be kept with its current parameters ($\mathcal{O}^{t_j}$) or also "restarted" ($\mathcal{O}^{t_0}$).

3 On the Usefulness of Fuzzy Valuations

The global behavior of *FANS* is a function of its components and of their interactions. Bearing in mind Figure 1, we will discuss how *FANS* may be adapted to behave or reflect the behavior of other local search techniques. This behavior has to be understand in qualitative terms. We will show how this adaptation may be mainly obtained through the appropriate definition of the fuzzy valuation.

First, we need a more precise definition for the neighborhood scheduler. For simplicity, we will use a *FirstFound* scheme with this simple definition: given a solution s, the fuzzy valuation μ, and the minimum level λ of required "quality", the scheduler will return the first solution $r \in \hat{\mathcal{N}}(s)$ found (within a maximum number of trials). Suppose the fuzzy valuation μ represents some notion of "Acceptability". We will also assume a minimization problem, with a current solution s and $\mathcal{N}(s)$ the neighborhood of s.

With the above elements, we are in position to indicate how *FANS* is able to capture the behavior of traditional local search algorithms.

3.1 FANS with Random Walks-like Behavior

In order to obtain this behavior, all we need is to consider any solution from the neighborhood as equally acceptable. Using a fuzzy valuation with $\mu(f(\hat{s})) = 1 \;\forall\; \hat{s} \in \mathcal{N}(s)$, any solution from the operational neighborhood will have the chance of being selected. In this way, the desired behavior is obtained.

3.2 FANS with Hill Climbing-like Behavior

Again, we could also obtain the required behavior manipulating the fuzzy valuation. Using any valuation where $\mu(f(\hat{s})) = 1$ if $f(\hat{s}) < f(s)$ and setting $\lambda = 1$, we will only consider those solutions improving the current cost as acceptable. The classical Hill Climbing method will stop when a local optimum is reached. In the context of *FANS*, the implementation of a multi-start version is straightforward.

3.3 FANS with Simulated Annealing-like Behavior

SA uses an external parameter, called "Temperature" to manage the acceptance of new solutions. Better solutions are always accepted, but initially, *SA* with high temperature may accept bad solutions with high probability. As the execution goes on, the temperature is lowered, decreasing the probability of acceptance of worse solutions. Towards the end of the run, only better solutions will be considered.

In order to achieve this behavior with *FANS*, it is enough to use the following definition for the fuzzy valuation "Acceptability":

$$\mu(q,s) = \begin{cases} 0.0 & \text{if } f(q) > \beta \\ (\beta - f(q))/(\beta - f(s)) & \text{if } f(s) \leq f(q) \leq \beta \\ 1.0 & \text{if } f(q) < f(s) \end{cases} \qquad (4)$$

with f the objective function, s the current solution, q an operational neighborhood solution, and β a parameter defining the limit for what is considered as acceptable.

It is easy to see that the key element is the definition of the β value which determines what solutions belongs to the neighborhood or not. Defining β as some function $h(s,q,t)$ where t is an external parameter representing, for example, the current number of iterations or the number of cost function evaluations performed, the deterioration limit may be reduced as the simulation progresses. Towards the end of the run, only those solutions better than the current one will be taken into account. For example, h may be defined as $h = f(s)*(1 + \frac{1}{1+exp((f(s)-f(q))/t)})$ where the second term of the sum is the logistic probability used within Boltzmann trials.

The reader should note that under this situation, the parameter β acts as a threshold value, so we have a simple way to represent a wide class of threshold algorithms with small effort.

3.4 *FANS* with Tabu Search-like Behavior

The basic scheme of Tabu Search, makes use of a history or memory H in order to constrain and guide the search. Typically, the history H is used to constrain or discard the generation of particular neighborhood solutions, which are included in a *tabu list*. In this way, cycling situations are reduced, and the search can be guided to promising or unexplored regions [8].

The use of such memory can not be captured by the fuzzy valuation, or at least not easily. One possible way is to define a fuzzy valuation in terms of the set of tabu solutions which may lead to a fuzzy set of tabu solutions.

Then the selection rule in the neighborhood scheduler will need to be redefined as, for example, *Return the Best Solution with a tabu membership less than* λ. This is an approach which is now under investigation and we hope to have some results soon.

3.5 *FANS* as a General Purpose Heuristic

In the previous paragraphs we showed how *FANS* may be tailored to reflect the behavior of other techniques, mainly through the definition of adequate fuzzy valuations. Up until this point, we have emphasized *FANS* as a local search framework, but we now wish to point out the potential of *FANS* as a general purpose optimization tool.

In order to apply *FANS* to a particular problem, a particular instance of *FANS* has to be implemented: particular definitions for the components must be provided, and some parameters have to be set.

General purpose definitions are easily obtained in some cases. For example, we may use a modification operator based on *k-exchange* and an operator scheduler modifying k in some way. As a fuzzy valuation, the "Acceptability" property could be used within a *FirstFound* scheme for the neighborhood scheduler. With such definitions, a canonical and very simple version of *FANS* is obtained. Of course, the more problem-dependent the definitions are, the better will be the results.

It should be highlighted that each instantiation of *FANS* leads to an algorithm with its particular, possibly novel, behavior which is mainly related with the definition of the fuzzy valuation and the value of the parameter λ.

4 Examples of Application

In this section, we will review the main published results obtained with *FANS* over a diverse set of problems. This set includes knapsack problems with single and multiple restrictions, the minimization of real functions, and an application of *FANS* to a problem from the bioinformatic area: the protein structure prediction problem.

For each application, we will briefly describe how the components of *FANS* were defined and we will present the main results. Pointers to the corresponding publications are given.

4.1 Standard Knapsack Problem

The Knapsack problem is one of the most studied in both the Operational Research and Computer Science areas. The corresponding mathematical formulation is:

$$\text{Max } \sum_{j=1}^{n} p_j * x_j \tag{5}$$

$$\text{s.t. } \sum_{j=1}^{n} w_j * x_j \leq C, \ x_j \in \{0,1\}, j = 1, \ldots, n$$

where n is the number of items, x_j indicates if the item j is included or not in the knapsack, p_j is the profit associated with the item j, $w_j \in [0, .., r]$ is the weight of item j, and C is the capacity of the knapsack. It is also assumed $w_j < C$, $\forall j$ (every item fits in the knapsack); and $\sum_{j=1}^{n} w_j > C$ (the whole set of items does not fit).

Although it seems very simple, knapsack problems are a real challenge for a variety of search algorithms (see [23] for recent developments). In this section, we describe the results presented in [5].

Description of *FANS* implementation The application of *FANS* to this problem is quite simple. The solutions are represented by binary vectors so the modification operator used is k-BitFlip. It simply flips the values of k bits randomly selected. The fuzzy valuation represents some notion of "Acceptability" which captures the following idea: solutions improving the current cost will have a higher degree of acceptability than those with lower cost. Solutions with cost below a certain threshold will not be considered as acceptable. Thus, given the objective function f, the current solution s, a new solution $q = \mathcal{O}(s)$, and the limit β for what is considered as acceptable, the following membership function comprises those ideas:

$$\mu(q,s) = \begin{cases} 0.0 & \text{if } f(q) < \beta \\ (f(q) - \beta)/(f(s) - \beta) & \text{if } \beta \leq f(q) \leq f(s) \\ 1.0 & \text{if } f(q) > f(s) \end{cases}$$

As a first approximation, $\beta = f(s) * (1 + \gamma)$ was used with $\gamma = 0.05$.

To adapt the modification operator, the operator scheduler changed the value of the parameter k. Each time the scheduler was called, the current value of k was replaced by a new value $\hat{k}$ obtained as a random integer value from $[1, 2*k]$. Also, if $\hat{k} > top = n/10$, where n is the number of items of the instance, then $\hat{k} = top$.

The last element to describe is the neighborhood scheduler. For its definition, a *Quality Based Grouping Scheme* [20] was used. This scheduler takes into account the membership values of the solutions provided by the fuzzy valuation in order to select the solution to return. The Quality Based Grouping Scheme or $R|S|T$ scheme tries to generate R "Acceptable" solutions with $\mathcal{O}$ in at most $maxTrials$ trials, then those solutions are grouped into S fuzzy sets based on their acceptability degree, and finally T solutions are returned. The second step may be viewed as a primitive clustering process.

A scheme with parameters $R = 5|S = 3|T = 1$ and $maxTrials = 25$ was implemented. The $S = 3$ fuzzy sets or clusters were represented by overlapped triangular membership functions and their boundaries were adjusted to fit the range $[\lambda, 1.0]$ being $\lambda = 0.98$ the minimum level of acceptability required. The sets represented the terms *Low, Medium, High* for the linguistic variable *Quality*. At the end of the process, $T = 1$ solution must be returned. The selection rule returned *any solution of the highest quality available*. If *High* quality solutions exist, any one of them was returned. If the set was empty, the same procedure was tried with the *Medium* quality set and if it was also empty, the *Low* level quality was used. If no acceptable solution was found, an exception condition was raised.

Experiments and Results The performance of *FANS* was assessed through comparisons against two implementations of genetic algorithms (*GA*, in what follows) and one of simulated annealing (*SA*). The *GA*'s varied in the crossover operator they used. One employed a one-point crossover and the other a uniform one (*GAop* and *GAux* respectively). The mutation operator was the same as the modification operator of *FANS* and given that no repair procedure is used, the following consideration was taken into account: if the solution obtained after mutation was infeasible, it was discarded and the operator was applied again over the original solution. This process was repeated at most four times. If no feasible solution could be obtained by mutation, the original individual was kept.

Table 1. Mean and variance of the errors taken over the 15 problems of each type of instance and its corresponding 30 runs' results.

Method	Uncorr Mean	Uncorr Var.	Weakly Corr Mean	Weakly Corr Var.	Strong Mean	Strong Var.
FANS	1.08	0.55	1.75	1.08	0.84	1.06
SA	1.61	3.17	1.80	1.37	1.32	1.25
GAop	2.06	0.80	2.74	1.18	1.32	1.26
GAux	1.32	0.77	2.63	1.25	1.37	1.25

The algorithms were compared over a test set of 45 randomly generated problems of size $n = 100$. Three types of correlations between weights and profits were considered giving raise to three groups: no correlated (*UN*), weakly correlated (*WC*), and strongly correlated (*SC*) instances. Each group consisted of 15 problems.

For the comparison, none or minimal knowledge of the problem was assumed (reflected by the use of very simple definitions for the operators in the 4 algorithms compared) and a fixed amount of resources (i.e., number of cost functions evaluations) was allocated for each algorithm.

Results were measured in terms of the error with respect to the Dantzig bound and they are shown in Table 1 where each value represents the mean and variance of the errors taken over the 15 problems of each type of instance and its corresponding 30 runs' results.

The results in Table 1 together with the t-test performed, showed that *FANS* consistently outperformed both *GA* in all three types of instances. *FANS* also presented better results in *UN* and *SC* than *SA*, achieving the same performance on *WC* instances. Both *GA* achieved the same performance over *WC* and *SC* instances, but *GAux* was better over *UN*. *SA* performance was better than both *GA* over *UN* and *WC* instances; the three algorithms were equally good on *SC*.

When the whole set of results over the 45 test instances was considered, *FANS* achieved significantly lower mean error values than *SA*, *GAop* and *GAux*. It was followed by *SA*, which outperformed both *GA*, and *GAux* which resulted better than *GAop*, mainly due to the results over *UN* instances.

Another interesting point was that *FANS*, *SA* and *GAop*, reached lower mean error values instances *SC*, contradicting the idea which relates higher correlation with higher difficulty.

4.2 Knapsack Problem with Multiple Restrictions

This version is the most general class of knapsack problems. It is an integer linear problem with the following formulation:

$$\text{Max} \sum_{i=1}^{n} p_i * x_i \qquad (6)$$

$$\text{s.t} \sum_{i=1}^{n} w_{ij} * x_i \leq C_j \text{ with } j = 1, \ldots, m$$

where n is the number of items, m the number of restrictions, $x_i \in \{0, 1\}$ indicates if the i item is included or not in the knapsack, p_j is the profit associated with the item i, and $w_{ij} \in [0, \ldots, r]$ is the weight of item i with respect to constraint j. Two additional restrictions are assumed: $w_{ij} < C_j \forall i, j$ (i.e. every item alone fits in the knapsack) and $\sum_{i=1}^{n} w_{ij} > C_j \ \forall j$ (i.e. the whole set of items do not fit in the knapsack).
The details of the application described in this section can be found in [21].

Description of FANS Implementation The characteristics of this implementation of *FANS* are quite similar to those used in the previous section. For the neighborhood scheduler, another definition was used besides that of the Quality Based Grouping Scheme. The new one is called *First* and its definition is very simple: given a current solution s and certain level of acceptability desired λ, the scheduler will use at most $maxTrials$ trials to obtain a solution $x \in \mathcal{N}(s)$ such that $\mu(x) \geq \lambda$.

Experiments and Results In order to analyze the performance of *FANS*, comparisons were made among the following five algorithms: F_{rst}, is *FANS* with scheduler $R|S|T$; F_{ff}, is *FANS* with scheduler $First$; *GAux*, is the *GA* with uniform crossover; *GAop*, is the *GA* with one-point crossover; and *SA*, the Simulated Annealing implementation.

Again, the performances of the algorithms were compared under none or minimal knowledge of the problem and when they are given a fixed an equal number of resources. The experiments were done over 9 instances selected from a standard set of problems available from [3]. The instances are named *pb5, pb7, Weing1, Weing3, Weing7, Weish10, Weish14, Weish18, Weish27*. The number of variables ranged from 20 to 105 with 2 to 30 restrictions.

For each problem and algorithm 30 runs were made; each one ending when $maxEvals = 15000$ cost function evaluations were done or when $maxEvals*4$ solutions were generated. This limit is needed because only feasible solutions are evaluated. The results were analyzed in terms of the error (with respect to the optimum value) for each problem and globally over the whole test set.

Table 2 (a) shows the mean of the errors over 30 runs for each test problem. The results indicated that both versions of *FANS* achieved the lower values of error for all problems, except for Weing3, Weing7 y Weish27. *SA* achieved the better value on Weing7 and *GAux* did it on the other two problems.

Table 2. Results for Knapsack with Multiple Restrictions. On (a), the mean of errors for each problem and algorithm. On (b), an x means the algorithm of the column reached the optimum of the problem on the row on any run.

	F_{rst}	F_{ff}	SA	GAop	GAux
pb5	1.04	0.92	6.52	3.37	3.07
pb7	0.94	1.19	4.13	3.83	4.23
weing1	0.20	0.19	8.07	0.92	1.37
weing3	1.54	1.32	22.04	1.85	0.91
weing7	0.50	0.51	0.48	1.13	0.93
weish10	0.27	0.14	1.22	1.34	1.18
weish14	0.85	0.78	1.93	1.85	0.91
weish18	0.73	0.71	1.39	0.95	0.89
weish27	3.02	2.89	2.85	3.21	1.18

(a)

	F_{rst}	F_{ff}	SA	GAop	GAux
pb5	x	x			
pb7	x	x			
weing1	x	x		x	x
weing3				x	x
weing7					
weish10	x	x	x	x	x
weish14	x	x		x	x
weish18	x	x		x	x
weish27					x
# Opt.	39	39	9	17	31

(b)

On part (b) of Table 2, an 'X' indicate if the algorithm on the column reached the optimum of the problem on the row in any of the 30 runs. It can be seen that F_{rst}, F_{ff} and $GAux$ obtained the optimum on 6 out of 9 problems, while $GAop$ on 5 out of 9 and SA just in 1 out of 9 (this optimum was reached by all algorithms). The last row (# Opt) in Table 2 (b), indicated the number of executions ending at the optimum for a total of 9*30=270 runs. F_{rst} y F_{ff} achieved the higher values followed by $GAux$. SA and $GAop$ were quite ineffective from this point of view.

The analysis of the mean and variance of the errors over the whole set of problems (results not shown) revealed that both versions of $FANS$ achieved the lowest values, followed by $GAux$. The mean error in $GAux$ and $GAop$ was almost twice of that in F_{rst} and F_{ff}, and SA values were 5 times higher.

To conclude the analysis, and to confirm if the differences in the mean of the errors over the whole test set were of statistical significance, t-tests were done with a confidence level of 95%. The results confirmed that both versions

Table 3. Test functions used for the experiments

$f_{sph}(\hat{x}) = \sum_{i=1}^{n} x_i^2$ [-5.12, 5.12]
$f_{ras}(\hat{x}) = 10.n + \sum_{i=1}^{n}[x_i^2 - 10 * cos(2\pi x_i)]$ [-5.12, 5.12]
$f_{ros}(\hat{x}) = \sum_{i=1}^{n-1}(100 * (x_{i+1} - x_i^2)^2 + (x_i - 1)^2)$ [-5.12, 5.12]
$f_{gri}(\hat{x}) = \frac{1}{4000}\sum_{i=1}^{n} x_i^2 - \prod_{i=1}^{n} cos(\frac{x_i}{\sqrt{i}}) + 1$ [-600,600]
$f_{sch}(\hat{x}) = \sum_{i=1}^{n}(\sum_{j=1}^{i} x_j)^2$ [-65.5, 65.5]
$ef_{10}(\hat{x}) = f_{10}(x_1, x_2) + \ldots + f_{10}(x_{i-1}, x_i) + \ldots + f_{10}(x_n, x_1)$
where $f_{10} = (x^2 + y^2)^{0.25} * [sin^2(50 * (x^2 + y^2)^{0.1}) + 1]$ [-100.0, 100.0]

of *FANS* outperformed *GAux*, *GAop* y *SA*. Both *GA* outperformed *SA*, and no significative differences were found among them, in spite of the apparent superiority of *GAux*.

4.3 Minimization of Real Functions

In this section we will describe the results presented in [24] where *FANS* was applied to the minimization of real functions $f : \mathcal{R}^n \rightarrow \mathcal{R}$ with n variables. The test set used covered a broad range of situations like multi modality, separability, etc. [2,16].

Description of *FANS* implementation The main differences between this application of *FANS* and the previous ones, lies in the representation of solutions, which in turns affects the definition of the modification operator and the operator scheduler. The fuzzy valuation "Acceptability" and the neighborhood scheduler $R|S|T$ were used again with the same definition as before but with different parameters.

The Modification Operator used, called r_k *Move*, randomly selected a certain number of variables, and performed a random positive or negative perturbation on each one. Given some value $m \in (0..Top]$, with $0 < Top \leq 100$, and a variable $x_i \in [min, max]$, a modification range r was calculated as $r = (max - min) * m/100$. The value m represented a percentage of the full range available to the variable. After that, the new value for each x_i was a random number in $[x_i - (r/k), x_i + (r/k)]$. The k value represented a factor of "compression" of the available search space.

The values m and Top were tunable parameters and it was clear that changes on them, modified the behavior of the operator. The adaptation of m and Top was done by the Operator Scheduler with the following strategy: given an initial value for $m = m_0$, then m was initially decremented by some small amount (0.1) each time the activation conditions were verified. When $m = 0.1$ was reached, the value started to be incremented until $m = Top$. After that, m was decremented/incremented between $[0 \ldots Top]$.

	Sphere				Rosen			
	AE	AB	SD	BF	AE	AB	SD	BF
BCGA	3.0E+05	2.6E-02	2.1E-02	6.7E-03	3.0E+05	1.1E+02	3.8E+01	2.6E+01
CHC	*	2.0E-32	9.0E-32	5.0E-32	*	2.0E+01	7.0E-01	2.0E+01
FANS	3.0E+05	2.6E-26	2.7E-26	3.3E-27	2.9E+05	8.7E+00	1.9E+01	1.8E-04
SA	3.0E+05	2.0E-08	5.4E-09	9.1E-09	3.0E+05	3.0E+00	4.6E+00	4.0E-03

	Rastringin				Schaffer			
	AE	AB	SD	BF	AE	AB	SD	BF
BCGA	3.0E+05	1.6E+01	4.4E+00	7.2E+00	3.0E+05	1.0E+03	3.2E+02	4.1E+02
CHC	*	0.0E+00	0.0E+00	100%	*	1.0E-01	3.0E-01	7.0E-12
FANS	1.3E+05	9.6E-01	9.8E-01	40%	3.0E+05	3.5E-05	4.8E-05	7.9E-07
SA	3.0E+05	2.4E-08	5.3E-09	1.6E-08	3.0E+05	8.7E-05	9.1E-05	4.3E-06

	Griewank				EF10			
	AE	AB	SD	BF	AE	AB	SD	BF
BCGA	3.0E+05	1.1E+00	1.1E-01	8.9E-01	3.0E+05	1.2E+00	3.6E-01	7.1E-01
CHC	*	0.0E+00	0.0E+00	100%	*	1.0E-07	2.0E-08	9.0E-08
FANS	2.0E+05	1.5E-02	1.6E-02	5.4E-20	3.0E+05	2.4E-11	1.7E-11	9.1E-12
SA	3.0E+05	2.1E-02	2.1E-02	1.9E-08	3.0E+05	1.0E-07	2.5E-08	5.9E-08

Table 4. Results obtained for *FANS*, *BCGA* and *SA*.

The *Top* limit was also adapted. Its value was reduced by 0.01 until some limit was reached. In that way, the upper limit for the variation for m was reduced in order to concentrate the search.

Experiments and Results The test set was composed of the functions shown in Table 3, where their definitions and the range for the variables $x_i \in [min, max]$. The functions have their optimum located at $x^* = (0, .., 0)$ with $f(x^*) = 0$ and $n = 25$. For ef_{10}, the dimension was set to $n = 10$.

FANS was compared against a public domain binary genetic algorithm (*BCGA*) [1] and an implementation of simulated annealing.

The results obtained are presented in Table 4, where AE is the average number of cost function evaluations done to reach the best value; AB is the average of the best values found on each run; SD is the standard deviation of the best values found; BF is the best value ever found. If a percentage appears, that means the percentage of the runs that reached the optimum.

It is easy to see from Table 4 that *FANS* clearly outperformed *BCGA* in all functions. This fact was confirmed by a t-test, with confidence 95%. When compared against *SA*, the t-test confirms that *FANS* is better than SA in ef_{10}, f_{sph} and f_{sch}. No statistical difference was found for f_{ros} and f_{gri}, although the corresponding BF values were better for *FANS*. For f_{ros}, authors thought that *FANS* was making sound progress but rather slow. For f_{gri}, *FANS* was unable to escape from the high number of local minimums. Just on 7 out of 25 runs, *FANS* obtained values around $1.e^{-20}$. While the t-test showed that *SA* outperformed *FANS* on average for f_{ras}, *SA* never found the optimum and *FANS* did it on the 40% of the runs. That value could be improved to 90% on f_{ras} using other parameters on the Neighborhood Scheduler.

Finally, in Table 4 are included some results obtained with the CHC algorithm in [10]. The CHC is a real coded genetic algorithm designed to overcome the premature convergence problem and it's usually taken as a reference point. The results shows that such simple version of *FANS* clearly outperformed CHC in f_{sch}, f_{ros} and ef_{10} in terms of the *AB* values. On f_{sph} the level of precision (10^{-26}) achieved by *FANS* was good enough for any practical application. The CHC performance was excellent on f_{ras} and f_{gri} where always achieved the corresponding optimum. *FANS* results on f_{gri} were rather bad, but they were similar to those obtained by some distributed GA also presented in [10].

4.4 The Protein Structure Prediction Problem

Protein Structure Prediction is one the most exciting problems that computational biology faces today. In simple terms, it can be formulated as follows: given a sequence of amino acids, which is the corresponding 3D structure of minimum energy?. This is a very important problem because the 3D structure of a protein determines its biological function. The determination of the sequence may be considered a solved problem, but the experimental determination of the structure via x-ray crystallography or nuclear magnetic resonance is still a time consuming and difficult task.

One of the most studied simple protein models is the hydrophobic - hydrophilic model (HP model) proposed by Dill [8]. It was shown NP-Hard on the square lattice [7] and cubic lattice [3]. Genetic Algorithms, Simulated Annealing, GRASP and also Cellular Automata were applied to this problem[13,14,17,26].

Under this model, a sequence s is represented as a string $s \in \{H, P\}^+$, where H represents a hydrophobic amino acid and P represents a hydrophilic one. The goal is to find a self avoiding embedding of maximum score (or minimum energy) in a given lattice. Figure 2 shows strings embedded in the square, and triangular lattices, with HH contacts highlighted with dotted lines. The conformation in Figure 2(a) has a score of -17 (seventeen contacts) and the conformation in Figure 2(b) has a score of -9 (nine contacts).

One possible representation for the structures is called Internal Coordinates, where two types arise: absolute and relative. Under the absolute encoding, the structures are represented as a list of absolute moves in the corresponding space. For example, if a 2D square lattice is used,a structure s is codified as a string $s = \{\textbf{U}p, \textbf{D}own, \textbf{L}eft, \textbf{R}ight\}^+$. When using a relative encoding, each move must be interpreted in terms of the previous one, like those of the LOGO turtle: a structure s is encoded as a string in the alphabet $s = \{\textbf{F}orward, \text{Turn}\textbf{R}ight, \text{Turn}\textbf{L}eft\}^+$.

Just recently, a direct comparison among both encodings was done for evolutionary algorithms [13]. Now, we will review two works [18,25] where

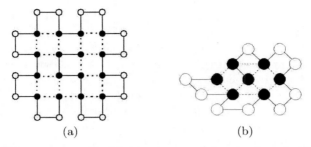

Fig. 2. HP sequences embedded in a square (a) and triangular (b) lattices.

FANS was applied to *PSP* with two objectives: first, to analyze how the codification affected the performance of *FANS*, and second, to overcome a theoretical problem arisen in the application of *GA*'s to this problem.

Description of *FANS* implementation Given that a new representation was used, new definitions were needed for the operator and its corresponding scheduler. The fuzzy valuation was kept with the definition of acceptability and the neighborhood scheduler implemented a *First* strategy.

The operator $\mathcal{O}$ used a parameter k representing the number of positions to change in the given solution. Two modes of operation were available: *Segment mode*, where k consecutive positions of the structure were modified; or *Flip mode* where k positions randomly selected were changed.

The operator scheduler modified the parameter k of the operator in the following manner: each time the scheduler was called, the value of k was decremented by one. In this way, the operator performed coarse modifications initially (corresponding to an exploration stage), making them finer as the simulation progress (like fine tuning the structures). The initial value was set to $k = n/4$, where n stands for the length of the sequence. When $k = 0$ ocurred, the doRestart() procedure was executed generating a new initial random solution and re-setting the k value of the operator to $n/4$. Then *FANS* restarted from that new solution.

Experiments and Results for Encoding Comparison For the experiments, three values of the parameter $\lambda = \{0.0, 0.9, 1.0\}$ were used, each one leading to a different behavior for *FANS* namely λ_0, $\lambda_{.9}$, and λ_1. When λ_0 was used, the behavior was that of a random walk procedure. Any solution from the neighborhood could be selected. When *FANS* operated in $\lambda_{.9}$ mode, some demeliorating moves were accepted, while the use of λ_1 led to a hill climber like behavior: only improving solutions could be selected.

The experiments were done for a set of instances in a square, triangular and cubic lattices. For each version of *FANS*, encoding (absolute or relative) and instance, 30 runs were made.

T-test analyses were performed to detect if the mean values of energy obtained for the various algorithms under both encodings were different or not. When a random walk type of search was used, λ_0, the relative encoding was superior in the three lattices. When $\lambda_{.9}$ was used, allowing demeliorating steps, then the relative encoding was superior in the square lattice and indistinguishable in the other two. Finally, when *FANS* performed as multi-start hill-climber, λ_1, the absolute encoding was better for the triangular lattice, both encodings achieved similar values for the cubic lattice and the relative encoding was superior in the square one.

These results agreed with those presented in [13], where the same kind of experiment was done for genetic algorithms. In this way, if new heuristics methods are to be applied to this problem, both results should be taken into account to decide the representation to use.

***FANS* vs. *GA*: Experiments and Results** It is known that from a practical point of view, genetic algorithms perform well on *PSP*. However there is a theoretical problem arising from the use of standard crossover operators with an internal coordinates representation. A couple of years ago, a *GA* was used for the *PSP* and it was shown that the crossover was not transferring information between individuals [15] (the *idea of crossover* was not working). That was verified with a*headless chicken test* [11]: this test compares the results of a *GA* using a standard crossover against a *GA* using a random one.

The results showed that the *GA* with random crossover performed equal or better than the one with standard crossover. So, given that there was no information exchange between individuals, it was then argued that the use of a population is not needed and it was claimed that the same results could be *potentially* obtained applying mutations over a unique individual.

In this section we will describe how *FANS* obtained as good results as those obtained by a *GA*, giving experimental evidence to confirm both, the claim described above and the power of this fuzzy sets-based heuristic.

The experiments were done over a set of instances on the square lattice. Based on the results of the previous section, the solutions were encoded using relative codification, and the version of *FANS* named $\lambda_{.9}$ was used. The *GA* implemented used the same encoding for solutions. As mutation operator, the modification operator of *FANS* was used. The value for the parameter k was randomly set each time the mutation was applied. A standard two point crossover was employed and the selection scheme was Tournament Selection.

In terms of the mean values, *FANS* obtained better results on 5 out of 7 cases. In general the standard deviation is less in *FANS* than in the *GA* and both algorithms achieved the optimal values for all instances excepting *I*5, where the *GA* reached a higher value. In other words, *FANS* obtained the same results as the *GA* using a quite simple scheme, thus verifying the claim that the use of a population is not needed if the *GA* uses a standard crossover and an internal coordinates representation.

5 Conclusions

In this chapter we reviewed the main characteristics of our Fuzzy Adaptive Neighborhood Search, *FANS*, with special emphasis on the fuzzy valuation. We showed how each combination of fuzzy valuation and level λ of minimum membership required for a solution to be selected, led to a particular behavior of *FANS*, which in turn, enabled as to consider the method as a framework of local search methods.

Then, we described a set of published results where *FANS* was compared against other heuristic optimization methods over a set of instances of four problems with very different characteristics. The main conclusion was that *FANS* was able to obtain similar or better results than the other algorithms.

We consider both aspects as a proof of the the benefits of merging basic ideas from the Fuzzy Sets and Systems area with simple optimization methods. *FANS* may be obtained upon request from dpelta@ugr.es

Acknowledgments

Research supported in part by Projects PB98-1305, TIC 99-0563. David Pelta is a grant holder from Consejo Nacional de Invest. Cientificas y Tecnicas CONICET, Republica Argentina.

References

1. T. Bäck. Genesys 1.0 user guide. Technical report, Univ. of Dortmund, Germany, 1992.
2. T. Bäck. *Evolutionary Algorithms in Theory and Practice*. Oxford Univ. Press, 1996.
3. J.E. Beasley. The or-library: a collection of test data sets. Technical report, Management School, Imperial College, London SW7 2AZ., 1997. http://mscmga.ms.ic.ac.uk/info.html.
4. B. Berger and T. Leighton. Protein folding in the hydrophobic-hydrophilic (HP) model is NP-complete. *Journal of Computational Biology*, 5(1):27–40, 1998.
5. A. Blanco, D. Pelta, and J.L. Verdegay. A fuzzy valuation-based local search framework for combinatorial problems. *Journal of Fuzzy Optimization and Decision Making*, 1(2):177–193, 2002.
6. P. Crescenzi, D. Goldman, C. Papadimitriou, A. Piccolboni, and M. Yannakakis. On the complexity of protein folding. *Journal of Computational Biology*, 5(3):409–422, 1998.
7. K. A. Dill. Dominant forces in protein folding. *Biochemistry*, 24:1501, 1985.
8. F. Glover and M. Laguna. *Tabu Search*. Kluwer Academic Publishers, London, 1997.
9. P. Hansen and N. Mladenović. An introduction to variable neighborhood search. In S. Voss, S. Martello, I. Osman, and C. Roucairol, editors, *Metaheuristics: Advances and Trends in Local Search Procedures for Optimization*, pages 433–458. Kluwer, 1999.

10. F. Herrera and M. Lozano. Gradual distributed real-coded genetic algorithms. *IEEE Trans. on Evolutionary Computation*, 4(1):43–63, 2000.
11. T. Jones. Crossover, macromutation, and population based search. In *Proceedings of the Sixth International Conference on Genetic Algorithms*, pages 73–80. Morgan Kauffman, 1995.
12. T. Jones. *Evolutionary Algorithms, Fitness Landscapes and Search*. PhD thesis, University of New Mexico, Albuquerque, 1995.
13. N. Krasnogor, W.E. Hart, J. Smith, and D. Pelta. Protein structure prediction with evolutionary algorithms. In W. Banzhaf, J. Daida, A.E. Eiben, M.H. Garzon, V. Honavar, M. Jakaiela, and R.E. Smith, editors, *GECCO-99: Proceedings of the Genetic and Evolutionary Computation Conference*, pages 1596–1601. Morgan Kaufman, 1999.
14. N. Krasnogor, D. H. Marcos, D. Pelta, and W. A. Risi. Protein structure prediction as a complex adaptive system. In Cezary Janikow, editor, *Frontiers in Evolutionary Algorithms (FEA98)*, pages 441–447, 1998.
15. N. Krasnogor, D. Pelta, P. Martinez Lopez, P. Mocciola, and E. de la Canal. Genetic algorithms for the protein folding problem: A critical view. In C. Fyfe E. Alpaydin, editor, *Proceedings of Engineering of Intelligent Systems, EIS'98*, pages 353–360. ICSC Academic Press, 1998.
16. Z. Michalewicz. *Genetic Algorithms + Data Structures = Evolution Programs*. Springer Verlag, 1999.
17. J. T. Pedersen and J. Moult. Genetic algorithms for protein structure prediction. *Current Opinion in Structural Biology*, (6):227–231, 1996.
18. D. Pelta, A. Blanco, and J. L. Verdegay. Applying a fuzzy sets-based heuristic for the protein structure prediction problem. *International Journal of Intelligent Systems*, 17(7):629–643, 2002.
19. D. Pelta, A. Blanco, and J. L. Verdegay. A fuzzy adaptive neighborhood search for function optimization. In *Fourth International Conference on Knowledge-Based Intelligent Engineering Systems & Allied Technologies, KES 2000*, volume 2, pages 594–597, 2000.
20. D. Pelta, A. Blanco, and J. L. Verdegay. Introducing fans: a fuzzy adaptive neighborhood search. In *Eigth International Conference on Information Processing and Management of Uncertainty in Knowledge-based Systems, IPMU 2000*, volume 3, pages 1349–1355, 2000.
21. D. Pelta, A. Blanco, and J. L. Verdegay. Learning about FANS behaviour: Knapsack problems as a test case. In *International Conference in Fuzzy Logic and Technology, EUSFLAT'2001*, volume 1, pages 17–21, 2001.
22. D. Pelta, N. Krasnogor, A. Blanco, and J. L. Verdegay. F.a.n.s. for the protein folding problem: Comparing encodings and search modes. In *Fourth International Metaheuristics Conference, MIC 2001*, pages 327–331, 2001.
23. D. Pissinger. *Algorithms for Knapsack Problems*. PhD thesis, Dept. of Computer Science,University of Copenhagen, Denmark, 1995.
24. A. Sancho-Royo, J.L. Verdegay, and E. Vergara-Moreno. Some practical problems in fuzzy sets-based decision support systems. *MathWare and Soft Computing*, 6(2-3):173–187, 1999.
25. J.L. Verdegay and E. Vergara-Moreno. Fuzzy termination criteria in knapsack problem algorithms. *MathWare and Soft Computing*, 7(2-3):89–97, 2000.
26. K. Yue, K. M. Fiebig, P. D. Thomas, C. H. Sun, E. I. Shakhnovich, and K. A. Dill. A test of lattice protein folding algorithms. *Proc. Natl. Acad. Sci. USA*, 92:325–329, 1995.

Branch-and-bound algorithms using fuzzy heuristics for solving large-scale flow-shop scheduling problems

Jinliang Cheng[1], Hiroshi Kise[2], George Steiner[3], and Paul Stephenson[4]

[1] J.D. Edwards & Company, Toronto, Ontario, Canada
[2] Kyoto Institute of Technology, Kyoto, Japan
[3] McMaster University, Hamilton, Ontario, Canada
[4] Acadia University, Wolfville, Nova Scotia, Canada

Abstract. The flow-shop scheduling problem has a long history since Johnson's seminal paper in 1954, and is still actively studied. The problem is strongly NP-hard when the number of machines is greater than 2. Thus we have to rely on implicit enumeration procedures, such as branch-and-bound, for exact solutions. One of the objectives for researchers in studying such intractable problems is to develop new methods which can solve larger-scale problems. We discuss branch-and-bound algorithms for finding a permutation schedule minimizing the makespan in an m-machine flow shop where the jobs may all be available before processing begins or may be dynamically released. A branch-and-bound algorithm's performance is helped by sharper lower bounds, good search methods in the tree and more effective dominance relations. Dominance relations are only sufficient but not necessary conditions for optimality, which suggests that optimality may hold with high probability even if a dominance condition is not satisfied exactly, but only approximately. Our algorithms exploit this fact by using a *fuzzy approximation* of dominance relationships for tie breaking in the search tree and for building initial incumbent schedules. These fuzzy heuristics together with an adaptive branching technique and a generalized decomposition method proved to be very effective for solving large problems. We review the results of large-scale computational experiments for two- and three-machine problems and report on the algorithms' performance for problems on up to ten machines. The algorithms have proved to be effective in solving a large number of problems with up to several hundred jobs. An interesting byproduct of our research is the rather surprising insight that the difficulty of the problems with release times seems to depend much more on the relative size of the release times than on the number of jobs involved.

1 Introduction

We consider permutation flow shop scheduling problems where the jobs may all be available before processing begins or may be dynamically released. The set $J = \{1, 2, \ldots, n\}$ of n jobs has to be processed without preemption on each of m sequential machines in the same order $M_1, M_2, \ldots, M_m$. Job j has processing time $P_l(j)$ on machine $M_l (l = 1, 2, \ldots, m)$. Each job becomes available for processing at its release time r_j or at the begining of the processing ($r_j = 0$). The objective is to find an optimal sequence for the n jobs so as to minimize the maximum completion time or makespan.

According to the traditional notation introduced by Graham et al. [12], the problem can be denoted by $Fm/r_j, perm/C_{max}$. If all jobs are available at time zero, the corresponding $Fm/perm/C_{max}$ problem is a classical flow-shop problem, which is known to be NP-hard in the strong sense [9] even for $m = 3$. If not all jobs are available at time zero, the corresponding $Fm/r_j, perm/C_{max}$ problem is NP-hard in the strong sense even in the two-machine case [17].

The classical flow-shop problem has attracted the attention of many researchers since Johnson [14] discovered the well-known polynomial time solution for $F2//C_{max}$. The complexity results mentioned above strongly suggest, however, that an enumerative approach is essentially unavoidable for the exact solution of problems with $m \geq 3$. Branch-and-bound algorithms proved to be one of the most powerful methods for solving such NP-hard problems. They are characterized by the following basic components: branching operation, bounding operation and search strategy. Most of the previously published branch-and-bound algorithms for the permutation flow-shop problem focus on improving lower bounds [16] and use the same branching scheme that successively fixes jobs from the beginning of the schedule. Potts [19] has proposed an adaptive branching rule in which jobs get fixed from the beginning and the end of the schedule, and has found this scheme superior to previously published algorithms. Carlier and Rebai [2] adapted the disjunctive graph model to flow shop problems.

There are relatively few papers on computational experiments to find optimal schedules for flow-shop problems with release times. Grabowski [10] and Grabowski et al. [11] have proposed branch-and-bound algorithms using a dichotomic branching scheme based on the critical path concept for the $F2/r_j, perm/L_{max}$ and the $Fm/r_j, perm/L_{max}$ problems. They have experimented with two- and three-machine problems on up to 50 jobs and some instances with up to $m = 8$ but small number of jobs. Tadei et al. [20] have developed several branch-and-bound procedures by applying different lower bounds, branching schemes and dominance properties for the $F2/r_j/C_{max}$ problem. They have classified problem instances as 'easy' or 'hard' according to the distribution of the release times. They tested their algorithms on 'easy' problem instances with up to 200 jobs and 'hard' problem instances with up to 80 jobs. They have also found Grabowski's branching scheme less effective for the $F2/r_j/C_{max}$ problem than their n-ary branching scheme.

This paper aims to summarize, review and extend our results [3–6] on developing new branch-and-bound algorithms using fuzzy heuristics for solving large flow-shop problems $Fm/perm/C_{max}$ and $Fm/r_j, perm/C_{max}$. We present a general scheme of branch-and-bound algorithms for solving these problems with the following salient features. Three different lower bounds are considered. Two of them are obtained by relaxing the release time and the capacity constraints on some machines and the third one is based on a special solvable case of the $F2/r_j, perm/C_{max}$ problem. Two dominance properties are used: the first one is based on the concept of fuzzy dominance and it is used to find good initial schedules and to guide the search to reach an optimal solution quickly in the branch-and-bound tree; the second one leads to a decomposition procedure that reduces the problem size by fixing some jobs at the beginning of the schedule. A variant of Potts' adaptive branching rule is combined with the fuzzy search strategy to narrow the search tree and direct the search to an optimal solution faster. The remainder of the paper is organized as follows: Section 2 develops the lower bounds used. Section 3 discusses the dominance rules and describes the fuzzy heuristic. Section 4 defines the remainder of the branch-and-bound algorithm, including the upper bound, the branching rule and the fuzzy search strategy. Section 5 reports on extensive numerical experiments. The last section gives our concluding remarks.

2 Lower bounds

Lageweg et al. [16] have developed a conceptual framework of lower bound calculation for the permutation flow-shop problem. We generalize some of the lower bounds from this framework for our problems, and also discuss a new lower bound which uses a special case of the two-machine problem with release times [5].

2.1 Two-machine 1-minimal flow shop with preemption

Consider the problem of finding a preemptive schedule which minimizes the makespan for jobs $J = \{1, 2, \cdots, n\}$ in a two-machine flow shop with release times. The flow shop is called 1-*minimal* if the processing times satisfy $P_1(j) \leq P_2(j)$ for each job $j \in J$ [1,15]. This problem is denoted by $F2/r_j, 1 - min, pmtn/C_{max}$. If the processing times $P_1(j)$ and $P_2(j)$ are arbitrary, it is known that the problem $F2/r_j, pmtn/C_{max}$ is NP-hard in the strong sense [7]. On the other hand, an optimal preemptive schedule for $F2/r_j, 1 - min, pmtn/C_{max}$ can be found in $O(n \log n)$ time [5]. We will use this fact for finding new lower bounds for the general problem.

2.2 Lower bound computation

Consider a partial sequence (σ_1, σ_2), where σ_1 and σ_2 correspond to an initial partial sequence and a final partial sequence, respectively. Let S_i denote the set of jobs in σ_i for $i = 1, 2$, that have already been scheduled, and let S be the set of jobs still to be scheduled, i.e. $S = J \backslash (S_1 \cup S_2)$. For σ_1 we look at the original problem and let $C_l(\sigma_1)$ be its completion time on machine

M_l ($1 \leq l \leq m$). For σ_2 we consider the reverse problem, with release times treated as delivery times, and define $C_l(\sigma_2)$ ($1 \leq l \leq m$) analogously. We also define $C_{max}(\sigma_2)$ to be the maximum delivery completion time for σ_2.

Following Lageweg et al. [16], we relax the capacity constraint that each machine can process at most one job at a time for all machines but at most two, say M_u and M_v ($1 \leq u \leq v \leq m$). Then we obtain a scheduling problem on five machines, N_{ru}, M_u, N_{uv}, M_v, $N_{v.}$, in this order. The machines N_{ru}, N_{uv}, $N_{v.}$ have infinite capacity and auxiliary processing times on them are defined by

$$P_{ru}(j) = r_j + \sum_{h=1}^{u-1} P_h(j), j = 1, 2, \ldots, n \qquad (1)$$

$$P_{uv}(j) = \sum_{h=u+1}^{v-1} P_h(j), j = 1, 2, \ldots, n \qquad (2)$$

$$P_{v.}(j) = \sum_{h=v+1}^{m} P_h(j), j = 1, 2, \ldots, n \qquad (3)$$

(It is assumed that the value of any summation is zero when its lower limit exceeds its upper limit). Machines M_u and M_v have capacity one and processing times $P_u(j)$ and $P_v(j)$, $j \in S$, respectively. A lower bound can be obtained by finding an optimal schedule for this problem.

The framework developed by Lageweg et al. [16] is characterized by a string Ω of at most five symbols from $\{\Box, \bigcirc, *\}$ and any lower bound is denoted by $LB(\sigma_1, \sigma_2, u, v, \Omega)$, where

- $\Box$: indicates a bottleneck machine (i.e. M_u or M_v);
- $\bigcirc$: indicates a non-bottleneck machine on which the auxiliary processing times and/or release times are taken into account;
- $*$: indicates a non-bottleneck machine that is to be removed.

When any of the machines N_{uv}, $N_{v.}$ is removed from the problem, then this will be partially offset by strengthening $C_u(\sigma_1)$ and $C_u(\sigma_2)$: The processing of any job in S on a machine M_u ($1 \leq u \leq m$) cannot be started before $\max\{\min_{j \in S} P_{ru}(j), \max_{1 \leq l \leq u}(C_l(\sigma_1) + \min_{j \in S} P_{l-1,u}(j))\}$; and $C_u(\sigma_2)$ ($1 \leq u \leq m$) can be similarly strengthened by using $\max_{u \leq l \leq m}(\min_{j \in S} P_{u,l+1}(j) + C_l(\sigma_2))$ instead.

Computations of some lower bounds, like $LB(\sigma_1, \sigma_2, u, u, (\bigcirc, \Box, \bigcirc))$ and $LB(\sigma_1, \sigma_2, u, u+1, (\bigcirc, \Box, \Box, *))$, in this framework correspond to NP-hard problems. However, relaxing some constraints and then solving the relaxed problems may provide good lower bounds. Such a relaxation could be allowing preemption and/or creating an auxiliary problem with $1-minimal$ processing times.

We shall now discuss in detail each lower bound we have used.

(a1) $LB(\sigma_1, \sigma_2, u, m, (*, \square, \bigcirc, \square)), 1 \leq u \leq m - 1$.

This lower bound is obtained by solving a special three-machine flow-shop problem in which M_u and M_m are seperated by a non-bottleneck machine N_{um}. To do this, we can apply Johnson's algorithm with processing times $P_u(j) + P_{um}(j)$ and $P_{um}(j) + P_m(j), j \in S$ to determine an optimal Johnson schedule on M_u, N_{um}, M_m [8]. Let $C_u^a(\sigma_1, S)$ and $C_m^a(\sigma_1, S)$ be the completion times on machine M_u and M_m, respectively, of this Johnson schedule on S, then

$$LB(\sigma_1, \sigma_2, u, m, (*, \square, \bigcirc, \square)) = \max\{C_u^a(\sigma_1, S) + C_u(\sigma_2), C_m^a(\sigma_1, S) + C_m(\sigma_2), C_{max}(\sigma_2)\} \quad (4)$$

is a lower bound.

(a2) $LB(\sigma_1, \sigma_2, m - 1, m, (*, \square, \square))$.

This lower bound is obtained by applying Johnson's algorithm with processing times $P_{m-1}(j)$ and $P_m(j), j \in S$. Let $C_{m-1}^b(\sigma_1, S)$ and $C_m^b(\sigma_1, S)$ be the completion times on machine M_{m-1} and M_m, respectively, of this Johnson schedule on S, then

$$LB(\sigma_1, \sigma_2, m - 1, m, (*, \square, \square)) = \max\{C_{m-1}^b(\sigma_1, S) + C_{m-1}(\sigma_2), C_m^b(\sigma_1, S) + C_m(\sigma_2), C_{max}(\sigma_2)\} \quad (5)$$

is a lower bound.

We select the best one of the lower bounds $LB(\sigma_1, \sigma_2, u, m, (*, \square, \bigcirc, \square))$, $1 \leq u \leq m - 1$ of case (a1) and $LB(\sigma_1, \sigma_2, m - 1, m, (*, \square, \square))$ of case (a2) and denote it by LB_1, i.e.,

$$LB_1 = \max\{\max_{1 \leq u \leq m-1} LB(\sigma_1, \sigma_2, u, m, (*, \square, \bigcirc, \square)), \\ LB(\sigma_1, \sigma_2, m - 1, m, (*, \square, \square))\}. \quad (6)$$

This lower bound LB_1 has been shown to be the best bound among the polynomially computable one- and two-machine bounds for the classical permutation flow-shop scheduling problem [4,16]. It requires $O(n \log n)$ computation time at the root of the search tree, but only $O(n)$ at other nodes.

(b) $LB(\sigma_1, \sigma_2, u, u + 1, (\bigcirc, \square, \square, *), 1 - min, pmtn), 1 \leq u \leq m - 1$.

The problem of minimizing the maximum delivery completion time in a two-machine flow shop corresponding to $LB(\sigma_1, \sigma_2, u, u + 1, (\bigcirc, \square, \square, *))$ is NP-hard in the strong sense even with preemption [7]. Therefore, we replace every job $j \in S$ violating the 1-*minimal* condition by two jobs, i.e., if $P_u(j) > P_{u+1}(j)$, we cut off the amount of processing by which $P_u(j)$ is larger and create a new single-machine job (the *cut-off job*) with release time $P_{ru}(j)$ and requiring $\overline{P}_u(j) = P_u(j) - P_{u+1}(j)$ processing on M_u and no processing on subsequent machines and replace $P_u(j)$ by $P'_u(j) = P_{u+1}(j)$ on M_u and $P_{ru}(j)$ by $P'_{ru}(j) = P_{ru}(j) + P_u(j) - P_{u+1}(j)$. Jobs which satisfy the 1 - *minimal* condition remain unchanged. We will refer to both the shortened and the unchanged jobs requiring processing on both bottleneck machines as the *residual jobs*. It is clear that the length of the optimal preemptive schedule for this auxiliary problem on S is a lower bound for $LB(\sigma_1, \sigma_2, u, u+1, (\bigcirc, \square, \square, *))$. Furthermore, since the cut-off jobs need processing only on M_u, we can always give priority to a residual job over a cut-off job when scheduling them, i.e., we can find the optimal preemptive schedule on the residual jobs using the $O(n \log n)$ algorithm in [5] (starting at time $C_u(\sigma_1)$) and then schedule preemptively the cut-off jobs on M_u as early as possible. More precisely, we process the cut-off jobs in nondecreasing order of their release times and preemptively fit them into the earliest time slots when M_u is not processing any of the residual jobs. Note that we can have at most $O(n)$ such slots, since these have to start at the time of completion for some residual job and end at the time of arrival of another residual job. Thus the complexity of finding this optimal preemptive schedule for the auxiliary problem will be $O(n \log n)$ too.

Let $C_u^c(\sigma_1, S)$ and $C_{u+1}^c(\sigma_1, S)$ be the completion times of the preemptive schedule for the auxiliary problem on M_u and M_{u+1}, respectively, then

$$LB(\sigma_1, \sigma_2, u, u+1, (\bigcirc, \square, \square, *), 1 - min, pmtn) =$$
$$\max\{ \max_{u \leq l \leq u+1} (C_l^c(\sigma_1, S) + C_l(\sigma_2)), C_{max}(\sigma_2),$$
$$C_{u+1}^c(\sigma_1, S) + \max_{u+2 \leq l \leq m} (\min_{j \in S} P_{u+1,l}(j) + C_l(\sigma_2))\} \qquad (7)$$

and our second lower bound is as follows:

$$LB_2 = \max_{1 \leq u \leq m-1} LB(\sigma_1, \sigma_2, u, u+1, (\bigcirc, \square, \square, *), 1 - min, pmtn). \qquad (8)$$

(c) $LB(\sigma_1, \sigma_2, u, u, (\bigcirc, \square, \bigcirc), pmtn)$, $1 \leq u \leq m - 1$.

Computation of $LB(\sigma_1, \sigma_2, u, u, (\bigcirc, \square, \bigcirc))$ corresponds to minimizing maximum lateness on M_u with respect to release times $P_{ru}(j)$ and delivery times $P_{u.}(j)$, i.e. $1(M_u)/r_j (= P_{ru}(j))/L_{max}$, which has been shown to be NP-hard in the strong sense [17]. The preemptive case, however, is solved in $O(n \log n)$ time by a modification of Jackson's rule due to Horn [13].

Let $C_u^d(\sigma_1, S)$ be the completion time on M_u and $C_m^d(\sigma_1, S)$ be the completion time on M_m (where each job j has delivery time $q_j = P_{u.}(j)$) of the optimal preemptive schedule for this one machine problem, then

$$LB(\sigma_1, \sigma_2, u, u, (\bigcirc, \square, \bigcirc), pmtn) =$$
$$\max\{C_u^d(\sigma_1, S) + C_u(\sigma_2), C_m^d(\sigma_1, S) + C_m(\sigma_2), C_{max}(\sigma_2)\}. \qquad (9)$$

We denote the best one of these $m - 1$ lower bounds by LB_3, i.e.,

$$LB_3 = \max_{1 \leq u \leq m-1} LB(\sigma_1, \sigma_2, u, u, (\bigcirc, \square, \bigcirc), pmtn). \qquad (10)$$

LB_3 can be computed in $O(n \log n)$ time for each node of the search tree.

In the algorithm for $Fm//C_{max}$, only LB_1 was used, which can be computed in $O(n)$ time except for the root node. In the algorithm for $Fm/r_j/C_{max}$, we use the best lower bound given by $LB = \max_{1 \leq i \leq 3} LB_i$.

3 Dominance rules and fuzzy approximation

Dominance rules are normally used to eliminate nodes in a branch-and-bound algorithm before their lower bound is calculated. The new concept of fuzzy approximation of dominance was introduced in [3–5] for looking for a good initial solution or directing the search in a branch-and-bound algorithm. Other types of dominance may decompose the original problem into two smaller problems, where one of these may be solved by fixing a subsequence of an incumbent schedule [20].

3.1 Fuzzy approximation of dominance

The following sufficient conditions extend well-known dominance criteria to the $Fm/r_j, q_j, perm/C_{max}$ problem, where q_j is the delivery time of job j. (We include delivery times here because when applying these conditions to reversed problems, the release times are treated as delivery times.)

Proposition 1. *Consider a partial sequence σ for $Fm/r_j, q_j, perm/C_{max}$. If there are unsequenced jobs $i, j \in S$ such that*

$$C_l(\sigma ij) \leq C_l(\sigma ji), \quad 1 \leq l \leq m, \tag{11}$$
$$C_{max}(\sigma ij) \leq C_{max}(\sigma ji), \tag{12}$$

then the partial sequence σij dominates σji, i.e., any completion of σij has a makespan no longer than the same completion of σji.

Proof. Obvious.

Corollary 1. *Consider a partial sequence σ for $Fm/r_j, q_j, perm/C_{max}$. If there is an $i \in S$ such that for every $j \in S$ (11) and (12) hold then σ has an optimal completion in which i is the next job scheduled.*

Proof. Obvious.

The corollary requires very strong conditions and it is rare that there is an i for which they all hold. On the other hand, if they are 'almost all' satisfied, then it suggests that job i may precede any other job $j \in S$ in an optimal completion of σ with 'high probability'. We refer to such a situation by saying that the conditions are *satisfied approximately*. We measure the closeness of this approximation by a fuzzy membership function, which will be used to *break ties* in the search tree when deciding to which node to branch next among several candidate nodes with the same minimal lower bound. We also use this fuzzy inference in heuristically searching for a good initial schedule. Ties among several nodes with the same best lower bound seem to occur very frequently in flow-shop scheduling problems, and different tie breaking rules tend to have a significant effect on the performance of the algorithms.

The fuzzy scheduling method has been proposed for a permutation flow-shop problem without release times in [4] and with release times in [5]. It yielded nearly optimal solutions for the initial schedule and it was also used very successfully as the tie breaker for branching decisions. Let

$$D^l(\sigma ij) = C_l(\sigma ij) - C_l(\sigma ji), \quad l = 1, 2, \ldots, m, \tag{13}$$
$$D^{m+1}(\sigma ij) = C_{max}(\sigma ij) - C_{max}(\sigma ji), \tag{14}$$

then the *fuzzy membership function* that represents the likelihood that job i precedes job j in the best completion of σ is given by

$$\mu_\sigma(i,j) = 0.5 - \frac{D(\sigma ij)}{2D_{max}(\sigma)}, \tag{15}$$

where $D(\sigma ij) = \sum_{l=1}^{m+1} \alpha_l D^l(\sigma ij), \alpha_1, \ldots, \alpha_{m+1} \in [0,1]$ with $\sum_{l=1}^{m+1} \alpha_l = 1$ are real numbers and $D_{max}(\sigma) = \max_{i,j} |D(\sigma ij)|$. Then the *likelihood* of job i dominating the remaining jobs after partial sequence σ is measured by

$$\mu_\sigma^*(i) = \min_{j \in S} \mu_\sigma(i,j), \tag{16}$$

and job i^* satisfying

$$\mu_\sigma^*(i^*) = \max_{i \in S} \mu_\sigma^*(i) \tag{17}$$

is identified as the job that should immediately follow σ.

The rule determining i^* this way is referred to as the *fuzzy rule* and the schedule derived by successively applying the fuzzy rule is referred to as the *fuzzy schedule*. To obtain a fuzzy schedule for the initial solution, we apply the fuzzy rule for the forward problem, which has release times but no delivery times, and so $\alpha_{m+1} = 0$.

3.2 Decomposition of the incumbent schedule

We generalize a dominance relation used by Tadei et.al. [20] for the $F2/r_i/C_{max}$ problem and apply it to the incumbent schedule of our branch-and-bound algorithm for problems with release times.

Proposition 2. *Given a complete sequence* $\sigma = (\sigma(1), \sigma(2), \ldots, \sigma(n))$, *a partial sequence* $\sigma^{k-1} = (\sigma(1), \sigma(2), \ldots, \sigma(k-1))$ *is optimal, if there exists a* k $(2 \leq k \leq n)$ *such that*

$$\min_{k \leq i \leq n} (r_{\sigma(i)} + \sum_{h=1}^{l} P_h(\sigma(i))) \geq C_{l+1}(\sigma(k-1)), \quad l = 0, 1, \ldots, m-1 \tag{18}$$

simultaneously hold.

Eqs.(18) mean that if no job $\sigma(i)$ ($k \le i \le n$) is available on any machine M_l ($1 \le l \le m$) before the completion time of the partial sequence σ^{k-1} on this machine, then the partial sequence σ^{k-1} can be fixed at the root node in the branch-and-bound algorithm. Applying this decomposition procedure with the largest possible k can drastically reduce the size of a problem if some of the release times are relatively large.

4 Branch-and-bound algorithm

In this section we describe the remaining basic components of the branch-and-bound algorithm used to solve $Fm//C_{max}$ and $Fm/r_j/C_{max}$ problems.

4.1 Upper bound and incumbent schedule

The upper bound is calculated at the root node of the search tree, after this the upper bound is evaluated only at leaf nodes. For $Fm//C_{max}$ the fuzzy schedule is used as the incumbent schedule to calculate the initial upper bound. For $Fm/r_j/C_{max}$ both the fuzzy schedule and the M_NEH (modified algorithm of Nawaz, Enscore and Ham [18]) schedule [5] are used to calculate upper bounds at the root node and the better one is selected.

Using the fuzzy schedule as incumbent for the decomposition procedure tends to fix the largest number of jobs. This is because the fuzzy rule always tries to schedule a job to be completed on each machine as soon as possible and so it tries to minimize the idling between jobs on each machine. Using the fuzzy schedule, we determine the largest k ($2 \le k \le n$) such that Eqs.(18) simultaneously hold, and fix $\sigma^{k-1} = (\sigma(1), \sigma(2), \ldots, \sigma(k-1))$ as the intial subsequence of an optimal schedule.

4.2 Branching rule and search strategy

For $Fm/r_j/C_{max}$ we use a variant of Potts' adaptive branching rule [19] that fixes jobs at both ends of the schedule. More precisely, each node of the search tree is represented by a pair of initial and final partial sequences (σ_1, σ_2). Subsequent branching from (σ_1, σ_2) will either be of *type 1* branching in which a job is added to the end of an initial partial sequence σ_1, or of *type 2* in which a job is added to the beginning of a final partial sequence σ_2. The branching types are fixed to be the same within a level of the tree and are fixed on the very first visit to a level k according to the following rule: branch in the direction of the fewest number of nodes with the minimum lower bound. Let n_1 and n_2 be the number of nodes with the minimum lower bound for type 1 and type 2 branching, respectively, at level k during our first visit. If $n_1 \le n_2$ the next branching is of type 1, while if $n_1 > n_2$ then the branching is of type 2.

A depth-first search strategy is used, which selects a node with the minimum lower bound among those most recently created, breaking ties by the fuzzy rule. For type 1 branching, we look at the forward problems (with release times if applicable) and select the next node among the tied ones by the fuzzy rule. For type 2 branching, we consider the reverse problem (with release times treated as delivery times) and break ties using the fuzzy rule for this problem.

5 Computational experiments

We evaluated the performance of the branch-and-bound algorithm through very extensive numerical experiments. For this purpose, the following input parameters were set up:

1. The processing times $P_l(j)$ ($l = 1, 2, ..., m$) were generated using uniformly distributed random integer numbers on [1,100].
2. The release times r_j were determined using uniform random integer numbers in the range $[0, n \cdot 50.5 \cdot R]$ for $R = 0.4, 0.8, 1.0, 1.2, 1.6$.
3. For weights $\alpha_i (i = 1, 2, \ldots, m)$ of the membership function $\mu_\sigma(i, j)$ in the fuzzy approximation (see Eq.(15)) the following three types of functions were applied to each problem instance, and the best schedule obtained was adopted as the fuzzy schedule:

    (a) arithmetic progression weights $\alpha_i = i/m(m-1)$ ($i = 1, 2, \cdots, m$);
    (b) equal weights $\alpha_i = 1/m$ ($i = 1, 2, \cdots, m$);
    (c) inverse arithmetic progression weights $\alpha_i = (m + 1 - i)/m(m-1)$ ($i = 1, 2, \cdots, m$).
    (Note that $\sum_{i=1}^{m} \alpha_i = 1$ in each case.)

    In the search strategy for breaking ties, we use equal weights, which means that $\alpha_i = 1/m$ ($i = 1, 2, \ldots, m$) for σ_1, as we look at the original problem, and $\alpha_i = 1/(m + 1)$ ($i = 1, 2, \ldots, m + 1$) for σ_2, because we consider the reverse problem with delivery times.
4. The CPU time was limited to 20 minutes, and a problem instance that could not be solved within this time limit was identified as unsolved.

We used the random number generator of Taillard [22] to generate the problem instances. This means that all our test problem instances are reproducable by running the problem generation program from the same seeds (the seeds used have been saved and are available from the authors on request). The branch-and-bound algorithm was coded in FORTRAN 77 and run on a DEC 3000 (35 MFLOPS) Workstation for $Fm//C_{max}$ and on a Sun Sparc5 Workstation for $Fm/r_j/C_{max}$.

Table 1. The percentage of $Fm||C_{\max}$ instances solved by the algorithms

	$m = 3$			$m = 4$		
n	A	A_1	A_2	A	A_1	A_2
10–20	95	93	92	93	92	90
30–40	97	97	93	83	70	70
50–60	97	97	92	88	58	58
70–80	97	97	90	90	65	65
90–100	95	87	83	95	75	75
110–120	100	90	85	88	55	53
130–140	98	93	83	88	63	62
150–160	98	97	88	95	67	67
170–180	100	92	78	93	70	70
190–200	98	87	73	90	72	70
Average	98	93	85	91	69	68

5.1 Computational results for $Fm//C_{max}$

To evaluate the performance of the algorithm in a large-scale computational study, we generated 30 problem instances for each pair(n, m), $n \in \{10, 20, \ldots, 100, 150, \ldots, 1000\}$ and $m \in \{3, 4\}$. The total number of problem instances tested was 6000.

To examine the effectiveness of fuzzy search, the following three algorithms were implemented and compared on instances with $n \leq 200$.

1. A: the branch-and-bound algorithm described here;
2. A_1: algorithm A with no fuzzy search (i.e., only using the traditional depth-first search);
3. A_2: algorithm A with no fuzzy search and no fuzzy initial schedule.

Table 1 gives the percentage of the problem instances solved by the three algorithms. It is evident that Algorithm A is superior to A_1 and A_2. The comparison of A and A_1 shows that using fuzzy search as a tie breaker leads to a much more effective search strategy than the traditional depth-first search. Algorithm A has solved a substantially larger percentage of the problem instances than A_1, especially for $m = 4$. This is due to the fact that there tend to be many nodes with the same lower bound at each level of the search tree and fuzzy search helps to find one leading to an optimal schedule. The comparison of A_1 and A_2 shows that using the fuzzy schedule for finding upper bounds also helps.

For details on how Algorithm A performed for larger problem instances with up to 1,000 jobs, we refer the reader to [4]. Overall, it has solved 99.4% of the problem instances with up to 1000 jobs for $m = 3$, and 97.4% of the instances with up to 900 jobs for $m = 4$ within the time limit. A benchmark test was also conducted using the benchmark problems presented by Taillard [22] and 40% of all the benchmark problem instances were solved.

Table 2. The number of $F3/r_j/C_{max}$ instances solved by the algorithms

n	Alg	$R = 0.4$	$R = 0.8$	$R = 1.0$	$R = 1.2$	$R = 1.6$
	A	50	50	50	50	50
20	A_1	45	49	45	48	49
	A_2	50	50	50	50	50
	A	50	45	42	47	50
40	A_1	47	44	31	40	49
	A_2	50	46	41	47	50
	A	50	43	41	46	50
60	A_1	48	39	36	36	47
	A_2	49	42	36	45	50
	A	50	46	37	40	50
80	A_1	49	43	32	38	48
	A_2	48	46	32	40	50
	A	50	43	32	38	50
100	A_1	48	43	29	36	48
	A_2	49	39	30	40	50
150	A	50	39	28	42	49
200	A	50	41	28	36	50

5.2 Computational results for $F3/r_j/C_{max}$

To evaluate the performance of the algorithm for $F3/r_j/C_{max}$, we generated 50 problem instances for each (n, R) pair, $n \in \{20, 40, 60, 80, 100, 150, 200\}$ and $R \in \{0.4, 0.8, 1.0, 1.2, 1.6\}$. The total number of problem instances tested was 1750.

The fuzzy schedule was better than the M_NEH schedule (in spite of the fact that the NEH algorithm has been accepted as the best heuristic for the classical permutation flow-shop problem [21]) in most cases and its makespan was usually within 2% of the optimum. The fuzzy schedule has also fixed a larger number of jobs by our decomposition rule.

To examine the relative performance of the branch-and-bound algorithm proposed here, the following three versions were implemented and compared.

1. A: the branch-and-bound algorithm described here;
2. A_1: algorithm A with no type 2 branching (i.e., only using type 1 branching and fuzzy search);
3. A_2: algorithm A without fuzzy search.

For each group of 50 instances, we report the number solved in Table 1. For more detail, we refer the reader to [5]. It is evident from Table 1 that Algorithm A is superior to algorithms A_1 and A_2 in terms of the number of problem instances solved. For this reason, we have tested only algorithm A for instances with $n = 150$ or 200. The comparison of algorithms A and

A_1 shows that building the schedule from both ends leads to a much more effective search strategy than using traditional forward search with type 1 branchings only. Comparing algorithms A and A_2 shows that A was able to solve substantially more instances than A_2 for some hard groups, e.g., $R = 1.0$, $n = 60, 80$. On average, A also tended to be significantly faster than A_2, usually requiring only a few seconds. This shows again that fuzzy search as a tie breaker helps in directing the search through the tree.

Table 1 also indicates that the difficulty in solving a problem depends much more on the value of R than on the number of jobs, and there are substantial differences in how hard problem instances are, depending on their R values. Problems with relatively small ($R \leq 0.4$) or relatively large ($R \geq 1.6$) R values are much easier to solve than those with intermediate R values. The most difficult problems are those with $R = 1.0$, which means that the range of release times is between 0 and the expected total processing time on the first machine. The closer R is to 1.0, the harder the problems are. For $R = 0.4$ and $R = 1.6$ algorithm A solved all of the test problem instances on up to 200 jobs except for one with $R = 1.6$ and $n = 150$. It was also very fast for all solved problems. For $R = 1.0$, the number of problem instances solved decreases with a clear trend as n increases. Overall, algorithm A has solved 89% of the problem instances generated.

5.3 Computational results for $Fm/r_j/C_{max}$, $m \geq 4$

Since Algorithm A proved to be the most effective for $m = 3$, we report here only on its performance. Again, we have randomly generated 50 instances for each (m, R) pair, $m = 4, 5, 6, 8$ and $R \in (0.4, 0.8, 1.0, 1.2, 1.6)$. We have also solved a large number of the two groups of 50 problem instances generated for $m = 10$ and $n \in \{15, 20\}$. The results summarized in Table 3 further support our surprising insight that problem difficulty depends more on the R value used than m and n. The difficult problems are those with $R \in [0.8, 1.2]$ and the most difficult ones were generated with $R = 1.0$. Although the effectiveness of Algorithm A decreases as m and/or n increase, we were still able to solve a relatively large number of large problem instances with $R \notin [0.8, 1.2]$. To the best of our knowledge, ours are the first tests where truly large-scale ($m \geq 4$ and $n \geq 40$) problems have been solved to optimum within reasonable time. For the instances where we have exceeded the 20 minutes CPU time, the algorithm has always found good solutions, which were on average within 5% of the optimum. The exact solution of the most difficult problems, however, still represents a challenge for further research.

6 Concluding remarks

This paper considered the permutation flow-shop scheduling problem with or without release times and described branch-and-bound algorithms for its

Table 3. The number of problem instances solved by Algorithm A

m	n	$R = 0.4$	$R = 0.8$	$R = 1.0$	$R = 1.2$	$R = 1.6$
4	20	50	50	48	49	50
	40	50	43	36	43	50
	60	50	40	23	37	48
	80	50	40	24	33	49
	100	50	29	20	31	50
5	20	50	50	50	50	50
	40	48	43	23	33	49
	60	43	34	20	32	47
	80	49	32	13	31	50
	100	49	25	17	28	49
6	20	49	49	50	49	50
	40	45	30	17	35	49
	60	39	18	10	18	47
	80	42	24	8	18	44
	100	42	12	3	19	45
8	20	44	45	45	43	49
	40	22	8	3	5	47
	60	17	6	4	2	47
	80	14	3	2	4	39
	100	21	3	1	4	39
10	15	50	50	50	50	50
	20	27	40	34	45	50

solution. Three different lower bounds were presented. Two dominance rules were generalized. One was used to formulate the fuzzy heuristic for scheduling and searching. The other one reduced the size of the problems to be solved by decomposition, especially with large release times. The adaptive branching rule narrowed the search tree while the fuzzy search strategy led the search to an optimal solution quickly. Large-scale computational results showed that the proposed branch-and-bound algorithms are very effective for the problems without release times or with relatively small or large release times. The most difficult problems were those where the range of the release times was around the expected makespan on the first machine. These problems still represent a serious challenge for algorithm designers and could be used as benchmarks in future research.

References

1. Achugbue, J. and Chin, F. (1982): Complexity and Solutions of Some Three-stage Flow Shop Scheduling Problems. Math. of Oper. Res. 7, 532-544.

2. Carlier, J., and Rebai, I. (1996): Two Branch and Bound Algorithms for the Permutation Flow Shop Problem. Euro. J. of Oper. Res. 90, 238-251.
3. Cheng, J., Kise, H. and Karuno, Y. (1997): Optimal Scheduling for an Automated m-Machine Flowshop, J. Oper. Res. Soc. Japan 40, 356-372.
4. Cheng, J., Kise, H. and Matsumoto, H. (1997): A Branch-and-bound Algorithm with Fuzzy Inference for a Permutation Flowshop Scheduling Problem. Euro. J. of Oper. Res. 96, 578-590.
5. Cheng, J., Steiner, G., and Stephenson, P. (2001): A Computational Study with a New Algorithm for the Three-machine Permutation Flow-shop Problem with Release Times. Euro. J. of Oper. Res. 130, 559-575.
6. Cheng, J., Steiner, G., and Stephenson, P. (2002): Fast Algorithms to Minimize the Makespan and the Maximum Lateness for the Two-machine Flow-shop Problem with Release Times. To appear in J. of Scheduling.
7. Cho, Y. and Sahni, S. (1981): Preemptive Scheduling of Independent Jobs with Release and Due Date Times on Open, Flow and Job Shops. Oper. Res. 29, 511-522.
8. Conway, R.W., Maxwell, W.D. and Miller, L.W. (1967): Theory of Scheduling. Addison-Wesley, Reading, Mass.
9. Garey, M.R., Johnson, D.S. and Sethi, R. (1976): The Complexity of Flowshop and Jobshop Scheduling. Math. of Oper. Res. 1, 117-129.
10. Grabowski, J. (1980): On Two-machine Scheduling with Release and Due Dates to Minimize Maximum Lateness. Opsearch 17, 133-154.
11. Grabowski, J., Skubalska, E. and Smutnicki, C. (1983) On Flow Shop Scheduling with Release and Due Dates to Minimize Maximum Lateness. J. of the Oper. Res. Soc. 34, 615-620.
12. Graham, R.L., Lawler, E.L., Lenstra, J.K. and Rinnooy Kan, A.H.G. (1979): Optimization and Approximation in Deterministic Sequencing and Scheduling Theory: a Survey. Ann. of Disc. Math. 5, 287-326.
13. Horn, W.A. (1974): Some Simple Scheduling Algorithms. Naval Res. Logistics Quarterly 21, 177-185.
14. Johnson, S.M. (1954): Optimal Two- and Three-stage Production Schedules with Setup Times Included. Naval Res. Logistics Quarterly 1, 61-68.
15. Koulamas, C. (1998): On the Complexity of Two-machine Flowshop Problems with Due Date Related Objectives. Euro. J. of Oper. Res. 106, 95-100.
16. Lageweg, B.J., Lenstra, J.K., and Rinnooy Kan, A.H.G.(1978): A General Bounding Scheme for the Permutation Flowshop Problem. Oper. Res. 26, 53-67.
17. Lenstra, J.K., Rinnooy Kan, A.H.G. and Bruker, P. (1977): Complexity of Machine Scheduling Problems. Ann. of Disc. Math. 1, 343-362.
18. Nawaz, M., Enscore, Jr. E., and Ham, I. (1983): A Heuristic Algorithm for the m-machine, n-job Flow-shop Sequencing Problem. OMEGA 11, 91-95.
19. Potts, C.N. (1980): An Adaptive Branching Rule for the Permutation Flow-shop Problem. Euro. J. of Oper. Res. 5, 19-25.
20. Tadei, R., Gupta, J.N.D., Della Croce, F. and Cortesi, M. (1998): Minimising Makespan in the Two-machine Flow-shop with Release Times. J. of the Oper. Res. Soc. 49, 77-85.
21. Taillard, E. (1990): Some Efficient Heuristic Methods for the Flow Shop Sequencing Problem. Euro. J. of Oper. Res. 47, 65-74.
22. Taillard, E. (1993): Benchmarks for Basic Scheduling Problems. Euro. J. of Oper. Res. **64**, 278-285.

A Fuzzy Adaptive Partitioning Algorithm (FAPA) for Global Optimization

Melek Basak Demirhan[1] and Linet Özdamar[2]

[1] Yeditepe University, Dept. of Systems Engineering, Kayisdagi, 81120 Istanbul, Turkey.
[2] Nanyang Technological University, School of Mechanical and Production Engineering, Systems and Engineering Management Division, 50 Nanyang Avenue, Singapore 639798

Abstract. Adaptive Partitioning Algorithms (APA) divide the feasible region into non-overlapping partitions (regions) in order to direct the search to the promising region(s) that are expected to contain the global optimum. APA usually collect data from pre-determined locations in each partition and use evaluation measures that are based on assumptions or function approximations. The proposed Fuzzy Adaptive Partitioning Algorithm (FAPA) is a novel approach that aims at locating the global optimum of multi-modal functions without using any assumptions or approximations. FAPA introduces two new features: it selects the locations of data randomly in each partition and it utilizes a fuzzy measure in assessing regions.

1 Introduction

In its general form, a global minimization problem is to locate the point $x^* \in S$ such that

$$f(x^*) \leq f(x), \ \forall x \in S \subset \Re^n \tag{1}$$

The algorithms proposed for global optimization (GO) problems can be classified referring to the data collection method they utilize. Deterministic approaches take samples from pre-specified locations assuming a certain model for $f(x)$ whereas probabilistic approaches select these locations randomly.

Adaptive Partitioning Algorithms (APA) take place among deterministic GO approaches and they seek the solution for the problem defined in (1) as follows:

The feasible region is partitioned into sub-regions in an iterative and adaptive manner so that the location of the global optimum is enclosed. Sub-regions are assessed using data (with predetermined locations) and selected to be re-partitioned in the next iteration according to their assessment values.

Interval estimation techniques are among the most widely used assessment techniques in APA. The interval estimation approach that is based on a model supposition of $f(x)$, assumes $f(x)$ to be a Lipshitzian function or a realization of a stochastic process. This prior model of $f(x)$ is integrated with sample information resulting in an adaptive posterior model of $f(x)$ [1,2]. Another interval estimation approach is based on inclusion functions [3–6] where the bounds on $f(x)$ within a specific interval are calculated by replacing the real operations in $f(x)$ by their pre-determined interval operations resulting in an interval on $f(x)$ for each operation. Then, when all the terms are aggregated, an enclosure of $f(x)$ is obtained. Thus, each partition is evaluated with the aid of an inclusion function defined on $f(x)$. There is a continuous effort to improve the convergence of these methods with different partitioning schemes [7] and accelerating devices that discard partitions [8,9].

The second category of solution approaches, namely, probabilistic search methods, include random search (RS), metaheuristics, e.g., simulated annealing (SA) [10,11], genetic algorithms (GA) [12] and clustering methods, e.g., Controlled Random Search [13,14], Multi Level Single Linkage [15], Topographical Optimization [16]. Among these, meta-heuristics are intelligent methods, guided by their inherent operators such as, the annealing temperature (SA), and reproduction and crossover operators (GA). Hence, they are sophisticated stochastic search methods. On the other hand, in clustering methods, the search space is subjected to uniform random search and number of steepest descent iterations are applied locally to every seed selected. GO solution approaches are reviewed extensively in [17–19].

2 Fuzzy Adaptive Partitioning Approach

FAPA is an Adaptive Partitioning Algorithm that aims at locating the global optimum of multi-modal functions without using any function approximations or assumptions. It incorporates stochastic search and operates on the principle of reducing the search space as other APA do. The search space (a hypercube denoted by $\alpha_0, \alpha_0 \in S$) is reduced by iterative re-partitioning and re-sampling in the most favorable (promising) sub-regions.

FAPA has three key features that make it more efficient:

- The approach of partitioning the search space and taking samples from each partition provides a more uniform sampling on α_0. The advantages of conducting a random search (RS) in a partitioned feasible space, rather than a non-partitioned one is demonstrated by [20]. Also, in [21] numerical results indicate that SA and GA embedded in FAPA achieve a much better performance as compared to their stand-alone versions.
- The iterative and adaptive assessment of regions, guides the search and intensifies it into sub-regions having high potential of enclosing the global optimum.
- The fuzzy assessment criteria utilized both in evaluating regions and in transforming the data provide more flexible and assumption-free measures.

Hence, FAPA provides an iterative global optimization approach that incorporates the convergence capabilities of APA [22] with the power of stochastic search and the novel fuzzy assessment criteria.

Definitions given below will be useful in subsequent discussions involving uncertainties in FAPA.

Definition 1: Let $\alpha(t)$ for $t \in I$, be a subset of $\Re^n$. A collection of sets

$$C = \{ \alpha_i(t+1) : t \in I \text{ and } i = 1, 2, ..., k \}$$

is called the *k-partition* of $\alpha(t)$ if,

$$\alpha(t) = \cup_{i=1..k} \alpha_i(t+1) \text{ and } \alpha_i(t+1) \cap \alpha_j(t+1)$$

are almost disjoint for $\forall\ i = 1, 2, ..., k$ and $j \neq i$ (i.e., they have no common points except their common boundaries). Here, t denotes the partitioning iteration index.

Definition 2: For any $h > 0$, the region, $I_h = \{x \in \alpha_0 : |f(x) - f(x^*)| < h\}$ is called the *best region* (of size h) and any solution $x \in I_h$ is called a *near optimum* solution and also denoted by x^*.

Definition 3: At iteration t, any that contains solutions $x^* \in I_h$ is called the *region containing the optimum* and is denoted by $\alpha^*(t)$.

2.1 Uncertainties in FAPA

There are two kinds of uncertainties involved in evaluating a set of sample points taken from a given sub-region. The first one is probabilistic in nature and is due to the random selection of sample locations. However the second one is related to the assessment of the sub-region (resulting in the selection of the most promising region(s) to be re-partitioned in the following iteration) and it is fuzzy.

Probabilistic Uncertainty

The purpose of FAPA is to increase the probability of finding a near optimal solution, x^*, in consecutive iterations. To achieve this goal, FAPA reduces the size of the search space by "re-partitioning" the selected (promising) sub-region. That is, it reduces the Lebesgue measure of any partition $\alpha_i(t+1)$ originating from a selected $\alpha_s(t)$ to $\alpha_s(t)/k$ due to the k-partitioning operation. Then, by carrying out "re-sampling" on $\alpha_i(t+1)$ it increases the probability of finding a near optimum solution, $p_s(t)$, provided that the k-partitioning takes place on $\alpha^*(t)$.

Hence it can be stated that if it were possible to identify $\alpha^*(t)$ with certainty, then, the only source of uncertainty would be probabilistic since the relation $\alpha_s(t) = \alpha^*(t)$ would be true and $p_s(t)$ would represent the total uncertainty. Unfortunately this is not the case and we also have to conceive methods of reducing the uncertainty related to the identification of $\alpha^*(t)$.

Fuzzy Uncertainty

FAPA utilizes a fuzzy approach in order to reduce the uncertainty related to the identification of $\alpha^*(t)$. The existing uncertainties are due to the incomplete information about the unknown behavior of $f(x)$ and include:

- The extreme values that $f(x)$ attains within each crisp set $\alpha_i(t)$,
- Range of $f(x)$ within the best region I_h,
- Boundaries of the best region I_h.

To reduce the uncertainty related to the identification of $\alpha^*(t)$ or to increase the probability of identifying $\alpha^*(t)$, expressed as, $P[\alpha_s(t) \cap I_h \neq \phi]$, FAPA proceeds as described below.

Definition 4: At any iteration t, consider the set of sample points $A_i(t)$ selected from $\alpha_i(t)$. For any sample point $x_{ij}(t) \in A_i(t)$, its *degree of membership* to the best region, I_h, is measured by a membership function $\mu_{ij}(t)$ that maps $x_{ij}(t)$ to the unit interval [0,1].

Although various choices of $\mu_{ij}(t)$ are possible [23], only a modified version of the Gaussian membership function (2) is considered for demonstration purpose.

$$\mu_{ij}(t) = exp(-[f(x_{ij}) - f(x^*(t))]^2)/\sigma^2(t) \qquad (2)$$

Here $f(x^*(t))$ denotes the best value obtained up to iteration t and $\sigma(t)$ is the standard deviation of the functional values in all samples gathered in the current iteration t.

Definition 5: The *potential* $r_i(t)$ of a sub-region $\alpha_i(t)$, is a mapping into the unit interval such that,

$$r_i(t) = r[\mu_{i1}(t), \mu_{i2}(t), \ldots, \mu_{im}(t)] \; for \; x_{ij} \in A_i(t) \qquad (3)$$

The potential is a global measure, it aggregates the sample information related to a region and it evaluates the potential that the region contains the global optimum.

Depending on the context of the problem, any monotone non-increasing function may serve as a potential for a minimization problem. Here, the expression in (4) is adopted for $r_i(t)$. This expression is adapted from [24].

$$r_i(t) = 1/m \sum_j \mu_{ij}(t) \; exp(1 - \mu_{ij}(t)) \qquad (4)$$

Furthermore, the potential of the complement of $\alpha_i(t)$, denoted by, $r'_i(t)$ is also defined as follows.

$$r'_i(t) = 1/m \sum_j (1 - \mu_{ij}(t)) \; exp(\mu_{ij}(t)) \qquad (5)$$

FAPA computes a fuzzy global measure for $\alpha_i(t)$ by aggregating membership values $\mu_{ij}(t)$ of each observation. Hence, it determines $\alpha_s(t)$ accordingly and contributes to the increase of $P[\alpha_s(t) \cap I_h \neq \phi]$ in a fuzzy manner.

Composite Uncertainty

Entropy measures estimate the average ambiguity in fuzzy sets [25]. The authors in [25] discuss composite measures which combine fuzzy and probabilistic uncertainties and provide arguments on the conformity of hybrid entropy measures (total uncertainty measures) to the five fuzzy measure axioms [26].

The hybrid entropy $H_i(t)$ given in (6) represents the total uncertainty related to $\alpha_i(t)$ and it is proposed in [24]. This expression is selected to demonstrate that FAPA's fuzzy assessment criterion is consistent with the convergence property of APA [22]. The convergence property of APA is based on using region selection operators that assign higher potentials to the regions having better function values.

$$H_i(t) = -p_i(t)\ log(r_i(t)) - (1 - p_i(t))\ log(r'_i(t)) \qquad (6)$$

If it can be shown that, the total uncertainty for a given region goes to zero in the limit using the potentials given in (4) and (5), then it can be concluded that FAPA is a convergent partitioning algorithm. Clearly, when $H_i(t)$ equals zero, there is no uncertainty in the assessment of the sub-region $\alpha_i(t)$.

Theorem. For any selected sub-region $\alpha_s(t)$ the total uncertainty related to that sub-region decreases as the number of consecutive partitioning iterations increases. That is, for any selected sub-region $\alpha_s(t)$ we have,

$$lim_{t \to \infty} H_s(t) = 0 \qquad (7)$$

If we only consider nested partitioning iterations then either $\alpha_s(t) \cap I_h \neq \phi$ or $\alpha_s(t) \cap I_h = \phi$. In the first case, where $\alpha_s(t) \cap I_h \neq \phi$, $(1 - p_s(t))$ decreases uniformly and becomes zero after t exceeds a certain value T, since the Lebesgue measure of a selected partition decreases as t increases and $p_s(t)$ converges to 1. Therefore, $H_s(t) = -p_s(t)\ log(r_s(t)) \forall t > T$. The potential, $r_s(t)$ goes to 1 as nested partitioning always results in $\alpha_s(t) = \alpha^*(t)$ and total uncertainty becomes zero. In the second case, $\alpha_s(t) \cap I_h = \phi$ implies $\alpha_s(t) \neq \alpha^*(t)$ and $p_s(t)$ is zero for all t, however $r'_s(t)$ goes to 1 as t increases. Hence in both cases (7) is satisfied. The proof for the general case, where the partitioning iterations do not necessarily take place in nested sub-regions, is given in [27].

Consequently, FAPA combines the tools of subdivision, probabilistic search and fuzzy logic to result in a powerful fuzzy set based search technique for GO. In Section 3, the computational results where FAPA and other various GO approaches are summarized.

3 Computational Results

The performance of FAPA is compared against statistical estimation and Bayesian interval estimation measures as well as clustering methods and SA.

In [28] an extensive performance comparison is provided both on basic methods and new hybrids. Here, only the basic approaches are discussed. Their descriptions and implementation details are found in the related reference. The pseudocode of FAPA is given in the Appendix.

3.1 Algorithms Used in the Experiments

Simulated Annealing Algorithms

Three basic SA algorithms are discussed here: Simulated Annealing (SA), Simulated Annealing with Local Search (SALS) [28] and Adaptive Simulated Annealing (ASA) [10]. In SA and SALS a neighbor solution is generated by selecting a dimension randomly and moving in decreasing or increasing direction in that dimension. The additional feature of SALS is that a local hill climbing procedure is applied at given intervals during the search. On the other hand, ASA generates neighbors by using an expression that depends on the annealing temperature corresponding to the dimension considered.

Clustering Methods

The clustering algorithms presented here are Topographical Optimization with Local Search (TOLS) [16], MLSL [15] and TMLSL [29]. The general idea behind TOLS and TMLSL is to achieve a cover of the feasible space as uniformly as possible. While samples are taken, locations which are too near to existing samples are rejected. All three algorithms select seeds for local search by identifying the samples that have better function values than their close neighbors. The difference between the algorithms lies in the definition of close neighbors.

Adaptive Partitioning Methods (APA)

The proposed Fuzzy Adaptive Partitioning Algorithm method and APA based on interval estimation are considered here. Further, an APA with a statistical measure also takes place in the experiments.

Fuzzy Adaptive Partitioning In three FAPA versions, FAPA-SA-I, FAPA-SA-II and FAPA-SA-III, evidence collection in sub-regions is accomplished using SALS. SALS starts from a random sample taken from the sub-region. The difference between the three FAPA versions is their partitioning scheme. In the first version, the bound of every dimension is bisected resulting in 2^n new sub-spaces. In the second version, the dimension with the longest bound is bisected, whereas in the third version, five out of n dimensions are selected according to descending bound length and bisected. The fourth FAPA version (FAPA-RS) executes Random Search to collect samples.

Adaptive Partitioning Algorithms Based on Interval Estimation
In [22] the properties of partition assessment operators that satisfy the convergence conditions for APA is described. Six of them are selected and embedded in the partitioning scheme of FAPA. Hence, the fuzzy selection criterion is replaced by these operators. While some of these operators are based on a Lipschitzian overestimate of $f(x^*)$ in $\alpha_i(t)$, others reflect the probability of finding an improved estimate of $f(x^*)$ on $\alpha_i(t)$. Thus, all partition measures considered in this class are based on a prior class of models for $f(x)$. The operators based on a Lipschitzian prior model of $f(x)$ are DPS [30] and STR [31]. KAB [32,33] assumes a Wiener process for $f(x)$, whereas MTZ [2], ZIL [34] and BOE [35] are based on a Gaussian model.

Adaptive Partitioning Algorithm with a Statistical Measure This measure (TNG) is proposed in [20] and it is based on the expected value of the largest improvement in $f(x^*(t))$ which is to take place in the considered sub-region in the next sampling iteration. This expected value is computed using the Pareto distribution whose parameters are estimated using $f(x_{ij})$, for all $x_{ij} \in A_i(t)$.

3.2 Details of the Experiments

The methods listed above are tested on a wide test bed of 74 test functions (up to 10 variables) collected from the literature. The properties of these functions and their references are found in [28].

All stochastic methods used in the experiments are terminated after the execution of *1000n* function evaluations. 20 independent runs are carried out for each test function each starting with a different random seed. For a fair comparison, the interval estimation based partitioning methods are permitted to run until *20x1000n* function evaluations are executed. In Table 1, solution quality is indicated in terms of the absolute deviation from the known global optimum. The average absolute deviation (standard deviation) and the maximum absolute deviation from global optima are indicated for all methods in Table 1.

3.3 Results

The results in Table 1 indicate that stochastic approaches outperform deterministic interval estimation approaches. Among the clustering methods, the original version of MLSL outperforms its topographical variant, TMLSL. On the other hand, TMLSL achieves better results as compared to topographical optimization, TOLS. The clear winners among stochastic optimization methods are SALS and ASA. When embedded in FAPA, the best SA approach, SALS, improves considerably, and becomes the best algorithm among all. The differences in the performance of FAPA-SA-I, FAPA-SA-II and FAPA-SA-III demonstrate the effects of the partitioning scheme in APA [7]. Other partitioning schemes may further enhance FAPA's performance. An interesting result is that when pure random search, RS, becomes the evidence collection tool in FAPA, it produces quite impressive results. These empirical results and those obtained previously [21] show that FAPA increases the efficiency of stochastic evidence collection techniques.

4 Conclusion

We propose a novel Adaptive Partitioning Algorithm (FAPA) with a fuzzy assessment measure that derives information from randomly collected samples. FAPA reduces the uncertainty involved in the search and converges in the limit. Numerical results show that FAPA outperforms deterministic adaptive partitioning algorithms and improves the convergence of the stochastic search algorithms that are embedded in it. Hence, FAPA integrates the convergence properties of partitioning methods with the power of stochastic search by utilizing fuzzy set theory. Furthermore, FAPA can be used as a black box approach, where, given a fixed number of samples in a site, it is required to characterize function behavior without any knowledge of the function involved.

Table 1. Results on 74 small to moderate size test functions up to 10 variables [28]. Results for TNG are taken from [36].

Method	SA	SALS	ASA	TOLS	MLSL	TMLSL
Avg	1.27	0.99	1.09	40.13	3.24	27.78
Stdev	3.81	3.06	3.22	316.69	8.40	209.82
Max	19.36	17.68	17.68	2745.17	46.88	1817.93
Method	DPS	STR	KAB	MTZ	ZIL	BOE
Avg	50.59	78.70	617.41	606.76	582.16	603.74
Stdev	326.60	426.40	4886.70	4886.80	4681.10	4487.00
Max	2640.10	3074.32	40008.4	40008.4	38325.6	40008.4
Method	FAPA SA-I	FAPA SA-II	FAPA SA-III	FAPA RS	TNG*	
Avg	1.061	1.123	0.47	7.70	22.58	
Stdev	3.22	3.403	2.03	46.55	164.4	
Max	17.68	17.68	16.80	400.89	646.70	

Appendix

Notation

$C(t)$: set of existing sub regions in iteration t. r^*: lower bound for $r_i(t)$ used in discarding regions.

Pseudocode of FAPA

Step 0. Initialization: Identify the hypercube α_0. Set $\alpha_s(0) = \alpha_0$, $C(0) = \alpha_0$, and let $t = 1$.

Step 1. Partitioning: Partition the sub-region $\alpha_s(t-1)$ into k sub-regions to form the set C. Let $C(t) = [C(t-1)\backslash\alpha_s(t-1)] \cup C$.

Step 2. Evidence collection: Take a random sample $A_i(t)$ from each sub-region $\alpha_i(t) \in C(t)$.

Step 3. Evaluation: Compute from each $A_i(t)$, the potential $r_i(t)$ of sub-region $\alpha_i(t)$ to contain x^*.

Step 4. Selection: $\alpha_s(t) = max_i r_i(t)$. Discard $\alpha_i(t)$: $r_i(t) < r^*$. Set $t = t+1$ and go to Step 1.

References

1. Mockus J.B. (1989) Bayesian Approach to Global Optimization. Kluwer Academic Publishers, Dordrecht
2. Mockus J.B., Thiesis V., Zilinskas A. (1978) The application of bayesian methods for seeking the extremum. In: Dixon L.C.W., Szego G.P. (eds), Towards Global Optimization. Vol. 2., North Holland, Amsterdam, 117-129
3. Moore, R.E., Ratschek, H. (1988) Inclusion functions and global optimization II. Mathematical Programming **41**, 341-356
4. Ratschek H., Rokne J. (1988) New Computer Methods for Global Optimization. Ellis Horwood, Chichester
5. Horst R., Tuy H. (1996) Global Optimization- Deterministic Approaches. Springer Verlag, 3rd Ed.
6. Kearfott R.B. (1996) Rigorous Global Search: Continuous Problems. Kluwer Academic Publishers, Dordrecht.
7. Csallner, A.E., Csendes, T., Markot, M.C. (2000) Multisection in interval branch and bound methods for global optimization I. Theoretical results. Journal of Global Optimization **16**, 371-392
8. Csendes, T., Pinter, J. (1993) The impact of accelerating tools on the interval subdivision algorithm for global optimization. European Journal of Operations Research **65**, 314-320.
9. Csendes, T. (2000) New subinterval selection criteria for interval global optimization. Abstracts of EURO XVII Conference, Budapest, 133
10. Ingber, L. (1989) Very fast simulated re-annealing. J. Mathematical Computer Modelling **12**, 967-973
11. Dekkers, A., Aarts, E. (1991) Global optimization and simulated annealing. Mathematical Programming **50**, 367-393.

12. Michalewicz Z. (1996) Genetic Algorithms + Data Structures = Evolution Programs. 3rd Edn,.Springer-Verlag, Berlin
13. Price, W.L. (1978) A controlled random search procedure for global optimization. In: Dixon L.C.W., Szegö G.P. (eds), Towards Global Optimization 2. North-Holland, Amsterdam
14. Ali, M.M., Törn, A., Viitanen, S. (1997) A numerical comparison of some modified controlled random search algorithms. Journal of Global Optimization **11**, 377-385
15. Kan Rinnooy, A.H.G., Timmer, G.T. (1984) Stochastic methods for global optimization. American J. of Mathematical Management Science 4,7-40
16. Törn, A., Viitanen, S. (1994) Topographical global optimization using pre-sampled points. Journal of Global Optimization **5**, 267-276
17. Pinter J. (1996) Global Optimization in Action. Kluwer Academic Publishers, Dordrecht
18. Bomze I.M., Csendes T., Horst R., Pardalos P.M. (1997) Developments in Global Optimization. Kluwer Academic Publishers, Dordrecht
19. Pardalos P.M., Romeijn E. (eds) (2001) Handbook of Global Optimization - Volume 2: Heuristic Approaches. Kluwer Academic Publishers, Dordrecht
20. Tang, Z.B. (1994) Adaptive partitioned random search to global optimization. IEEE Transactions on Automatic Control **39**, 2235-2244
21. Demirhan, M., Özdamar, L., Helvacioglu, L., Birbil, S.I. (1999) FRACTOP: A Geometric partitioning metaheuristic for global optimization. Journal of Global Optimization **14**, 415-436
22. Pinter, J. (1992) Convergence qualification of adaptive partitioning algorithms in global optimization. Mathematical Programming **56**, 343-360
23. Ross, T.J. (1995) Fuzzy Logic with Engineering Applications. McGraw-Hill, New York.
24. Pal, N.R., Pal, S.K. (1989) Object-background segmentation using new definitions of entropy. IEE Proceedings **136**, part E, 284-295
25. Pal, N.R., Bezdek, J.C. (1994) Measuring fuzzy uncertainty. IEEE Transactions on Fuzzy Systems **2**, 107-118
26. Ebanks, B.R. (1983) On measures of fuzziness and their representations. J. of Math Anal. and Appl. **94**, 24-37
27. Demirhan, M., Özdamar, L. (1999) A note on the use of a fuzzy approach in adaptive partitioning algorithms for global optimization. IEEE Transactions on Fuzzy Systems **7**, 468-475
28. Özdamar, L., Demirhan, M. (2000) Experiments with new probabilistic search methods in global optimization. Computers and Operations Research **27**, 841-865
29. Ali, M.M., Storey, C. (1994) Topographical multi level single linkage. Journal of Global Optimization **5**, 349-358
30. Danilin, Y.M., Piyavskii, S.A. (1967) On an algorithm for finding the absolute minimum. In: Theory of Optimal Solutions. Institute of Cybernetics, Kiev, 25-37
31. Strongin R.G. (1978) Numerical Methods for Multiextremal Problems. Nauka, Moscow
32. Archetti, F., Betro, B. (1979) A probabilistic algorithm for global optimization. Calcolo **16**, 335-343
33. Kushner, H.J. (1964) A new method of locating the maximum point of an arbitrary multi-peak curve in the presence of noise. Transactions of ASME, Series D, Journal of Basic Engineering **86**, 97-105

34. Zilinskas, A. (1981) Two algorithms for one-dimensional multimodal minimization. Optimization **12**, 53-63
35. Boender, C.G.E. (1984) The generalized multinomial distribution: A Bayesian analysis and applications. Ph.D. Dissertation, Erasmus University, Rotterdam
36. Demirhan, M., Özdamar, L., (2000) A note on a partitioning algorithm for global optimization with reference to Z.B.Tang's statistical promise measure. IEEE Transactions on Automatic Control **45**, 510-515

Fuzzy Memes in Multimeme Algorithms: a Fuzzy-Evolutionary Hybrid

Natalio Krasnogor[1] and David A. Pelta[2]

[1] Computational Biophysics and Chemistry Group, School of Chemistry and
   Automatic Scheduling, Optimisation and Planning Group
   School of Computer Science and IT
   University of Nottingham, U.K.
[2] Departamento de Ciencias de la Computación e Inteligencia Artificial
   Universidad de Granada, Spain

Abstract. In this chapter we propose a robust Fuzzy-Evolutionary hybrid. Our approach successfully integrates the so called Fuzzy Adaptive Neighborhood Search with Multimeme Algorithms, blending together the expressiveness of fuzzy sets based optimization with the robustness of evolutionary search. We exemplify our approach on a hard combinatorial problem called Protein Structure Prediction.

1 Introduction

Fuzzy Adaptive Neighborhood Search (*FANS*) was introduced in [24,6]. Building upon local search, a classical method often used in optimization and operational research, and some basic elements of Fuzzy Sets theory, *FANS* was shown to be a robust optimization tool. This was noted for a variety of domains like knapsack problems [6], real function minimization [24] and more recently [25,23,13] in the protein structure prediction problem. See [26] for a review of these applications.

In our previous work, *FANS* was compared against a genetic algorithm and it was verified that both algorithms have similar performance for the range of problems studied. However, one of the advantages of using *FANS* instead of using a Genetic Algorithm is that the implementation and tuning of the former is simpler than the one required for the later. On the other hand, *FANS* performs its search by sampling one solution at a time which in some cases can be a drawback not shared with the Genetic Algorithm as the later keeps a population of solutions.

In this chapter we hybridize a Multimeme Algorithm[18,13] with a simplified version of *FANS* in order to implement the pool of local searchers that the Multimeme algorithm will use. We will demonstrate how *FANS*, and in turn fuzzy sets and systems ideas, can be used to design a wide range of memes' behaviors. Moreover, we will show some benefits of using our Fuzzy-Evolutionary hybrid to tackle the Protein Structure Prediction problem.

The problem of predicting the three-dimensional structure of a protein from its amino acid sequence, is one of the most important open problems that computational biology faces today even after more than 30 years of research. In words of John Maynard Smith [30]:

> " Although we understand how genes specify the sequence of amino acids in a protein, there remains the problem of how the one dimensional string of amino acids folds up to form a three-dimensional protein... it would be extremely useful to be able to deduce the three-dimensional form of a protein from the base sequence of the genes coding for it; but this is still beyond us."

Because "all-atom" simulations are extremely expensive, researchers often resort to simplified models of the problems. We use here one of such models.

The chapter is organized as follows: in section 2 the protein structure prediction problem is introduced. Then in section 3 a brief descriptions of Memetic and Multimeme algorithms are presented. The reader may find in another chapter of this volume a complete description of *FANS*. Consequently, in this chapter we will concentrate only on *tailoring FANS* to the Protein Structure Prediction Problem. The hybrid approach we propose, i.e. a Multimeme Algorithm that includes *FANS* as local searchers, is described in Section 4. In order to assess the usefulness of the approach, several computational experiments were performed. Their description appears in Section 5 and the results discussed there. Finally, Section 6 is devoted to conclusions.

2 The Protein Structure Prediction Problem

Modern molecular biology has made it possible to determine and store huge amounts of information about DNA sequences, including the amino acid sequences of a diverse set of proteins. Analyzing these proteins is a challenging, but crucial task because proteins play two important yet complementary roles: they are the "building blocks" and "architects" of life. A protein's structure determines its biological function, as a consequence, the problem of determining that structure from the sequence of amino acids alone becomes a central component of proteomics analysis. Solving this problem will facilitate the design of new products such as diagnostic tests, drugs, etc, and will also help in the correct understanding of gene expression with the ultimate goal of enabling gene therapy.

A protein is a chain of amino acid residues that folds into a specific *native* tertiary structure under certain physiological conditions. Proteins unfold when folding conditions provided by the environment are disrupted, and many proteins spontaneously re-fold to their native structures when physiological conditions are restored. This observation is the basis for the belief that prediction of the native structure of a protein can be done *computationally* from the information contained in the amino acid sequence alone.

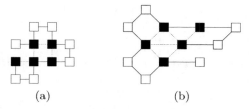

Fig. 1. HP sequence embedded in the square lattice and triangular lattice.

The Protein Structure Prediction Problem, predicting the native structure of a protein from its amino acid sequence, is one the most important open problems that computational biology faces today even after more than 30 years of intense research.

Because "all-atom" simulations are extremely expensive researchers, often resort to simplified models of the problem. In this chapter we will demonstrate the use of our metaheuristic with instances, i.e. proteins' sequences, of one of such models: the HP model [8].

HP models abstract the hydrophobic interaction process in protein folding by reducing a protein to a heteropolymer that represents a predetermined pattern of hydrophobicity in the protein; non-polar amino acids are classified as hydrophobes and polar amino acids are classified as hydrophilics. A sequence is $s \in \{H, P\}^+$, where H represents a hydrophobic amino acid and P represents a hydrophilic amino acid.

The HP model restricts the space of conformations to self-avoiding paths on a lattice in which vertices are labeled by the amino acids. The energy potential in the HP model reflects the fact that hydrophobic amino acids have a propensity to form a hydrophobic core. To capture this feature of protein structures, the HP model adds a value ϵ for every pair of hydrophobes that form a topological contact; a topological contact is formed by a pair of amino acids that are adjacent on the lattice and not consecutive in the sequence. The value of ϵ is typically taken to be -1.

Figure 1 shows a sequence embedded in the square and the triangular lattice, with hydrophobic-hydrophobic contacts (HH contacts) highlighted with dotted lines. The conformation in Fig. 1(a) embedded in a square lattice, has an energy of -4, while the embedding in the triangular lattice (b) has an energy of -6 (there are 4 and 6 dotted lines, i.e. contacts, in the figure).

The particular version of the problem that we are going to tackle in this chapter is given by:

Maximum Protein Structure Prediction
Instance: A protein, i.e. a string over the alphabet $\{H, P\}$ ($s \in \{H, P\}^*$).
Solution: A self avoiding embedding of s into a two dimensional square lattice.
Measure: The number of Hs that are topological neighbors in the embedding (neighbors in the lattice but not consecutive in s)

Protein structure prediction has been shown to be NP-complete for a variety of simple lattice models (see Atkins and Hart [1] for a recent review), including the version on the square lattice [7] and cubic lattice [3] of the HP model. A wide variety of global optimization techniques have been applied to modelizations of the problem, e.g. see the papers in Biegler et.al. [4], Pardalos, Shalloway and Xue [21]) and Pelta et.al. [25]. Genetic algorithms, and their variations, have proven as particularly robust and effective global optimization techniques. In particular, evolutionary methods have been used by several researchers [29,32,31,22,12,19,14,15,27,28,9] engaged in proteomics related activities.

3 Memetic Algorithms

Memetic Algorithms are metaheuristics designed to find solutions to complex and difficult optimization problems. They are evolutionary algorithms that include a stage of individual optimization or learning as part of their search strategy. Memetic Algorithms are also called hybrid genetic algorithms, genetic local search, etc. A simple Memetic Algorithm scheme is shown in Fig. 2.

The inclusion of a local search stage into the traditional evolutionary cycle of crossover-mutation-selection is not a naive or a minor change of the evolutionary algorithm architecture, on the contrary, it is a crucial deviation that affects how local and global search is performed. The reader should also note that the pseudocode shown in Fig. 2 is just one possible way to hybridize a genetic algorithm with local search. In fact, a great number of distinct memetic algorithms' architectures has been presented in the literature and even integrated into a formal model [13],[16].

One step beyond memetic algorithms are the so called MultiMeme Algorithms (*MMA*) as introduced in [18,13]. *MMA* may be regarded as memetic algorithms where several types of local searchers, called memes, are available to the evolutionary process during the local optimization phase.

An individual in a *MMA* is composed of a genetic part representing the solution to the problem being solved, and a memetic part encoding a meme, that is a local searcher, to employ during the individual optimization stage.

The set of memes available to the algorithm is called the *memepool* and its design is a critical aspect for the success of the metaheuristic. Several

```
Memetic_Algorithm():
Begin
  t = 0;
  /* We put the evolutionary clock (generations), to null */
  Randomly generate an initial population P(t);
  Repeat Until ( Termination Criterion Fulfilled ) Do
    Compute the fitness f(p) ∀p ∈ P(t) ;
    Accordingly to f(p) choose a subset of P(T), store them in M(t);
    Recombine and variate individuals in M(t), store result in M'(t);
    Improve_by_local_search( M'(t));
    Compute the fitness f(p) ∀p ∈ M'(t) ;
    Generate P(t + 1) selecting some individuals from P(t) and M'(t);
    t = t + 1;
  endDo
  Return best p ∈ P(t − 1);
End.
```

Fig. 2. A basic version of a memetic algorithm.

design criteria for the memepool are described in [13]. In general we can say that two design factors must be carefully addressed:

1. **the definition of memes:** which is related with the design of the appropriate local searchers (memes), that are going to be available in the memepool. The more diverse and robust the memepool, the more robust the local search stage is going to be.
2. **the scheduling of memes:** which is related to which meme to use, when to use it, to which point in space the meme must be applied, with how much intensity (i.e. is the local search executed until a local optimum is found?), etc. All of these issues has been studied elsewhere [10,20,13].

4 Fuzzy Memes for Multimeme Algorithms

As it was mentioned before, memes, as local search operators, are one of the key elements required for a successful application of *MMA*. Memes perform improvement steps where the selection or transition between the alternative solutions the local searcher is considering is decided in terms of cost; this decision often takes the form of a crisp criterion.

This criterion could be linguistically defined as: *accept the new solution if it is good enough for the established criterion*. We distinguish two main elements here: the *good enough*, and the *criterion* itself. In general, the most widely used definition for the criterion is *cost* and the *good enough* expression is usually taken as *lower than* or *greater than*. Thus, in the realm of local search (as an isolated heuristic or as part of a Multimeme Algorithm) this is translated to: *accept the neighbor solution if its objective function value is lower/greater than the current solution*.

FANS, and in turn fuzzy sets, aids to empower the local search performed by the memes in an elegant and compact manner. This is achieved by the use of rules that not only allow transitions to "Better" solutions but also to "Different" solutions, "Similar" solutions, "Potentially good" solutions, or any other criteria which can be linguistically defined by means of fuzzy sets. Such rules are easily represented in *FANS* by adequate combinations of the fuzzy valuation and the neighborhood scheduler. Each rule, representing a different behavior of *FANS*, will be mapped to a meme.

In this section we describe the core of our proposal: a metaheuristic where the "macro-architecture" is implemented as a Multimeme Algorithm while the "micro-architecture", that is the memes the *MMA* uses, is based on *FANS*. Such memes will be called *Fuzzy memes*.

We will first describe a stand-alone implementation of *FANS* for the protein structure prediction problem which will serve us for two objectives: as a baseline for algorithmic comparison purposes, and as a template for the design of memes.

4.1   *FANS* Implementation

In order to apply *FANS* to a particular problem, one needs to provide definitions for the *fuzzy valuation*, the *neighborhood scheduler*, the *operator scheduler* and the *basic operator moves*. The proposed definitions are described below.

Being S the set of solutions, i.e. candidate protein structures, the fuzzy valuation reflects the idea of "Solution Acceptability" which is based on the energy of a given solution. In this way, solutions that are neighbors (by means of the basic move operator) of the current solution and that improve over the cost of the later will have the highest degree of acceptability. Those neighbor solutions with slightly worse values than the current solution will be considered as acceptable but in a lower degree. Finally, neighbor solutions of bad quality (i.e. they are much worse than the current) will not be acceptable, i.e. their membership degree to the set of acceptable solutions will be zero. Thus, given the objective function f, the current solution $s \in S$, a new solution q obtained through the application of a basic move operator to s, the following membership function can be defined:

$$\mu(q,s) = \begin{cases} 0.0 & \text{if } f(q) < \beta \\ (f(q) - \beta)/(f(s) - \beta) & \text{if } \beta \leq f(q) \leq f(s) \\ 1.0 & \text{if } f(q) > f(s) \end{cases}$$

where β is a threshold specifying what is, and what is not, considered an acceptable deterioration in solution quality.

Given that the energy of a structure can take negative values (e.g. when the structure is not self-avoiding), the parameter β has two definitions[1]: when $f > 0$ then $\beta = f * 0.5$ (a deterioration in cost of 50% is allowed); when $f < 0$ then $\beta = f * 1.2$ (a deterioration in cost of 20% is allowed).

The neighborhood scheduler implements a first improvement strategy in terms of "Acceptability". That is, as soon as a neighbor solution deemed acceptable by the valuation function is found it will be accepted as the next current solution. Using at most 450 trials ($maxTrials = 450$), the scheduler tries to find an acceptable solution sampling the search space with the basic operators. The minimum level of acceptability is set to $\lambda = 0.9$, which implies that a candidate solution q will be deemed acceptable with respect to s if $\mu(q, s) \geq 0.9$. Four operators to generate the search space are used:

0. $Reflex(i, k)$: This operator reflects the protein structure across one of its symmetry axes. The change takes place between residues i and $i + k$.
1. $Shuffle(i, k)$: This operator performs a random re-positioning of the residues i^{th} to $(i + k)^{th}$.
2. $Stretch(i, k)$: The stretch operator unfolds a substructure of length k starting from residue i.
3. $Pivot(k)$: The pivot operator represents a rigid rotation. In this case, k random residues are selected and rigid rotations are performed sequentially on each one of them.

In general, it is not trivial to define and justify a particular order of application for the operators along the runs. To overcome this problem, the previous basic moves are encapsulated in a macro operator in such a way that each time this macro operator is called by the neighborhood scheduler, any sub operator (i.e. operators 0,1,2 or 3) will be randomly selected and applied.

The reader must note that this problem will not be present in the *MMA* as the scheduling of memes is handled automatically.

The value of the parameter k, which is initialized to 7 in this paper, specifies the radius of action of the basic operator. k is decremented by 1 each time the operator scheduler is called. Consequently, a broad search is performed at the beginning of the run and subsequently it is narrowed and focused on promising areas of the search space.

When the neighborhood scheduler fails to retrieve an acceptable solution for $k = 1$ then a cycle or optimization trajectory is finished. At this point, the value of k is reset to 7. Then, in order to continue the search from a different point in the search space, a new solution is generated and *FANS* is restarted from this newly created solution.

---

[1] Although in Protein Structure Prediction one tries to minimize the energy of the conformation, in this chapter we recast the problem to a maximization one by simply multiplying the energies by -1.

4.2 Multimeme Algorithm Implementation

We describe next the Multimeme Algorithm.

The memes employed by the Multimeme Algorithm are based on a simplified version of the *FANS* algorithm described before. The difference between *FANS* (as a stand-alone optimization heuristic) and when it is used as a meme, is found in the specifics of the operator used and the λ value employed to determine the minimum level of acceptability.

Each meme is identified by a 2-tuple:

$$\{< basic\ operator >< value\ of\ \lambda >\}. \tag{1}$$

where *basic operator* is one of the basic four operators described in 4.1 and $\lambda \in \{0.4, 0.8, 1.0\}$. Every combination maps to a meme which implies that the memepool will have 12 different memes (i.e. 12 pairs).

Each value of λ induces a particular behavior for the meme. Memes using $\lambda = 1$ will perform hill climbing with the corresponding operator, i.e. only augmenting paths are considered, while those using $\lambda < 1$, will accept non-improving moves. As $\lambda \to 0$, the meme may accept rather bad solutions, inducing a diversification behavior.

The value for the parameter k (the radius) of the operator is kept by each individual and it is inherited by its offsprings following the simple inheritance mechanism that is described below. The value, as in *FANS*, is modified by the operator scheduler following the same adaptation scheme presented before. Thus, each meme performs the search in the same way as *FANS* except that it has a fixed operator and take the value of k from the individual to be optimized.

To address the scheduling of memes, that is, which meme will be used in which individual and at what stage of the search, the algorithm uses the so called *Simple Inheritance Mechanism(SIM)*[18]. During the crossover stage, an offspring receives a meme (local searcher) L if such meme is shared by both of its ancestors. If they have different memes, then the fittest ancestor will pass its meme onto the offspring. Otherwise, if the ancestors have comparable fitnesses but each holds a different meme, a random selection will be done between their memes.

As we previously said, individuals will have access to a set of memes that represent different search strategies and then the simple inheritance mechanism will ensure that memes that are useful will be selected and spread in the population alongside the genes they help to improve.

As an additional mechanism of innovation, the meme inherited by an individual can be overridden (and the individual assigned a new meme chosen at random from the memepool) with certain probability. This probability is governed by the *innovation rate*. An innovation rate bigger than zero ensures the continuation of the search from a new neighborhood in those cases where the search is stagnated in a local optimum.

Table 1. HP model test Instances for the 2D Square Lattice.

Instance	Sequence	Opt	Length
I 1	PPHPPHHPPHHPPPPPHHHHHHHH HHPPPPPPHHPPHHPPHPPHHHHH	-22	48
I 2	HHPHPHPHPHHHHPHPPPHPPPHPP PPHPPPHPPPHPHHHHPHPHPHPHH	-21	50

5 Experiments setup and results

The blending of fuzzy memes within a multimeme approach gives raise to a novel metaheuristic of general application for which encouraging results have been obtained. We will presented below (for the two dimensional square HP model of the protein structure prediction problem) some of them. The candidate structures will be represented using the relative encoding as recommended in [14] and [25].

The experiments were done with the instances shown in Table 1. As mentioned before, *FANS* is used as a base line comparison; we perform 30 runs for each instance, each one ending when 200 cycles or optimization trajectories were completed.

In the case of the *MMA*, we perform 8 experiments per instance which arise from the combination of the following factors:

1. Replacement Strategy: $(\mu = 350, \lambda = 350), (\mu = 350 + \lambda = 350)$.
2. Depth of the local search, i.e. number of iterations performed by each meme application: 1,3.
3. Length of the local search, i.e. max. number of trials allocated in the neighborhood scheduler: $n/6, n/2$, with n the length of the sequence.

The first factor corresponds to the *MMA* itself while the last two are options for the memes. For each one of the eight combinations of parameters we perform 30 runs of the *MMA*. Each run was allocated 200 generations. The initial population was generated randomly and consisted of 350 individuals. Two-point crossover and two(consecutive)-point mutations were employed with 0.8 and 0.2 probability respectively. In the case of mutation the probability was per individual. The innovation rate was set to $IR = 0.2$. Tournament selection was used to select mating parents and a tournament size of 2 individuals was used.

Two variables are recorded at the end of every run: *bestF* which is the fitness of the best solution found, and *e2b* which is the number of fitness evaluations done to obtain the best solution. Tables 2 and 3 shows the average, standard deviation and minimum and maximum values over 30 runs for the variables *bestF* and *e2b*.

A given Multimeme algorithm was named as $iXdYtWS$ where the X can be either 1 or 2 accordingly to the instance being solved, Y can be either

Table 2. Statistics for instance 1

	bestF				e2b			
Algorithm	Avg	Sd	Min	Max	Avg	Sd	Min	Max
i1d1t6(350,350)	18.11	1.30	16.01	21.02	249432	105089	116404	444850
i1d1t6(350+350)	17.01	1.76	13.01	21.02	95556	88273	33240	485277
i1d1t2(350,350)	18.61	1.16	16.01	21.02	460148	211314	182490	928516
i1d1t2(350+350)	17.35	1.59	14.01	20.02	159085	199422	22519	1191330
i1d3t6(350,350)	18.65	1.56	16.01	22.02	704104	276819	270578	1245600
i1d3t6(350+350)	18.15	1.81	15.01	21.02	286994	363372	52399	1590160
i1d3t2(350,350)	19.48	1.36	17.02	22.02	1734563	743426	495304	3063660
i1d3t2(350+350)	18.82	1.24	16.02	21.02	1260605	1448553	142877	4876390
FANS	18.58	0.77	17.01	20.02	376614	235594	10501	732125

Table 3. Statistics for instance 2

	bestF				e2b			
Algorithm	Avg	Sd	Min	Max	Avg	Sd	Min	Max
i2d1t6(350,350)	18.18	1.23	16.01	20.01	225175	90445	82527	427009
i2d1t6(350+350)	15.71	1.66	13.01	19.01	91408	88494	27724	419319
i2d1t2(350,350)	18.28	1.36	16.01	21.01	438950	246878	136491	956606
i2d1t2(350+350)	16.11	1.54	13.01	20.01	227415	292371	36983	1676360
i2d3t6(350,350)	18.71	1.34	16.01	21.01	623300	281631	157267	1159780
i2d3t6(350+350)	16.91	1.06	14.01	19.01	257192	320362	73709	1467830
i2d3t2(350,350)	19.68	1.24	17.01	21.01	1508471	503728	793757	2878740
i2d3t2(350+350)	17.81	1.27	15.01	21.01	730525	842957	145596	3746700
FANS	18.31	0.70	17.01	20.01	501269	377920	7365	1017960

1 or 3 accordingly to the number of iterations allowed for the memes, W represents the maximum number of trials allocated to each iteration and can be either $n/2$ or $n/6$ (n is the size of the sequence). Finally S represents the replacement strategy; a (μ, λ) strategy or a $(\mu+\lambda)$. In both cases $\mu = \lambda = 350$.

In terms of *bestF*, (350,350) schemes achieved better values than those using (350+350) and, in almost all cases, with a smaller standard deviation. In terms of *e2b* the situation reverses: on average (350+350) schemes used fewer evaluations than those with (350,350). This is something expected since the (350+350) strategy is more exploitative than exploratory, producing a faster convergence although to worst solutions.

It is also interesting to note the large range of *e2b* values among the different algorithms and how these values are related with *bestF*. Although it can be consider relatively "cheap" to obtain rather good solutions, the effort required to obtain high quality solutions increases considerably. For example, the best algorithm for both instances is d3t2(350,350) which on average uses 1.5 millions of energy evaluations to obtain the best individual, but runs with about 3 millions also occurred (see columns Max). This is an observation that

Table 4. Number of runs (out of 30) that achieved a given energy value.

Algorithm	Energy Instance 1					Energy Instance 2				
	≤ 18	19	20	21	22	≤ 17	18	19	20	21
FANS	16	10	4			1	22	4	3	
d1t6(350+350)	25	2	2	1		27		3		
d1t6(350,350)	22	3	3	2		9	9	7	5	
d1t2(350+350)	25	2	3			25	4		1	
d1t2(350,350)	12	12	5	1		8	11	5	4	2
d3t6(350+350)	19	4	2	5		24	4	2		
d3t6(350,350)	15	6	5	3	1	8	3	10	7	2
d3t2(350+350)	11	10	7	2		12	10	6	1	1
d3t2(350,350)	10	4	9	5	2	1	6	5	8	10

can have an important impact in other models of protein structure prediction, e.g. the Functional Model Proteins [11,5], where an energy gap between the native state of the protein and its first excited state exists. The tables also suggest that *FANS* by itself represents a good compromise between quality and effort of search.

Table 4 shows the number of runs (out of 30 per algorithm and instance) that achieved a given energy value. Several versions of the *MMA* achieved optimal configurations for instance 2, namely $i2d1t2(350, 350)$, $i2d3t6(350, 350)$, $i2d3t2(350, 350)$, and $i2d3t2(350 + 350)$. It is evident that the *MMA* benefits from either a deep local search ($d3$) or a large number of trials for its memes ($t2$). There was no case when the shorter local searchers ($d1t6$) reported optimal results for instance 2.

For instance 1, just two algorithmic combinations were capable of achieving optimal structural conformations: $i1d3t6(350, 350)$ and $i1d3t2(350, 350)$ both with a $(350, 350)$ strategy and allowing a large number of iterations ($d3$) for the memes. It can also be concluded that instance 1 is harder than 2: this fact becomes clear if we add up the last 2 columns of each instance. The number of runs that ended in the optimum or the best sub-optimum is 22 for instance 1 while it is 40 for instance 2.

FANS acting alone achieved only 20 bonds for both instances although it used just a fraction of the energy evaluations required by the *MMA*. For the relevant energy values, i.e. the optimum minus one or two contacts, the *MMA* using *FANS* based memes achieves such values more frequently than *FANS* alone.

In Fig. 3 we show the evolution of protein structures' energy as a function of generations. Instance 1 is plotted in (a) and instance 2 in (b). In the case of instance 1 we see that the algorithms that reach the better (highest) energy values are $i1d3t2(350, 350)$ and $i1d3t2(350 + 350)$. In this case they are the ones that employ the deepest search (3 iterations) and the highest number of trials ($n/2$ trials) per local search phase.

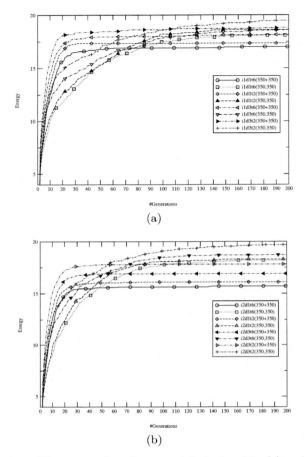

Fig. 3. Evolution of fitness vs. time. Instance 1 is depicted in (a) and instance 2 in (b). Several algorithms compared.

The worst algorithmic combinations are those using just one local search iteration and a very aggressive replacement strategy: $i1d1t6(350+350)$ and $i1d1t2(350+350)$. In general one can observe that for instance 1 the $(350, 350)$ strategy achieves better final values, although it takes longer to converge.

The graph for instance 2, Fig. 3 (b), presents similar general trends. However, in this case the best algorithmic combinations are $i2d3t2(350, 350)$ and $i2d3t6(350, 350)$. It is clear from these graphs that in general it is better to have a moderate selection pressure $(350, 350)$ rather than a large one. It is also important to note that the more relevant parameter governing the computational effort of the local search is the number of iterations rather than the number of trials per iterations. That is, the algorithms benefits more from sampling points that are deep within a basin of attraction (several iterations of local search) than from sampling points on the outskirts of several basins of attractions (several trials per local search).

5.1 Inner Workings of the Multimeme Algorithm

In order to analyze how the *MMA* performs its search we can look at the concentration of memes or their evolutionary activity [2]. This analysis has been done before for memetic algorithms in [13,18].

The concentration c of meme i at time t, c_i^t, is the number of individuals that carry meme i in the population at time (generation) t. The value of c_i^t is a measure of how much the Multimeme Algorithm is employing meme i at that time, hence plotting the evolution of this values for the different memes can give us insights on how an effective search is produced. The concentration can vary from 0 (i.e. no individual is using this meme at time t) to 350 (i.e every individual in the population is using the meme). The values are re-scaled here to the range $[0, 100]$, so $\sum_{i=0}^{12} c_i^t = 100$.

In Fig. 4 and 5 we plot the average concentration of memes for instance 1 under the two selection schemes $(350, 350)$ and $(350 + 350)$ respectively. Each panel in Fig. 4 shows the variation in concentration for memes using a particular move operator but with different λ values.

The first thing to notice is that although both algorithms ($i1d1t6(350, 350)$ and $i1d1t6(350 + 350)$) achieves the same sub-optima (see Table 4) they do so by using the memes in very different ways.

In the *plus* strategy plotted in Fig. 5 there are three memes that clearly dominates the local search phase, namely $M1L0.4, M1L0.8, M1L1.0$. All the other memes remains with an average concentration below 5. The three memes mentioned used the same operator, a shuffle move, with different λ values for the fuzzy valuation function. In contrast, the plots in Fig. 4 show that, while there is a small selective advantage for memes $M2L1.0$ and $M1L1.0$ at the beginnings of the run, they soon became neutral to each other. All memes have concentrations above 5.

In Fig. 6, we analyze how algorithms $i2d3t2(350, 350)$ and $i2d3t2(350 + 350)$ achieved the optimum configuration for instance 2. We can see that although the difference between the dominant meme in 6(b) and the other memes is bigger than that in (a), the rest of the meme concentrations seems to follow similar patterns. That is, for this instance and version of the algorithm, the simple inheritance mechanism is detecting that the best memes to use are those that employ the basic move 1 (shuffle) for both selection strategies.

Also, from inspection of the graphs in Figs. 4, 5 and 6 it is possible to note that those memes that allows deterioration ($\lambda = 0.4$ or $\lambda = 0.8$) are beneficial to the algorithms. Results similar to these are reported in [17,13].

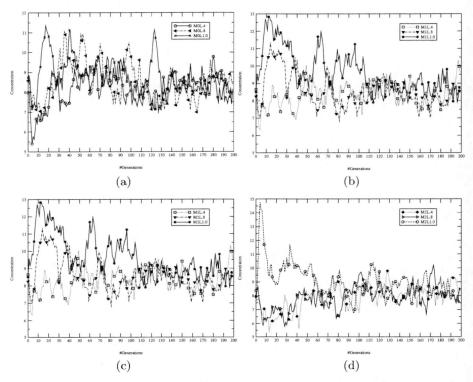

Fig. 4. Meme concentrations for instance 1, algorithm $i1d1t6(350, 350)$. Each panel shows the variation in concentration for memes using a particular move operator but different λ value.

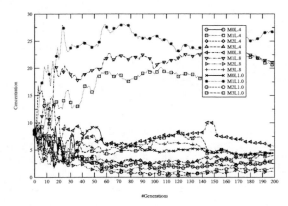

Fig. 5. Meme concentration for instance 1, algorithm $i1d1t6(350 + 350)$.

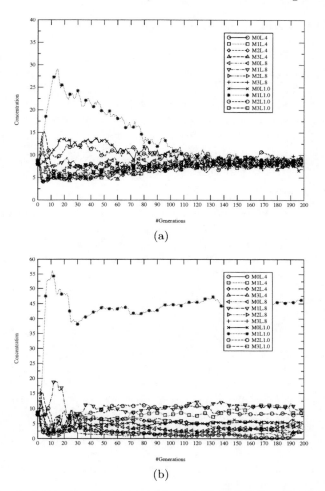

Fig. 6. Meme concentration for instance 2, algorithm $i2d3t2$ under a $(350, 350)$(a) and a $(350 + 350)$(b) selection scheme.

6 Conclusions

A hybridization strategy between a fuzzy sets-based heuristic, *FANS*, and a Multimeme algorithm was proposed and tested.

The construction of the memepool using simplified versions of *FANS*, enabled us to obtain a wide range of fuzzy memes, each one with its particular behavior, in a simple, compact manner and unified fashion. The advantage of using *FANS* as the memes for a *MMA* over using adaptive helpers as in [17,13] is that it is much easier to tune the search of the memes in the former than in the later. Moreover, human knowledge or instance specific knowledge (e.g. secondary structure information) can be readily incorporated into *FANS* based memes.

The scheduling of memes by the simple inheritance mechanism, was proven successful in the detection of the most suitable fuzzy meme for different stages of the search. This has been verified in other domains [18,13] which deems Multimeme Algorithms a very robust metaheuristic.

The coupled effect of both elements, lead to a robust and general purpose metaheuristic which is able to obtain better results than *FANS* alone in the protein structure prediction problem.

Acknowledgments

Research partially supported by Fundación Antorchas, Argentina, and the British research council BBSRC (Bioinformatics Initiative, ref:42BIO14458). D.A. Pelta is a Grant Holder from Consejo Nacional de Investigaciones Cientificas y Técnicas CONICET, Argentina.

References

1. J. Atkins and W. E. Hart. On the intractability of protein folding with a finite alphabet. *Algorithmica*, pages 279–294, 1997.
2. M.A. Bedau and N.H.Packard. Measurement of evolutionary activity, teleology and life. In C.G. Langton, C. Taylor, D. Farmer, and S. Rasmussen, editors, *Artificial Life II*, volume 98-03-023, pages 431–461. Addison-Wesley, 1992.
3. B. Berger and T. Leight. Protein folding in the hydrophobic-hydrophilic (HP) model is NP-complete. In *Proceedings of The Second Annual International Conference on Computational Molecular Biology, RECOMB 98*, 1998.
4. L. T. Biegler, T. F. Coleman, A. R. Conn, and F. N. Santosa, editors. *Large-Scale optimization with applications. Part III: Molecular structure and optimization*, volume 94 of *The IMA Volumes in Mathematics and its Applications*. Springer-Verlag, New York, 1997.
5. B.P. Blackburne and J.D. Hirst. Evolution of functional model proteins. *Journal of Chemical Physics*, 115(4):1935–1942, 2001.
6. A. Blanco, D.A. Pelta, and J.L. Verdegay. A fuzzy valuation-based local search framework for combinatorial problems. *Journal of Fuzzy Optimization and Decision Making*, 1(2):177–193, 2002.
7. P. Crescenzi, D. Goldman, C. Papadimitriou, A. Piccolboni, and M. Yannakakis. On the complexity of protein folding. In *Proceedings of The Second Annual International Conference on Computational Molecular Biology, RECOMB 98*, 1998.
8. Ken A. Dill. Theory for the folding and stability of globular proteins. *Biochemistry*, 24:1501, 1985.
9. G.W. Greenwood, B. Lee, J. Shin, and G.B. Fogel. A survey of recent work on evolutionary approaches to the protein folding problem. In *Proceedings of the Congress of Evolutionary Computation (CEC)*, pages 488–495. IEEE, 1999.
10. W. E. Hart. Adaptive global optimization with local search. *Ph.D. Thesis*, University of California, San Diego, 1994.

11. J.D. Hirst. *Protein Engineering*, 12:721–726, 1999.
12. M. Khimasia and P. Coveney. Protein structure prediction as a hard optimization problem: The genetic algorithm approach. In *Molecular Simulation*, volume 19, pages 205–226, 1997.
13. N. Krasnogor. *Studies on the Theory and Design Space of Memetic Algorithms*. Ph.D. Thesis, Faculty of Computing, Mathematics and Engineering, University of the West of England, Bristol, United Kingdom., 2002.
14. N. Krasnogor, W.E. Hart, J. Smith, and D. Pelta. Protein structure prediction with evolutionary algorithms. In W. Banzhaf, J. Daida, A.E. Eiben, M.H. Garzon, V. Honavar, M. Jakaiela, and R.E. Smith, editors, *GECCO-99: Proceedings of the Genetic and Evolutionary Computation Conference*. Morgan Kaufman, 1999.
15. N. Krasnogor, D. Pelta, P. Martinez Lopez, P. Mocciola, and E. de la Canal. Genetic algorithms for the protein folding problem: A critical view. In C. Fyfe E. Alpaydin, editor, *Proceedings of Engineering of Intelligent Systems*. ICSC Academic Press, 1998.
16. N. Krasnogor and J.E. Smith. Memetic algorithms: Syntactic model and taxonomy. submitted to The Journal of Heuristics. Available from the authors.
17. N. Krasnogor and J.E. Smith. A memetic algorithm with self-adaptive local search: Tsp as a case study. In *Proceedings of the 2000 Genetic and Evolutionary Computation Conference*. Morgan Kaufmann, 2000.
18. N. Krasnogor and J.E. Smith. Emergence of profitable search strategies based on a simple inheritance mechanism. In *Proceedings of the 2001 Genetic and Evolutionary Computation Conference*. Morgan Kaufmann, 2001.
19. N. Krasnogor, D. Pelta, D. H. Marcos, and W. A. Risi. Protein structure prediction as a complex adaptive system. In *Proceedings of Frontiers in Evolutionary Algorithms 1998*, 1998.
20. M.W.S. Land. Evolutionary algorithms with local search for combinatorial optimization. *Ph.D. Thesis, University of California, San Diego*, 1998.
21. P. M. Pardalos, D. Shalloway, and G. L. Xue, editors. *Global minimization of nonconvex energy functions: Molecular conformation and protein folding*, volume 23 of *DIMACS Series in Discrete Mathematics and Theoretical Computer Science*. American Mathematical Society, Providence, Rhode Island, 1996.
22. A. L. Patton. A standard ga approach to native protein conformation prediction. In *Proceedings of the Sixth International Conference on Genetic Algorithms*, pages 574–581. Morgan Kauffman, 1995.
23. D. Pelta, A. Blanco, and J. L. Verdegay. Applying a fuzzy sets-based heuristic for the protein structure prediction problem. *International Journal of Intelligent Systems*, 17(7):629–643, 2002.
24. D. Pelta, A. Blanco, and J. L. Verdegay. A fuzzy adaptive neighborhood search for function optimization. In *Fourth International Conference on Knowledge-Based Intelligent Engineering Systems & Allied Technologies, KES 2000*, volume 2, pages 594–597, 2000.
25. D. Pelta,N. Krasnogor, A. Blanco, and J. L. Verdegay. F.a.n.s. for the protein folding problem: Comparing encodings and search modes. In *Proceedings of the Fourth International Metaheuristics Conference, MIC 2001*, 2001.
26. D. Pelta, A. Blanco, J.L. Verdegay, (2002) *Fuzzy Adaptive Neighborhood Search: Examples of Applications* In this same volume

27. A. Piccolboni and G. Mauri. Protein structure prediction as a hard optimization problem: The genetic algorithm approach. In N. et al. Kasabov, editor, *Proceedings of ICONIP '97*. Springer, 1998.
28. A. A. Rabow and H. A. Scheraga. Improved genetic algorithm for the protein folding problem by use of a cartesian combination operator. *Protein Science*, 5:1800–1815, 1996.
29. S. Schulze-Kremer. Genetic algorithms for protein tertiary structure prediction. In *Parallel Problem Solving from Nature - PPSN II*. North-Holland, 1992.
30. J. Maynard Smith. *Shaping Life: Genes, Embryos and Evolution*. Weidenfeld and Nicolson, 1998.
31. R. Unger and J. Moult. A genetic algorithm for three dimensional protein folding simulations. In *Proceedings of the 5th International Conference on Genetic Algorithms (ICGA-93)*, pages 581–588. Morgan Kaufmann, 1993.
32. R. Unger and J. Moult. Genetic algorithms for protein folding simulations. *Journal of Molecular Biology*, 231(1):75–81, 1993.

Vehicle Routing Problem With Uncertain Demand at Nodes: The Bee System and Fuzzy Logic Approach

Panta Lučić and Dušan Teodorović

The Charles E. Via Jr. Department of Civil and Environmental Engineering
Virginia Polytechnic Institute and State University
Northern Virginia Center, 7054 Haycock Road
Falls Church, VA 22043-2311, U.S.A.

Abstract. This paper describes an "intelligent" system designed to make "real-time" decisions regarding route shapes for situations in which locations of the depot, nodes to be served and vehicle capacity are known, and the demand at the nodes is only approximated. The model developed is based on the combination of the new computational paradigm - the Bee System and Fuzzy Logic. The results are found to be very close to the best solution assuming that the future node demand pattern is known. The system applications appear to be very promising.

1 Introduction

Vehicle routing problems that appear in various transportation activities certainly belong to the class of complex transportation problems. Vehicles leave the depot, serve nodes in the network, and on completion of their routes, return to the depot. Every node is described by a certain demand (the amount to be delivered to the node or the amount to be picked up from the node). Other known values include the locations of the depot and nodes, the distance between all pairs of nodes, and the capacity of the vehicles providing service. The classical vehicle routing problem consists of finding the set of routes that minimizes transport costs.

During last two decades, papers have started to appear [13], [12], [38], [21], [10], [11], [3], [28], [29], [16], [39], [46], [34], [31], [41] in which demand at nodes is treated as a random variable and actual demand value is known only after the visit to the node. The problem of routing vehicles in the case of stochastic demand at nodes is known in the literature as the stochastic vehicle routing problem . The basic characteristic of the stochastic vehicle routing problem is that the real value of demand at a node is only known when the vehicle reaches the node. Due to the uncertainty of demand at the nodes, a vehicle might not be able to service a node once it arrives there due to an insufficient capacity. Such situation is known as a "route failure".

This paper describes the development of an *"intelligent system"* designed to make real-time decisions regarding route shapes for situations in which locations of the depot, nodes to be served and vehicle capacity are known, and the demand at the nodes is only approximated (represented by probability density functions). The model developed is based on the combination of the new computational paradigm - the Bee System and Fuzzy Logic.

The paper is organized in the following way. Statement of the problem is given in section 2. Proposed solution of the problem is given in section 3. Numerical experiments are shown in section 4. Section 5 contains conclusions and directions for further research.

2 Statement of the problem

Let us assume that there are n nodes in the network to be served and that service is provided by vehicles of the same capacity. We will denote vehicle capacity by C. Vehicles leave depot D, serve a number of nodes, and on completion of their service, return to the depot. Let us note depot D and n nodes shown in Fig. 1.

Let's further assume that the demand at each node is only approximately known. Such demand can be represented by a probability density function or in the case of a subjective estimate by the appropriate fuzzy number. Due to the uncertainty of demand at the nodes, a vehicle might not be able to service a node once it arrives there due to insufficient capacity. Teodorović and Pavković [39] and Teodorović and Lučić [34] assumed in such situations that the vehicle returns to the depot, empties what it has picked up thus far, returns to the node where it had a "failure," and continues service along the rest of the planned route (Fig. 1). The same assumption is made in this paper. When evaluating the planned route, the additional distance that the vehicle makes due to "failure" arising at some nodes must be taken into consideration. The problem logically arises of designing such a set of routes, which will result in the least total sum of planned route lengths and additional distance covered by vehicles due to failure. The assumption central in the conduct of this research is that it is possible to develop an "*intelligent system*" capable of making on line decisions regarding sending a vehicle to the next node or to the depot.

3 A Proposed solution of the problem

3.1 Solving Combinatorial Optimization problems by Swarm Intelligence

In this paper, the Traveling Salesman problem will first be solved. Following this, the authors will take into account all existing constraints including stochastic demand at nodes, and vehicle capacity. One of the goals of this paper is also to explore the possible applications of collective bee intelligence in solving complex transportation engineering problems. Lučić and Teodorović

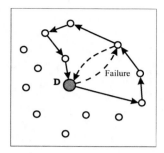

Fig. 1. "Failure" at a node of the planned route

[24], [25] have developed the Bee System - the new heuristic algorithm for the complex combinatorial optimization problems. Using the improved Bee System we can create a Traveling Salesman tour through the n nodes on the network to be served.

3.2 Bee system: the new computational paradigm

Bees in the nature and artificial bees

Self-organization of the bees is based on a few relatively simple rules of individual insect's behavior ([17], [18], [1], [14], [19], [27], [32], [15], [20], [7], [8], [9], [43], [45]). In spite of the existence of a large number of different social insect species, and variation in their behavioral patterns, it is possible to describe individual insects' as capable of performing a variety of complex tasks [6]: Each bee decides to reach the nectar source by following a nestmate who has already discovered a patch of flowers. Each hive has a so-called dance floor area in which the bees that have discovered nectar sources dance, in that way trying to convince their nestmates to follow them. If a bee decides to leave the hive to get nectar, she follows one of the bee dancers to one of the nectar areas. Upon arrival, the foraging bee takes a load of nectar and returns to the hive relinquishing the nectar to a food storer bee. After she relinquishes the food, the bee can (a) abandon the food source and become again uncommitted follower, (b) continue to forage at the food source without recruiting the nestmates, or (c) dance and thus recruit the nestmates before the return to the food source. The bee opts for one of the above alternatives with a certain probability. Within the dance area the bee dancers "advertise" different food areas. The mechanisms by which the bee decides to follow a specific dancer are not well understood, but it is considered that "the recruitment among bees is always a function of the quality of the food source" [6]. It is also noted that not all bees start foraging simultaneously. The experiments confirmed that "new bees begin foraging at a rate proportional to the difference between the eventual total and the number presently foraging".

Bees' behavior in nature has been a great inspiration for the authors of this paper to create artificial bee colony. Our artificial bee colony behaves partially alike, and partially differently from bee colonies in nature.

The bee system described in this paper represents the improvement of our previous work [24]. Our artificial system is the system composed of a number of precisely defined elements (individuals). We define the behavior rules of our artificial bees (agents) and simulate the interaction between them. In this way, different scenarios with different rules can be simulated and explored.

Solving The Traveling Salesman Problem With The Bee System

The well known Traveling Salesman problem is defined in the following way: Given n nodes, find the shortest itinerary that starts in a specific node, goes through all other nodes exactly once and finishes in the starting node. We define an artificial bee environment. This environment is an artificial representation of the space. Our artificial bees (agents) perform only activities defined by our model and a corresponding computer program. We define the way in which artificial bees communicate with each other. In other words, we perform Multi-agent simulation. Let us denote by $G = (N, A)$ the network in which the bees are collecting nectar (the graph in which the traveling salesman route should be discovered). Let us also randomly locate the hive in one of the nodes. When foraging, the bees are trying to collect as much nectar as possible. Let us also assume that the nectar quantity that is possible to collect flying along a certain link is inversely proportional to the link length. In other words, the shorter the link, the higher the nectar quantity collected along that link. This means that the greatest possible nectar quantity could be collected when flying along the shortest traveling salesman route. Our artificial bees will collect the nectar during the certain prescribed time interval. After that, we will randomly change the hive position. The bees will start to collect the nectar from the new location. We will then again randomly change the hive location etc. The iteration in searching process represents one change of the hive position. We assume that our artificial bees live in an environment characterized by the discrete time. Each iteration is composed of a certain number of stages. The stage is an elementary time unit in bees' environment. During one stage the bee will visit s nodes, create partial traveling salesman tour, and after that return to the hive (the number of nodes s to be visited within one stage is prescribed by the analyst at the beginning of the search process). In the hive the bee will participate in a decision making process. The bee will decide whether to abandon the food source and become again uncommitted follower, continue to forage at the food source without recruiting the nestmates, or dance and thus recruit the nestmates before returning to the food source. Let us denote by B the total number of bees in the hive and by $B(u, z)$ the total number of bees collecting nectar during stage u ($u = 0, 1, 2, ..., \left\lceil \frac{|N|-1}{s} \right\rceil$) in iteration z. During any stage, bees are choosing nodes to be visited in a random manner. Logit model is one of the most successful and widely accepted discrete choice model. We have assumed that the probability of choosing node j by the k-th bee, located in the node i (during stage $u + 1$ and iteration z) equals:

$$p_{ij}^k(u+1,z) = \begin{cases} \dfrac{exp\left(-ad_{ij}\,z/\sum_{r=\max(z-b,1)}^{z-1} n_{ij}(r)\right)}{\sum_{l \in N_k(u,z)} exp\left(-ad_{il}\,z/\sum_{r=\max(z-b,1)}^{z-1} n_{il}(r)\right)}, & \begin{array}{l} i = g_k(u,z), \\ j \in N_k(u,z), \\ \forall k, u, z \end{array} \\ 0, & \text{otherwise} \end{cases} \quad (1)$$

where:

$i, j-$	node indexes $(i, j = 1, 2, ..., \|N\|,)$
$d_{ij}-$	length of link (i, j),
$k-$	bee index $(k = 1, 2, ..., B)$,
$z-$	iteration index $(z = 1, 2, ..., M)$,
$n_{il}(r)-$	total number of bees that visited link (i, l) in r-th iteration,
$b-$	"memory length" [1],
$g_k(u, z)-$	last node that bee k visits at the end of stage u in iteration z,
$N_k(u, z)-$	set of unvisited nodes for bee k at stage u in iteration z (in one stage bee will visits s nodes; we have $\|N_k(u, z)\| = \|N\| - us$),
$a-$	input parameter.

The greater the distance between node i and node j, the lower the probability that the k-th bee located in the node i will choose node j during stage u and iteration z. The distance d_{ij} is obviously a very important factor influencing the bee's choice of the next node. The influence of the distance is lower at the beginning of the search process. The greater the number of iterations z, the higher the influence of the distance. This is expressed by the term z in the nominator of the exponent (relation (1)). In other words, at the beginning of the search process bees have more freedom of flight. They have more freedom to search the solution space. The more iterations we make, the bees have less freedom to explore the solution space. The more we are approaching the end of the search process, the more focused the bees are on the flowers (nodes) in the neighborhood. Our artificial bees have memory and they can remember how many bees visited a certain link during last b iterations. The greater the total number of bees that visited a certain link in the past, the higher the probability of choosing that link in the future. This represents the interaction between individual bees in the colony.

---

[1] While foraging in stage u, every artificial bee has the ability to notice the total number of bees in every link. Every artificial bee has the same ability for previous stages as well. That is, artificial bees have the capacity to remember former bee assignments in the network. However, bee recollection is limited. The maximum number of stages that bee can recall represents memory length.

After relinquishing the food, the bee is making the decision about abandoning the food source or continuing the foraging at the food source. It is assumed that every artificial bee (agent) can obtain the information about nectar quantity collected by every other artificial bee. The probability that the bee k will at beginning of the stage $u + 1$ use the same partial tour that is defined in stage u in iteration z equals:

$$p_k(u+1,z) = exp\left(-\frac{L_k(u,z) - \min_{r \in w(u,z)}(L_r(u,z))}{uz}\right) \quad (2)$$

where $L_k(u,z)$ is the length of partial route that is discovered by bee k in stage u in iteration z.

When bee decides not to abandon the food source she can: (a) continue of foraging at the food source without recruiting the nestmates; (b) fly to the dance floor area and start dancing, thus recruiting the nestmates before the return to the food source. The bee opts for one of the above alternatives with a certain probability. Within the dance area the bee dancers "advertise" different food areas. Since the bees are, before all, social insects (the interaction between individual bees in the colony has been well documented), it is assumed in this paper that the probability p^* of an event that the bee will continue foraging at the food source without recruiting the nestmates is very low:

$$p^* << 1 \quad (3)$$

After relinquishing the food, and after making the decision to continue foraging at the food source, the bee flies to the dance floor and starts dancing with the probability equal to (1- p^*). Bee dancing represents the interaction between individual bees in the colony. This kind of communication between individual bees contributes to the formation of the collective intelligence of the bee colony.

In the case when at the beginning of stage $u + 1$, bee does not use the same partial traveling salesman tour, the bee will go to the dancing area and will follow another bee(s). We assume that every partial traveling salesman tour x that is being advertised in the dance area has two main attributes: (a) the total length, and (b) the number of bees that are advertising the partial route. We introduce the normalized value of the total length of the partial traveling salesman tour and the normalized value of the number of bees advertising the partial tour. Both normalized values are defined in the following way: (a) Both normalized values can take any value between 0 and 1; (b) The smaller the total length normalized value the better the partial tour; (c) The bigger the number of bees normalized value the better the partial tour.

Let us denoted by $Y(u,z)$ - the set of partial tours that were visited by at least one bee and by $B_\xi(u,z)$ - the number of bees that discovered partial route ξ.

We have assumed in this paper that the probability that the partial route ξ will be chosen by any bee that decided to choose the new route equals:

$$p_\xi(u,z) = \frac{exp\left(\rho\beta_\xi(u,z) - \theta\alpha_\xi(u,z)\right)}{\sum_{r \in Y(u,z)} exp\left(\rho\beta_r(u,z) - \theta\alpha_r(u,z)\right)}, \quad \xi \in Y(u,z), \forall u, z \quad (4)$$

where:

$\alpha_\xi(u,z)-$ the normalized value of the partial route length,
$\beta_\xi(u,z)-$ the normalized value of the number of bees advertising the partial tour,
$\rho, \theta-$ parameters given by the analyst.

Before relocating the hive to the next location we tried to improve the solution obtained by the bees in current iteration by applying different tour improvement algorithms. The most frequently used tour improvement algorithms are based on k-opt procedure. In this paper we used 2-opt algorithm, 3-opt algorithm [23] and modified 3-opt algorithm.

Experimental study of the Bee System

The proposed Bee System was tested on a large number of numerical examples. In all the cases one of tour improving algorithms has been employed. The benchmark problems were taken from the following Internet address: http://www.iwr.uni-heidelberg.de/iwr/comopt/software/TSPLIB95/tsp/.
The following problem instances were considered: Eil51.tsp, Berlin52.tsp, St70.tsp, Pr76.tsp, Kroa100.tsp, Eil101.tsp, Tsp225.tsp, A280.tsp, Pcb442.tsp and Pr1002.tsp. Problem instances Pcb442.tsp and Pr1002.tsp were considered only with modified 3opt ("short version") tour improving algorithm. All tests were run on an IBM compatible PC with PIII processor (533MHz). Table 1 presents the results obtained by Bee System when search is limited on 100 cycles.

We can see from the Table 1 that the proposed Bee System produced results of a very high quality.

The crucial assumption: We can predict future without a mistake

The system behaves "intelligently" if it "emits" similar output results for similar input variables. Artificial neural networks and fuzzy systems are "intelligent" systems since they have the ability to "learn from experience". Recognition without definition is a characteristic of intelligent behavior. The initial assumption in this paper is that it is possible to develop an "intelligent" vehicle routing systems that makes real-time decisions of high quality. In other words, the paper assumes that it is possible to develop a vehicle routing system that will recognize different situations. As in other intelligent

Table 1. The results obtained by the Bee System enriched with 3-opt heuristic

Problem (Number of nodes)	Optimal Value (O)	The best value obtained by the Bee System (B)	(B-O)/O (%)	Time required to find the best solution (sec)	Average value obtained by the Bee System over 20 runs (A)	(A-O)/O (%)
Eil51 (51)	428.87	428.87	0	37	428.87	0
Berlin52 (52)	7544.366	7544.366	0	1	7544.366	0
St70 (70)	677.11	677.11	0	22	677.11	0
Pr76 (76)	108159	108159	0	11	108159	0
Kroa100 (100)	21285.4	21285.4	0	10	21285.4	0
Eil101 (101)	640.21	640.21	0	1741	643.05	0.44
Tsp225 (225)	3859	3876.05	0.44	5153	3905.32	1.2
A280 (280)	2586.77	2600.34	0.53	13465	2627.45	1.57

systems, the "intelligent" vehicle routing system should be able to generalize, adapt and learn based on new knowledge and new information.

The assumption can be made that future could be predicted without a mistake. In our case this means that we are able to exactly predict demand values at all nodes in the transportation network. In case of perfect prediction we must be able to make optimal decisions. In other words, we know the random demand values at all nodes and we are trying to develop the best vehicle routes.

The problem considered could be denoted by (P). For a known "scenario" (known the random demand values at all nodes) the problem (P) can be solved using a particular optimization technique or heuristic algorithm. In this paper, we applied the developed Bee System, solved Traveling Salesman Problem and created "Giant route". In the next step, we "walked" along the created giant route and we have decided when to finish with one vehicle route and when to start with the next vehicle route. These decisions were easily made since we knew demand at every node and vehicle capacity. The problem (P) can be solved many times for different scenarios. If one can precisely predict the future then he/she can get the optimal or close to optimal solution. Instead of predicting it though, one can simulate future events. In other words, one can simulate realizations of the random variables representing demand at nodes. In the next step, after solving problem (P), the optimal or close to the optimal solution will be obtained. One can repeat the simulation, and again, after solving the problem (P) one can get the optimal or close to the optimal solution. After third simulation one will get the third optimal or close to the optimal solution, etc. In this way one can get the "best" solution for every simulated "scenario". The similar methodological approach was also used by Teodorović and Lučić [34], and Teodorović et al. [37]. This statistical material enables the generation of a fuzzy rule base. In this paper the fuzzy rule bases are generated from numerical examples.

The Algorithm to Create Intelligent Vehicle Routing System

In order to generate fuzzy rule bases from numerical examples, the well-known procedure proposed by Wang and Mendel [44] was used. We could establish fuzzy sets for all the antecedents and the consequences. We will do it in such a way that, at the very beginning, we will establish the domain intervals for all input and output variables. Let us assume that we have the following set of the input-output data pairs:

(x_1^1, y^1), (x_1^2, y^2), (x_1^3, y^3), ...

where x_1, is input and y is output. We will generate the fuzzy rule base from this set of input-output data pairs. The values of x_1, and y belong to the domain intervals (x_{1min}, x_{1max}), (y_{min}, y_{max}) respectively. From every input-output data pair we can eventually generate one fuzzy rule.

We applied the developed Bee System, solved Traveling Salesman Problem and created "Giant route". In the next step, we "walked" along the created giant route and we have decided when to finish with one vehicle route and when to start with the next vehicle route. These decisions were easily made since we knew demand at every node and vehicle capacity. After serving the first k nodes, the available capacity of vehicle B_k will equal:

$$\boldsymbol{B}_k = C - \sum_{i=1}^{k} D_i \qquad (5)$$

It is clear that the "strength" of our preference for the vehicle to serve the next node after it has served k nodes depends on the available capacity B_k, as well as on expected demand in the next node. In other words, the bigger the available capacity and the smaller the expected demand in the next node, the higher our expectation that the route will contain more nodes. We can expect that at a certain time point the route will have "small," "medium," or "big" number of nodes. We will denote by n_k the expected number of new nodes in the route after vehicle already has served k nodes. The linguistic expressions "small number of new nodes," "medium number of new nodes," and "big number of new nodes" can be represented by corresponding fuzzy sets. Available capacity can also be subjectively estimated, for example, as "small," "medium," and "large." Let us denote respectively by X_1, X_2 and X_3 the following variables:

$$\boldsymbol{X}_{1k} = \frac{B_k}{C}; \quad \boldsymbol{X}_{2k} = \frac{\mu_{k+1}}{C}; \quad \boldsymbol{X}_{3k} = \frac{\sigma_{k+1}}{C}; \quad (6)$$

The first variable represents relative available capacity after serving the first k nodes. The second variable represents relative expected demand in the next node, while the third one describes relative variability of the demand in the next node. The typical rule in the approximate reasoning algorithm to determine the expected number of the new nodes in the route can be the following one:

    If   Relative available capacity is **Large** and Relative expected demand in the next node is **Small** and Relative variability of the demand in the next node is **Small**

    Then Expected number of the new nodes in the route is **Big**.

We can see that the antecedent of the rules contains remaining vehicle capacity and the expected demand in the next node. The consequence contains the expected number of new nodes in the route. As we already explained, in this paper the fuzzy rule base is generated from numerical examples using the procedure proposed by Wang and Mendel [44]. For known available capacity B_k that remains after serving k nodes, and for known characteristics of the demand in the next node it is possible to use the approximate reasoning rules to determine the expected number of new nodes in the route.

We are now able to answer the following question:
Should we send the vehicle to the next node or return it to the depot after completing service to k nodes?

Let the expected number of the new nodes in the route equal n_k^*. Based on this value, a decision must be made whether to send the vehicle to the next node or return it to the depot. The vehicle should be sent to the next node if the following relation is fulfilled: $n_k^* \geq 1$. When $n_k^* < 1$, the vehicle should be returned to the depot.

The algorithm to create the fuzzy system developed in this paper consists of the following steps:

Algorithm 1:

Step 1: Generate many different "scenarios" (collect demand values for all nodes that are based on a large number of simulations).
Step 2: Using developed heuristic algorithm based on the Bee System, find the "good" solution for each generated "scenario".)
Step 3: Based on the statistical data resulting from Steps 1 and 2 use the Wang Mendel's algorithm to generate the fuzzy rule base.

The generated fuzzy rule base enables "on line" developing of the vehicle routes. The vehicle routes are created in the following way:

Algorithm 2:

Step 1: Using Bee System (or some other "good" heuristic algorithm) solve the Traveling Salesman Problem.

Step 2: First include the depot in a route. Then include nodes in the route in the same order as they appear in the Traveling Salesman Route. Before deciding to include a node into the route, first use generated fuzzy rule base to calculate the expected number of new nodes in the route. If the calculated expected number of the new nodes is greater than or equal to one, include the node in the route. Otherwise, this node becomes a first point of the new vehicle route. Finish with the algorithm when all nodes are included in the routes.

4 Results obtained using the "intelligent" vehicle routing system

The developed model was tested on ten TSP examples introduced in chapter 3.2.3. In order to convert the original TSP problems into the corresponding Vehicle Routing Problems, the first node was treated as a depot in all considered examples. The vehicle capacity equals 1,000 (C = 1,000). Demand D_i at any node i has been represented by the Normal distribution with mean μ_i and standard deviation σ_i, i.e.:

$$\boldsymbol{D}_i \sim N(\mu_i, \sigma_i); \qquad i = 1, 2, ..., n. \tag{7}$$

The characteristics of the node demand (mean, standard deviation, and "real demand values") are generated randomly for every considered node. When randomly generating mean and standard deviation of the demand in every node the following relations are fulfilled:

$$0 \leq \mu_i \leq 200; \qquad 0 \leq \sigma_i \leq 60; \qquad i = 1, 2, ..., n. \tag{8}$$

The vehicle routes were generated using heuristic algorithm based on the Bee System. Generated routes enabled us to "walk" along the routes and to "read" the available capacity in every node, as well as demand characteristics of the next node on the route. In this way, we obtained statistical material for generating fuzzy rule base. When generating fuzzy rule base, we have used triangular fuzzy numbers uniformly distributed over entire domain interval of input and ouput variables. The domain intervals of input variables x_1, x_2 and x_3 have been divided into 9 regions. The domain intervals of output variables y has been divided into 17 regions.

We have compared the results obtained by the proposed process above with the "best" solution obtained by the heuristic algorithm based on Bee System (a priori known solution).

Because the Bee System result was the *retrospectively* derived best solution for a given demand pattern, the performance associated with it is considered as the target or reference for evaluation. In the past, one of the difficulties of evaluating fuzzy systems was the lack of a reference against which the result could be checked. The criterion used to compare the two cases (our result vs. Bee System result) is the total distance traveled by all vehicles. Demand at each node is a deterministic amount that is obtained by simulation. By moving along the route designed by the approximate reasoning algorithm and accumulating the amounts picked up at each node, it was easy to determine the nodes where failures occurred and to calculate the additional distance that the vehicles had to make.

First, our results and the Bee System results were compared for the same set of demand patterns that were used to develop the rules and relative errors were calculated . Second, for a different set of demand pattern (test set), the best set of routes was developed by the Bee System method and the performance was measured . This performance was then compared with the one obtained using the previously developed fuzzy rules (relative errors were calculated). This last step made sure that the comparison was not biased. All computer experiments were done on a PC computer (PIII processor). Practically negligible CPU times were achieved, and were thus absolutely acceptable for the "real time" application of the developed algorithm. Maximum and average relative error values of all considered numerical examples are given in Table 2.

Table 2. Maximum and average relative error values

Examples	Maximum relative error [%]	Average relative error [%]
Eil51	5.83	0.607
Berlin52	5.98	1.44
St70	4.05	1.63
Pr76	5.83	1.75
Kroa100	5.75	2.42
Eil101	5.44	2.18
Tsp225	4.70	1.49
A280	3.48	1.26
Pcb442	4.32	1.15
Pr1002	2.38	0.93

Each performed numerical experiment is represented by the following two solutions: (a) the solution obtained when future demand is known in advance;

(b) the solution obtained by the proposed "Intelligent" system that makes real-time decisions. The obtained solution pair is shown in Fig. 2 (Example: Berlin52).

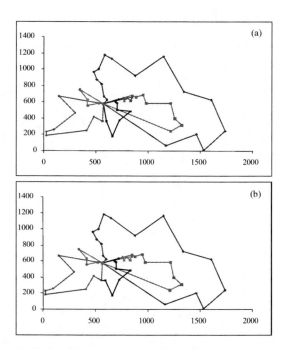

Fig. 2. Example Berlin52; (a) the solution obtained when future demand is known in advance;(b) the solution obtained by the proposed "Intelligent" system

5 Conclusion

In this paper, an "intelligent" vehicle routing system is developed. Applications of the system are considered for the vehicle routing problem when demand at nodes is uncertain (actual demand value is known only after the visit to the node). The developed "intelligent" system makes "real time" decisions as to whether to reject or include a new node into vehicle route. The results are found to be very close to the best solution assuming that the future node demand pattern is known. The system applications appear to be very promising. The model developed is based on the combination of the Bee System and Fuzzy Logic. The proposed process learns from the solutions obtained assuming that the future situations are known. Combining many solutions, a set of rules is developed. All pairs ("demand scenario- appropriate set of vehicle routes") were used to produce a fuzzy rule base. Evaluating the performance of the fuzzy rules developed by this process is also noble. Because the best solution is known for a particular demand pattern (the solution from the proposed heuristic algorithm), the performance of the proposed rules can easily be checked against the result of the best solution. Many tests show that the outcome (the total expected distance that the vehicles were to cover (the total distance is the sum of the total lengths of the planned routes and the additional distances that are covered due to failures at the nodes)) of the proposed rules is nearly equal to the best solution. While in this paper the heuristic algorithm based on Bee System is used to derive the "good" solution for a hypothetical demand patterns, other optimization or heuristics techniques can be used instead. The proposed method is based on the concept of learning from the best example solution. Practically negligible CPU times were achieved, and were thus absolutely acceptable for the "real time" application of the developed algorithm.

Acknowledgments

We wish to thank the National Science Foundation for this research was supported by a grant #CMS-0085430.

References

1. Baschbach, V.S., Waddington, K.D. (1994) Risk-sensitive Foraging in Honey Bees: No Consensus Among Individuals and No Effect of Colony Honey Stores, Animal Behavior, **47**, 933-941
2. Beckers, R., Deneubourg, J.L., Goss, S. (1992) Trails and U-turns in the Selection of a Path by the Ant Lasius niger, Journal of Theoretical Biology, **159**, 397-415
3. Bertsimas, D., Chervi, P., Peterson, M., (1995) Computational approaches to stochastic vehicle routing problems, Transportation Science, **29**, 342-352
4. Biesmeijer, J.C., van Nieuwstadt, M.G.L., Lukacs, S., Sommeijer, M.J. (1998) The Role of Internal and External Information in Foraging Decisions of Melipona Workers (Hymenoptera: Meliponinae), Behavior Ecology Sociobiology, **42**, 107-116
5. Bonabeau, E., Dorigo, M., Theraulaz, G. (1999) Swarm Intelligence, Oxford University Press, Oxford
6. Camazine, S., Sneyd, J. (1991) A Model of Collective Nectar Source by Honey Bees: Self-organization Through Simple Rules, Journal of Theoretical Biology, **149**, 547-571
7. Chittka, L., Gumbert, A., Kunze, J. (1997) Foraging Dynamics of Bumble Bees: Correlates of Movements Within and Between Plant Species, Behavioral Ecology, **8**, 239-249.
8. Chittka, L., Thompson, J.D. (1997) Sensori-motor Learning and its Relevance for Task Specialization in Bumble Bees, Behaviour Ecology Sociobiology, **41**, 385-398
9. Collevatti, R.G., Campos, L.A.O., Schoereder, J.H. (1997) Foraging Behaviour of Bee Pollinators on the Tropical Weed Triumfetta semitriloba: Departure Rules from Flower Patches, Insectes Sociaux, **44**, 345-352

10. Dror, M. (1993) Modeling Vehicle Routing with Uncertain Demands as a Stochastic Program: Properties of the Coresponding Solution, European Journal of Operational Research, **64**, 432-441
11. Dror, M., Laporte, G., Louveaux, F. (1993) Vehicle Routing with Stochastic Demands and Restricted Failures, Operations Research, **37**, 273-283
12. Dror, M., Laporte, G., Trudeau, P. (1989) Vehicle Routing with Stochastic Demands: Properties and Solution Frameworks, Transportation Science, **23**, 166-176
13. Dror, M., Trudeau, P. (1986) Stochastic Vehicle Routing with Modified Savings Algorithm, European Journal of Operational Research, **23**, 228-235
14. Dukas, R., Real, L.A., (1991) Learning Foraging by Bees: a Comparison Between Social and Solitary Species, Animal Behaviour, **42**, 269-276
15. Dukas, R., Visscher, P.K. (1994) Lifetime Learning by Foraging Honey Bees, Animal Behavior, **48**, 1007-1012
16. Gendreau, M., Laporte, G., Seguin, R. (1996) Stochastic Vehicle Routing, European Journal of Operational Research, **88**, 3-12
17. Gould, J.L. (1987) Landmark Learning by Honey Bees, Animal Behaviour, **35**, 26-34
18. Hill, P.S., Wells, P.H., Wells, H. (1997) Spontaneous Flower Constancy and Learning in Honey Bees as a Function of Colour, Animal Behaviour, **54**, 615-627
19. Kadmoon, R., Shmida, A. (1992) Departure Rules Used by Bees Foraging for Nectar: a Field Test, Evolutionary Ecology, **6**, 142-151
20. Keasar, T., Shmida, A., Motro, U. (1996) Innate Movement Rules in Foraging Bees: Flight Distances are Affected by Recent Rewards and are Correlated with Choice of Flower Type, Behaviour Ecology Sociobiology, **39**, 381-388
21. Lambert, V., Laporte, G., Louveaux, F.V. (1993) Designing Collection Routes through Bank Branches, Computers and Operations Research, **20**, 783-791
22. Laporte, G. (1992) The Vehicle Routing Problem - An Overview of Exact and Approximate Algorithms, European Journal of Operational Research, **59**, 345-358
23. Lin, S., Kernigham, B.W. (1973) An Effective Heuristic Algorithm for the Traveling Salesman Problem, Operational Research, **21**, 498-516
24. Lučić, P., Teodorović, D. (2001) Bee System: Modeling Combinatorial Optimization Transportation Engineering Problems by Swarm Intelligence, TRISTAN IV - Triennial Symposium on Transportation Analysis (preprints), 441-445, Sao Miguel, Azores Islands, Portugal, June 13-19
25. Lučić, P., Teodorović, D. Computing with Bees: Attacking Complex Transportation Engineering Problems, paper under the review in the Transportation Research (Part C) journal
26. Malmborg, C.J. (1996) A Genetic Algorithm for Service Level Based Vehicle Scheduling, Eurpoean Journal of Operational Research, **93**, 121-134
27. Peleg, B., Shmida, A., Ellner, S. (1992) Foraging Graphs: Constraint Rules on Matching Between Bees and Flowers in a Two-sided Pollination Market, Journal of Theoretical Biology, **157**, 191-201
28. Popović, J. (1995) Vehicle Routing in the Case of Uncertain demand: A Bayesian Approach, Transportation Planning and Technology, **19**, 19-29
29. Potvin, J.Y., Duhamel, C., Guertin F. (1996) A Genetic Algorithm for Vehicle Routing with Backhauling, Applied Intelligence, **6**, 345-355

30. Powell, W.B. (1987) An Operational Planning Model for the Dynamic Vehicle Allocation Problem with Uncertain Demands, Transportation Research, **21B**, 217-232
31. Secomandi, N. (2000) Comparing neuro-dynamic programming algorithms for the vehicle routing problem with stochastic demands, Computers and Operations Research, **27**, 1201-1225
32. Seeley, T.D. (1992) The Tremble Dance of the Honey Bee: Message and Meanings, Behavior Ecology Sociobiology, **31**, 375-383
33. Seeley, T.D., Visscher, P.K. (1988) Assesing the Benefits of Cooperation in Honeybee Foraging: Search Costs, Forage Quality, and Competitive Ability, Behavior Ecology Sociobiology, **22**, 229-237
34. Teodorović, D., Lučić, P. (2000) Intelligent Vehicle Routing System, Proceedings of the 3rd IEEE Conference on Intelligent Transportation Systems, October Dearborn, U.S.A., 482-487
35. Teodorović, D., Lučić, P. The Combined Ant System-Fuzzy Logic approach to the vehicle routing problem when demand at nodes is uncertain, paper under the review in the Fuzzy Sets and Systems
36. Teodorović, D., Lučić, P. Schedule Synchronization in Public Transit by Fuzzy Ant System, paper under the review in the European Journal of Operational Research
37. Teodorović, D., Lučić, P., Popović, J., Kikuchi, S., Stanić, B. (2001) Intelligent Isolated Intersection, in Proceedings of the 10th International IEEE Conference on Fuzzy Systems, December, Melbourne, Australia
38. Teodorović, D. , Pavković, G. (1992) A Simulated Annealing Technique Approach to the Vehicle Routing Problem in the Case of Stochastic Demand, Transportation Planning and Technology, **16**, 261-273
39. Teodorović, D. , Pavković, G. (1996) The fuzzy set theory approach to the vehicle routing problem when demand at nodes is uncertain, Fuzzy Sets and Systems **82**, 307-317
40. Teodorović, D., Vukadinović, K. (1998) Traffic Control and Transport Planning: A Fuzzy Sets and Neural Networks Approach, Kluwer Academic Publishers, Boston/ Dordrecht/London
41. Van Breedam, A. (2001) Comparing descent heuristics and metaheuristics for the vehicle routing problem, Computers and Operations Research, **28**, 289-315
42. Vienne, K., Erard, C., Lenoir, A. (1998) Influence of the Queen on Worker Behaviour and Queen Recognition Behaviour in Ants, Ethology, **104**, 431-446
43. Waddington, K.D., Nelson, C.M., Page, R.E.Jr. (1998) Effects of Pollen Quality and Genotype on the Dance of Foraging Honey Bees, Animal Behaviour, **56**, 35-39
44. Wang, L-X., Mendel, J. (1992) Generating Fuzzy Rules by Learning from Examples, IEEE Transactions on Systems, Man and Cybernetics, **22**, 1414-1427
45. Williams, N.M., Thompson, J.D. (1998) Trapline Foraging by Bumble Bees: III. Temporal Patterns of Visitation and Foraging Success at Single Plants, Behavioral Ecology, **9**, 612-621
46. Yang, W.-H., Mathur, K., Ballou, R.H. (2000) Stochastic Vehicle Routing Problem with Restocking, Transportation Science, **34**, 99-112
47. Zadeh, L. (1965) Fuzzy Sets, Information and Control, **8**, 338-353
48. Zimmermann, H.J. (1991) Fuzzy Set Theory and Its Applications, Boston: Kluwer

Fuzzy Constructive Heuristics

José A. Moreno Pérez and J. Marcos Moreno Vega

Dpto. de E.I.O. y Computación
Centro Superior de Informática
Universidad de La Laguna
38271 La Laguna, Santa Cruz de Tenerife, Spain

Abstract. We consider the design of fuzzy constructive heuristic algorithms for solving combinatorial problems. The fuzzy technology is used in constructive heuristics to select the item to be included in the solution and to stop the search. This is done by considering fuzzy sets of promising elements and satisfactory solutions. The procedures that implement several of these algorithms were tested on moderate and large size instances of a well known cutting problem. These procedures show better performance than the best known heuristic for them.

1 Introduction

In a constructive method an element is iteratively added to an initially empty structure until a solution of the problem is obtained. The choice of the item to be included in the partial solution is based on one or several heuristic evaluations that measure the convenience of considering the item as belonging to the solution. The heuristic functions depend on the problem and also on the knowledge of the decision maker about the problem.

The heuristic functions are used to intelligently guide the process of searching for solutions with high quality. If the evaluation of an element depends on the items already in the solution, the function and the method are adaptive. In addition to the heuristic function, it is necessary a strategy to select the elements. One of the most known strategy is the *greedy* rule, which selects the element that optimizes the heuristic function. However, this strategy shows poor performance in most cases.

In some cases, after knowing the evaluation of an element, the expert provides comments like: *high recommendable element*, *quite good item*, *acceptable enough element*, etc. In those cases, it is possible to construct fuzzy sets of elements from which to choose the element to be included in the solution.

One of the most relevant questions, with a high effect on the quality of the solutions, is the stopping criteria applied. As indicated above, the usual expressions are: *acceptable solution*, *good enough solution*, etc. The construction of a fuzzy set from the desirable situation, allows us to design appropriate stopping criteria.

In this paper we introduce the fuzzy constructive methods and design fuzzy stopping rules. Moreover, we show the performance of the proposed method on a well known cut problem.

2 Fuzzy Constructive Methods

Let $E = \{e_1, e_2, ..., e_n\}$ be the set of elements to be included in the solution and f an adaptive heuristic function. The functions that evaluate the elements are such that better elements have smaller values. Let X denote the structure that stores the obtained partial solution so far, and then $T = E - X$ is the set of elements not included in the partial solution.

The pseudocode of a classical constructive method with the greedy strategy is shown in figure 1. The main drawback of the greedy strategy is that,

Procedure Greedy_Constructive_Method
    begin
        $X = \emptyset$;
        $T = E$;
        repeat
            Evaluate($f(e) : e \in T$);
            $e^* = \arg\min_{e \in T}\{f(e)\}$;
            $X = X \cup \{e^*\}$;
            $T = T - \{e^*\}$;
        until (X is feasible);
        return X;
    end.

Fig. 1. Greedy Constructive Method

in general, it provides low quality solutions. It is true even if an improvement procedure (a local search) is applied to the obtained solution.

Let us assume that there is a membership function $\mu(.)$ that evaluates the degree of belonging of an element $e \in E$ to the set of *best elements*. The membership function $\mu(e)$ of an element e depends on its heuristic evaluation, $f(e)$. Then the set of *best elements* E^* of E is constructed by:

$$E^* = \{e \in E / \mu(e) \geq \alpha\},$$

i.e., E^* is the α-cut of μ ($\alpha \in [0, 1]$ is fixed by the decision maker).

Therefore, the pseudocode of the fuzzy constructive method that we propose is as shown in figure 2.

In this pseudocode, by $Choose(e \in E^*)$ we mean the method used to determine the element $e \in E^*$ that is included in the partial solution X.

Procedure Fuzzy_Constructive_Method
begin
   $X = \emptyset$;
   $T = E$;
   repeat
      Evaluate($f(e) : e \in T$);
      Generate(E^*);
      Choose($e^* : e^* \in E$);
      $X = X \cup \{e^*\}$;
      $T = T - \{e^*\}$;
   until (X is feasible);
   return X;
end.

Fig. 2. Fuzzy Constructive Method

Some alternatives for this method are the equiprobable sampling in E^* or the proportional sampling in E^*. They are:

Equiprobable sample:
$$\Pr(choose\ e) = \frac{1}{|E^*|}.$$

Proportional sample:
$$\Pr(choose\ e) = \frac{\mu(e)}{\sum_{e \in E^*} \mu(e)}.$$

Note that, in contrast to the classical constructive method, several executions of the fuzzy constructive method will provide different solutions. In that way, the probability of getting high quality solutions increases.

Figure 3 shows the iterated fuzzy constructive method. It includes an improvement method after each execution of the fuzzy constructive procedure. In this pseudocode, $Improving\_Method(X, X')$ returns an improved solution X' obtained from X.

The improving method applied in this procedure can be, for example, Fuzzy Adaptive Neighbourhood Search (FANS) [7], but can also be a crisp procedure like any local search.

3 Fuzzy Stopping Rules

Ideally, the search process must finish when the optimal solution of the problem is found. However, this stopping criterion is not applicable in real situations, since the optimal solution is unknown. Therefore, a real stopping criterion is to finish the search when a high quality solution that satisfies the preference of the decision maker is met.

Procedure Iterated_Fuzzy_Constructive_Method
begin
  Fuzzy_Constructive_Method;
  Improving_Method(X, X');
  $X^* = X'$;
  repeat
    Fuzzy_Constructive_Method;
    Improving_Method(X, X');
    if (X' is better than X^*) **then**
      $X^* = X'$;
  until (stopping_rule);
  return(X^*);
end.

Fig. 3. Iterated Fuzzy Constructive Method

For many problems, the quality of the solution can be measured by some characteristics. These characteristics indicate that the reached situation is *acceptable, good enough, difficult to improve, ...* These situations are determined by a set of variables that characterize the quality of the solution. Using these variables, some fuzzy sets of solutions are used to provide appropriated fuzzy stopping criteria. The criteria consist of stopping the search when a solution of these fuzzy set is met.

4 The non-guillotine rectangular two-dimensional cutting problem

The two-dimensional cutting problem consists of determining the optimal cut of a single sheet into a set of small pieces of given shapes and sizes. In the rectangular two-dimensional cutting problems, the stock sheet and the small pieces have rectangular shape. The optimality of the cut is given by the amount of space used or the waste material. In other problems, given the number of small pieces with economical values, the solution consists of selecting pieces to cut in order to maximize the total value. These problems appear in relevant commercial and industrial application areas, where one or several big sheets of wood, cloth, paper or metal have to be cut in a large number of small pieces (see, for example, [1] and [3]).

The non-guillotine rectangular two-dimensional cutting problem [4], [5] is formulated as follows. Given a rectangular sheet of a fixed width w and unlimited high, and a set of small rectangles

$$\mathcal{R} = \{R(w_1, h_1), R(w_2, h_2), \ldots, R(w_n, h_n)\},$$

given by the lengths of their sides, determine the way to cut all the pieces in a sheet with minimum total height. In this problem the cuts can be of

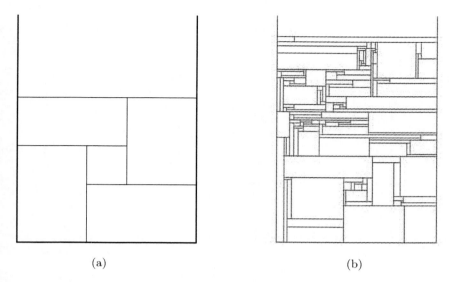

Fig. 4. Rectangular cutting problems. **(a)** Non guillotine cutting. **(b)** Non-guillotine rectangular two-dimensional cutting problem

non-guillotine type and the objects can be rotated. The guillotine cuttings are those that go from one edge of the stock rectangular sheet to the opposite edge. In a non guillotine cutting, this is not true (see figure 4(a)). Obviously, each small piece has to be with one of its sides parallel to the fixed edge of the stock sheet in the optimal cut. Thus, the values w_i and h_i are named the width and high of the pieces and can be rotated 90^0. At least one of the two sides, w_i or h_i, of each rectangle is assumed to be smaller than the width w.

4.1 Contour

A partial solution is given by the position of the bottom-left corner of each small rectangle (and the boolean variable to know if the piece is rotated or not). The solution is feasible if there is not an overlap between two rectangles. The rectangles are obviously shifted down in order to minimize the total height wasted. The partial solution determines the waste of space determined by the fixed width side, the unlimited edges and an upper piecewise rectilinear contour as shown in figure 5. In addition, not useful areas called trim loss can be generated (see the partial solution obtained after the insertion of the rectangle 4 in figure 5).

Let us assume that $\mathcal{R} = \mathcal{R}_1 \cup \mathcal{R}_2$, where $\mathcal{R}_1$ is the set of rectangles already packed in the partial solution and $\mathcal{R}_2 = \mathcal{R} \setminus \mathcal{R}_1$ the set of remainder rectangles to be packed.

Consider the upper envelope of the set of rectangles already included in the partial solution. This envelope is given by a sequence of segments,

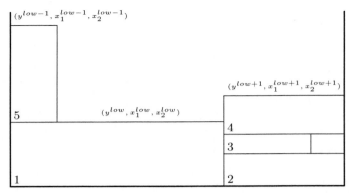

Fig. 5. Contour

$\mathcal{C} = (s^1, s^2, ..., s^c)$, where each segment s^i is given by the height and the position of the extremes, i.e., $s^i = (y^i, x_1^i, x_2^i)$. The contour $\mathcal{C}$ verifies $x_1^1 = 0$, $x_2^c = w$ and, for $1 \leq i < c-1$, $x_2^i = x_1^{i+1}$. Then,

$$\mathcal{C} = \{(y^1, x_1^1, x_2^1), (y^2, x_1^2, x_2^2), ..., (y^c, x_1^c, x_2^c)\}.$$

Let $s^{low} = (y^{low}, x_1^{low}, x_2^{low})$ be the segment with small height; i.e. such that $y^{low} = \min_{i=1,...,c}\{y^i\}$.

Let us assume that the width of any not packed rectangle is the side closer to the size of the lowest segment $w^{low} = |x_2^{low} - x_1^{low}|$ of the upper contour. We can do this assumption since otherwise a simple rotation of the object will provide this hypothesis.

Since we want to obtain the packing with smallest height, we can design several natural constructive ways in which the rectangles are packed in the segment with smallest height C. Among all the possible alternatives, we describe three of them that also take into account other partial objectives.

4.2 Sets of best rectangles

Consider the situation shown in figure 5 where the lowest segment is $s^{low} = (y^{low}, x_1^{low}, x_2^{low})$. The next and before segments are

$$s^{low+1} = (y^{low+1}, x_1^{low+1}, x_2^{low+1})$$

and

$$s^{low-1} = (y^{low-1}, x_1^{low-1}, x_2^{low-1}).$$

Let us assume, without loss of generality, that the heights of these segments are such that $y^{low} \leq y^{low+1} \leq y^{low-1}$.

¿From the segment $s^{low} = (y^{low}, x_1^{low}, x_2^{low})$, there are several ideal situations that could be achieved by the inclusion of a new rectangle. A very good situation appears if we can include a rectangle with width equal to s^{low}.

Another very good situation appears if the height of the included rectangle is close to $y^{low+1} - y^{low}$ or to $y^{low-1} - y^{low}$ (see figure 5). Therefore, there are three notions of ideal situations that are used to build fuzzy sets of best rectangles. The membership functions of these sets are as follows.

The first fuzzy set of best rectangles corresponds to those that most fit to the width w^{low}. Let $\alpha_1 \in [0,1]$ be a fit threshold given by the decision maker. Then the membership function of any object $e = R(w,h) \in \mathcal{R}_2$ is given by:

$$\mu^1(R(w,h), \alpha_1) = \begin{cases} \frac{1}{\alpha_1}(\alpha_1 - w^{low} + w) & \text{if } w^{low} - \alpha_1 \leq w \leq w^{low} \\ 0 & \text{otherwise} \end{cases}$$

This membership function defines the first fuzzy set $\mathcal{E}_1$ of the *best elements* on the universe of the not packed rectangles.

A second fuzzy set of best rectangles, $\mathcal{E}_2$, is obtained by considering the set of not packed rectangles such that, if one of them is packed on s^{low} it will provide a segment with height close to y^{low+1} (see figure 5).

Let $\alpha_2 \in [0,1]$ be a second fit threshold provided by the decision maker. Let the second membership function of an object $e = R(w,h)$ be given by:

$$\mu^2(R(w,h), \alpha_2) = \begin{cases} \frac{1}{\alpha_2}|\alpha_2 - y^{low+1} + y^{low} - h| & \text{if } |y^{low+1} - y^{low} + h| \leq \alpha_2 \\ 0 & \text{otherwise} \end{cases}$$

Similarly, consider the set $\mathcal{E}_3$ of the not packed rectangles such that, if one of them is packed on s^{low} the corresponding segment has height very close to y^{low-1}.

Let $\alpha_3 \in [0,1]$ be a third fit threshold provided by the decision maker. Let a third membership function of an object $e = R(w,h)$ be given by:

$$\mu^3(R(w,h), \alpha_3) = \begin{cases} \frac{1}{\alpha_3}|\alpha_3 - y^{low-1} + y^{low} - h| & \text{if } |y^{low-1} - y^{low} - h| \leq \alpha_3 \\ 0 & \text{otherwise} \end{cases}$$

We have three rules for selecting the new rectangle to be included in the partial solution. These rules consist in selecting a rectangle from $\mathcal{E}_1$, from $\mathcal{E}_1 \cap \mathcal{E}_2$ and from $\mathcal{E}_1 \cap \mathcal{E}_3$. With the first rule, from the list

$$\mathcal{L}_1 = \{R(w,h) \in \mathcal{R}_2 : \mu^1(R(w,h), \alpha_1) \geq 0\},$$

an element is chosen to be packed in the partial solution. For the second rule, from the list

$$\mathcal{L}_2 = \{R(w,h) \in \mathcal{L}_1 : \mu^2(R(w,h), \alpha_2) \geq 0\},$$

an element is chosen to be packed in the partial solution. Finally, for the third rule, from the list

$$\mathcal{L}_3 = \{R(w,h) \in \mathcal{L}_1 : \mu^3(R(w,h), \alpha_3) \geq 0\},$$

an element is chosen to be packed in the partial solution.

In any case, if the corresponding list $\mathcal{L}_i (i = 1, 2, 3)$ is empty, we take from $\mathcal{R}_2$ the rectangle that better fits to s^{low}. If such rectangle does not exist, we rebuild the contour C replacing the segments s^{low} and s^{low+1} by a new segment $s = (y^{low+1}, x_1^{low}, x_2^{low+1})$. The area between x_1^{low}, x_2^{low}, y^{low} and y^{low+1} is wasted.

4.3 Fuzzy stopping rules for the cutting problem

Besides the objective value, there are other values that can be used to evaluate the quality of a solution for the non-guillotine rectangular two-dimensional cutting problem. Some of them are the total area of trim loss and the shape of superior contour.

Ideally, we want to pack the rectangles in such a way that the total area of trim loss is zero and the upper contour is smooth enough.

Given a solution X with objective value $f(X)$, let $TrimLoss(X)$ be the total area of trim loss of X and $Shape(X)$ be the average distance of the upper contour to the total height of X. This average distance is computed by:

$$Shape(X) = \frac{1}{c} \sum_{i=1}^{c} |f(X) - y^i|.$$

Let $\beta_1 \in [0,1]$ a value fixed by the decision maker. The set of solutions with an acceptable trim loss is given by the following membership function μ^4. Let X be a solution, then:

$$\mu^4(X, \beta_1) = \begin{cases} \frac{1}{\beta_1}(\beta_1 - TrimLoss(X)) & \text{if } 0 \leq TrimLoss(X) \leq \beta_1 \\ 0 & \text{otherwise} \end{cases}$$

Let

$$S_1 = \{X : \mu^4(X, \beta_1) \geq 0\}.$$

Similarly, let $\beta_2 \in [0,1]$ be another value fixed by the decision maker. The set of solutions with a smooth enough contour is given by the membership function μ^5. This membership function for a solution X is given by:

$$\mu^5(X, \beta_2) = \begin{cases} \frac{1}{\beta_2}(\beta_2 - Shape(X)) & \text{if } 0 \leq Shape(X) \leq \beta_2 \\ 0 & \text{otherwise} \end{cases}$$

Let
$$S_2 = \{X : \mu^5(X, \beta_2) \geq 0\}.$$

The proposed stopping criterion combines the above sets of solutions. The search stops when a solution of $S_1 \cap S_2$ is reached.

5 Computational experiments

The computational experiments were performed in two phases. In the first phase, the values of the parameters that define the fitness of the rectangles were determined. Then, the proposed constructive methods were compared with the best known procedure for solving the problem. In a second phase, the performance of the three constructive methods to solve moderate and large size instances of the problem was analyzed.

5.1 Tuning the parameters

We have three constructive methods obtained by using the lists of best rectangles $\mathcal{L}_1$, $\mathcal{L}_2$ and $\mathcal{L}_3$. Let FCM_1, FCM_2 y FCM_3 denote these methods. The stopping rule used (see figure 3) was to fix the number of iterations of the loop to $n_{iter} = 40$.

To fix the parameters $(\alpha_i,\ i = 1, 2, 3)$ that control the fitness of the rectangles to the lowest segment, randomly generated packing problems were solved. Several values for α_i, $i = 1, 2, 3$ where fixed and each method was run 5 times for the instances generated. The output variable was the average objective value in the n_{iter} executions of the loop.

For each parameter three levels were taken: $\alpha_1 = 0, 0.1, 0.2$, $\alpha_2 = 0, 0.1, 0.2$ and $\alpha_3 = 0, 0.1, 0.2$ (with $\alpha_2 = \alpha_3$). Therefore we considered the treatments T_{α_1}, T_{α_1,α_2} and $T_{\alpha_1,\alpha_2,\alpha_3}$.

We applied the Friedman nonparametric test to analyze the data (see [2]), since the previous normality and variance equality tests were negative. When the null hypothesis of equality between treatment was rejected, we applied the Friedman multiple comparison tests ([2], page 274) to obtain the significative difference between treatments.

In tables 1 and 2 we show the p-value associated with the Friedman statistic for the problem. In addition, for those problems where the null hypothesis of the equality between treatments was rejected, we show the treatment with the smallest average value and, when the test did not give significative differences, we also show the treatment with the second smallest average value (between round brackets).

From the results of the tests we conclude:

1. FCM_1: the appropriated value for α_1 is 0; i.e., that corresponding to the best fit;

2. FCM_2: the values of the parameters depend on the size of the problem. For problems with 200 or more rectangles, the treatment that gives the best performance is $T_{0,0}$, with the only exception $n = 200$, $w = 60$ and $h_{opt} = 100$. For problems with 100 rectangles, there is not a clear conclusion, since it could be chosen between the treatments $T_{0,0}$ and $T_{0,0.2}$. We selected $T_{0,0.2}$ taking into account the number of times that both $T_{0,0}$ and $T_{0,0.2}$ appear as the best or second best treatment. For $n = 50$, the ties is even more clear if we compare the number of times that a treatment appears as the best one. Then, taking into account the second best treatment, we selected $T_{0,0.1}$.
3. FCM_3: using the above arguments it follows that, for problems with 200 or less rectangles, the recommended choices are $\alpha_1 = 0$ and $\alpha_2 = \alpha_3 = 0.2$. For problems with 300 rectangles, the choice is $\alpha_1 = \alpha_2 = \alpha_3 = 0$.

Table 1. Significance levels (p-value) for the equality among treatments and best treatments (between brackets, second best one)

n	w	h_{opt}		FCM_1	FCM_2	FCM_3
50	30	45	p-value	0.008	0.000	0.000
			treatment	T_0	$T_{0,0.1}$	$T_{0,0}$ ($T_{0,0.2}$)
		60	p-value	0.022	0.005	0.244
			treatment	T_0	$T_{0.1,0.2}$ ($T_{0,0.1}$)	=
	60	90	p-value	0.022	0.026	0.001
			treatment	T_0	$T_{0.1,0.2}$ ($T_{0,0}$)	$T_{0.1,0}$ ($T_{0,0.2}$)
		100	p-value	0.819	0.000	0.000
			treatment	=	$T_{0,0.2}$ ($T_{0,0.1}$)	$T_{0.1,0}$ ($T_{0,0.2}$)
	90	120	p-value	0.007	0.002	0.010
			treatment	T_0	$T_{0.1,0}$ ($T_{0,0}$)	$T_{0,0.2}$
		150	p-value	0.022	0.001	0.156
			treatment	T_0	$T_{0,0.1}$	=
100	50	70	p-value	0.007	0.001	0.000
			treatment	T_0	$T_{0,0}$ ($T_{0,0.2}$)	$T_{0,0}$ ($T_{0,0.2}$)
		110	p-value	0.007	0.000	0.000
			treatment	T_0	$T_{0,0.2}$	$T_{0,0.2}$ ($T_{0,0}$)
	80	90	p-value	0.022	0.003	0.032
			treatment	T_0	$T_{0,0.1}$ ($T_{0,0.2}$)	$T_{0,0}$ ($T_{0,0.2}$)
		140	p-value	0.007	0.001	0.009
			treatment	T_0	$T_{0,0.2}$	$T_{0,0.1}$ ($T_{0,0.2}$)
	100	180	p-value	0.022	0.003	0.000
			treatment	T_0	$T_{0,0}$	$T_{0,0}$ ($T_{0,0.2}$)
		200	p-value	0.247	0.000	0.000
			treatment	=	$T_{0,0}$	$T_{0.1,0}$ ($T_{0,0.2}$)

Table 2. Significance levels (p-value) for the equality among treatments and best treatments (between brackets, second best one)

n	w	h_{opt}		FCM_1	FCM_2	FCM_3
200	60	100	p-value	0.074	0.004	0.001
			treatment	T_0	$T_{0.2,0.2}$	$T_{0.1,0.2}$ ($T_{0.2,0.2}$)
		130	p-value	0.007	0.000	0.000
			treatment	T_0	$T_{0,0.2}$ ($T_{0,0}$)	$T_{0,0}$ ($T_{0,0.2}$)
	100	90	p-value	0.007	0.000	0.000
			treatment	T_0	$T_{0,0}$	$T_{0.1,0.2}$ ($T_{0,0.2}$)
		140	p-value	0.022	0.000	0.000
			treatment	T_0	$T_{0,0}$	$T_{0,0.1}$ ($T_{0,0.2}$)
	200	150	p-value	0.015	0.000	0.000
			treatment	T_0	$T_{0,0}$	$T_{0.1,0}$ ($T_{0,0.2}$)
		200	p-value	0.022	0.000	0.000
			treatment	T_0	$T_{0,0}$	$T_{0,0}$ ($T_{0,0.2}$)
300	80	150	p-value	0.022	0.000	0.000
			treatment	T_0	$T_{0,0}$	$T_{0,0}$
		200	p-value	0.007	0.000	0.000
			treatment	T_0	$T_{0,0}$	$T_{0,0}$
	140	120	p-value	0.022	0.000	0.000
			treatment	T_0	$T_{0,0}$	$T_{0,0}$
		200	p-value	0.022	0.000	0.000
			treatment	T_0	$T_{0,0}$	$T_{0,0.1}$ ($T_{0,0}$)
	200	200	p-value	0.007	0.000	0.000
			treatment	T_0	$T_{0,0}$	$T_{0,0}$
		280	p-value	0.022	0.000	0.000
			treatment	T_0	$T_{0,0}$	$T_{0,0}$

5.2 Comparative

To compare the efficiency and the efficacy of our proposals with respect to Simulated Annealing that uses the Bottom-Left ($SA + BL$) strategy (see [4]), we used the test bed described in [4] (the data of these instances can be found in [6]). It is a bed of 21 instances arranged in 7 categories of 3 instances.

In the first four columns of table 3 we show the characteristics of these problems. The fifth column shows the best objective value and the computational time (in minutes) needed by $SA + LB$. The values have been taken from [4]. From columns 6 to 8 we show the best objective values obtained with FCM_1, FCM_2 and FCM_3. For these methods we do not show the computational time because it was insignificant.

The first notable improvement in the performance is shown by the considerable decreasing in the running time when we use any of our proposals. For instance, to the category C_7 it goes from 4181 minutes to a insignificant time. In addition, the efficacy of FCM_1, FCM_2 and FCM_3 is comparable or better than that of $SA + BL$. Among the three proposals, the best performance is, in general terms, FCM_3. This behavior can also be seen in the results obtained for problems with bigger sizes (see table 4).

Table 3. Best objective values and required times in minutes (average values for category)

Category	n	w	h_{opt}		$SA+BLF$	FCM_1	FCM_2	FCM_3
C_1	16 or 17	20	20	objective	20.8	22.6	22	22
				time	0.7			
C_2	25	40	15	objective	15.9	17	17	16.33
				time	2.4			
C_3	28 or 29	60	30	objective	31.5	33.66	35.33	33.66
				time	4			
C_4	49	60	60	objective	61.8	62.66	64.33	63
				time	33			
C_5	72 or 73	60	90	objective	92.7	94.33	94	93
				time	115			
C_6	97	80	120	objective	123.6	125.33	124.33	124
				time	382			
C_7	196 or 197	160	240	objective	249.6	247	245	246
				time	4181			

5.3 Results for large instances

Table 4 shows the results obtained on random instances with large sizes. The three first columns show the number of rectangles, the width of the sheet and the optimal objective value. The last three columns show the best average objective value in 5 runs of FCM_1, FCM_2 and FCM_3. ¿From these results we conclude that FCM_3 is slightly better than FCM_1 and FCM_2, and the good performance of FCM_1, FCM_2 and FCM_3. As indicated above, the time required is not significant.

5.4 Stopping rule

To evaluate the quality of the proposed stopping rule, randomly generated instances of the cutting problem were solved using FCM_3. Each instance was solved 5 times. Table 5 shows the best average objective values and the

Table 4. Random instances: average objective values

n	w	h_{opt}	FCM_1	FCM_2	FCM_3
50	50	50	52	52	51.8
50	40	60	62.6	63.4	62
100	50	50	52	51.4	51.2
100	50	75	76.8	77	77
200	100	100	101.4	102	101.8
200	120	160	163	162.4	163
500	100	200	202.2	202.8	202.6
500	150	200	202.2	202	202
700	250	320	322.6	323	322.6
700	250	400	403.2	403.6	403

average number of iterations for those runs. Three values for the parameter β_1 and two values for β_2 were fixed. For β_1 we use $0.05A$, $0.03A$ and $0.02A$ where A is the total area of the rectangles

$$A = \sum_{i=1}^{n} w_i h_i.$$

The two values for β_2 are $\beta_2 = 3$ and $\beta_2 = 1.5$. From the results obtained we conclude that:

1. For a fixed value of β_1, better solutions were obtained by decreasing β_2.
2. The best objective values were obtained for small values of the parameters β_1 y β_2.
3. The number of required iterations is small enough. This indicates that the constructive method FCM_3 provides solutions of high quality.

References

1. Bischoff, E.E., Wäscher, G. (1995) Cutting and Packing. European Journal of Operational Research **84** , 503-505
2. Daniel, W.W. (1990) Applied Nonparametric Statistics. PWS-Kent Publishing Company, Boston
3. Dyckhoff, H. (1990) A Typology of Cutting and Packing Problems. European Journal of Operational Research **44** , 145-159
4. Hopper, E., Turton, B. (2001) An Empirical Investigation of Meta-heuristic and Heuristic Algorithms for a 2D Packing Problem. European Journal of Operational Research **128** , 34-57

Table 5. Stopping rule: average objective values and iterations

| | | | | $\beta_1 = 0.05$ | | $\beta_1 = 0.03$ | | $\beta_1 = 0.02$ | |
| | | | | $\beta_2 = 3$ | $\beta_2 = 1.5$ | $\beta_2 = 3$ | $\beta_2 = 1.5$ | $\beta_2 = 3$ | $\beta_2 = 1.5$ |
n	w	h_{opt}							
50	40	60	objective	63.2	62	63.4	62.2	62.6	62
			iteration	3.6	16.6	3	21.8	4	10.2
50	50	50	objective	51.8	51.8	52	51.4	52	51.6
			iteration	1.4	4.2	1.8	11.8	9.8	16.8
100	50	50	objective	52.4	51.8	51.8	52	51.6	51.8
			iteration	1	1	1	1.8	1	1.8
100	50	75	objective	77.2	77	77.4	77	77.4	76.8
			iteration	1.2	2	1	1.4	2	2
200	100	100	objective	102	102	102.2	102	102	102
			iteration	1.2	1.4	1	1.2	1	1.4
200	120	160	objective	163.4	62.4	162.8	162	162.8	162
			iteration	1.8	10.6	2.2	13.2	7.6	10.8

5. Hopper, E., Turton, B. (2001) A Review of the Application of Meta-Heuristic Algorithms to 2D Strip Packing Problems. *Artificial Intelligence Review* **16**, 257-300
6. *OR Library*. http://mscmga.ms.ic.ac.uk/jeb/orlib/stripinfo.html)
7. Pelta, D., Blanco, A., Verdegay, J. L., (2002) *Fuzzy Adaptive Neighborhood Search: Examples of Applications* In this same volume

Heuristics for Optimization: two approaches for problem resolution

J.M. Cadenas, M.C. Garrido, and F. Jimenez

Dpto. Ingeniería de la Información y las Comunicaciones
Universidad de Murcia. 30100-Espinardo. Murcia. Spain

Abstract. In this paper we describe two approaches for a sample problem. One way is to adapt our problem to procedures and techniques which have already been designed, and obtain results in line with the technique applied. The other involves moving away from techniques and concentrating purely on to statement the problem correctly.

For the first approach we propose a heuristic based factoring method (MFGN). For the second, we propose Fuzzy Mathematical Programming (FMP) in order to model the problem. Any technique which can solve the mathematical programming models will serve to solve those obtained. We will use specifically the Optimization Toolbox extends the MATLAB environment (the language of technical computing) and a evolutionary algorithm (EA). This paper describes these approaches to solving different problems within the fuzzy rule learning context.

1 Introduction

When we try to solve a problem, from a conventional viewpoint, we generally use procedures or techniques which have already been designed. This supposes adapting the problem to the requirements of the technique and, hence, consideration is limited to those aspects of the problem that the technique is able to take into account. Furthermore, when we try to compare results obtained by the application of different techniques to a particular problem, it may occur that these are not directly comparable, on account of each technique's taking into consideration different aspects of the problem, i.e., we are not using a common framework.

However, we could ignore well known, already designed procedures or techniques. In this case, in order to solve our problem, we would characterize it and highlight the relevant aspects or characteristics to be taken into account for the solution of the problem. These characteristics would enable us to model the problem.

In this paper we will describe both approaches to the same problem. We will show the process to be followed when applied to the problem of learning fuzzy rules. This process can be understood as the obtaining of a reduced rule set to model a system as simply as possible.

For the first approach we will use the technique known as MFGN (mixture of factorized generalized normals). We will apply it directly to the problem, only some of whose aspects or characteristic we will be able to take into account.

In the second approach we will use Fuzzy Mathematical Programming (FMP) to model the problem. Thus we will be able to move away from the technique used and to concentrate exclusively on to statement the problem correctly.

A large number of data sets have been used to evaluate the accuracy of these methods. We will show the rule learning process when applied to the classical Iris data set (IRIS).

Classical Iris data form a set of examples which have been used extensively to illustrate clustering and classifiers designs. The set of examples contains three classes with 50 examples in each, where each class refers to a type of Iris plant. The three types are Iris Setosa, Iris Versicolour and Iris Virginica. Each example contains four features. The first refers to the length of the sepal, the second to the width of the sepal, the third to the length of the petal and the fourth to the width of the petal (all measurements are in cms).

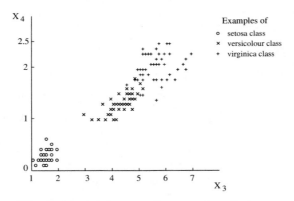

Fig. 1. Iris database on the x_3 and x_4 features

The rest of the paper is organized as follows:

Sections 2 and 3 present both approaches. Firstly, we propose the modelling of the problem through FMP. Once modelled, we solve the model using appropriate technique. Secondly, we propose the MFGN technique as a means to solve the problem. This technique finds solutions within its own constraints.

In the section 4 we analyze the results obtained through the different approaches. Section 5 gives conclusions and comments on the work performed in the paper.

2 Fuzzy rule learning in FMP

2.1 Introduction

The need to find the optimal, or best available, solution to a correctly formulated problem is what gives rise to the proposals of suitably adapted methodologies. From a more specific, although very general, viewpoint, one very important class of problems is the so called optimization problems, usually associated with the search for the maximum o minimum value that a function can attain in a previously specified set. Everything that is related to these problems is situated within the corpus known as Mathematical Programming, which includes a large variety of situations, be they linear cases, non-linear, random, a single or several decision maker, etc.

We have developed the modelling of the fuzzy rules learning problem through (non-linear) mathematical programming. Mathematical programming separates the activity relating to the statement of the problem from the techniques which can be used to solve it. Hence a "user" of mathematical programming may move completely away from the technique used to solve the problem, and concentrate exclusively on the correct statement. This offers a level of abstraction which is of great importance with respect to other techniques.

Once the mathematical programming model has been stated, any of the many techniques which have been studied for solving non-linear programming models can be used (e.g. descending gradient, genetic algorithms, ..).

Let us suppose that a set of M examples (of n characteristics) are classified in N classes. In the fuzzy rule learning process, we will first solve the problem of selecting features. This selection of features aims to simplify the initial set of examples by seeking to override some of the features which form it.

Once the set of examples has been simplified, we apply the clustering method. This method solves the problem of making a fuzzy partition. Once we have found the clusters we use them to generate the fuzzy rules. The rules will be like: "If Antecedent Then Conclusion", where the antecedent will be the fuzzy cluster and the conclusion will be the class with which the cluster is labelled.

Finally, we apply the last method proposed, which is able to simplify rules and to remove rules which are not important in the classification task.

We now describe each one of the problems.

2.2 Feature selection from examples

The mathematical programming model we propose here is based on the model proposed in [9]. In this paper the problem of feature selection is solved (for two classes) by a mathematical programming model with a concave objective function and linear constraints. We will extend the model to solve the feature selection problem for the classification of N classes. The model proposed is a non-linear biobjective one.

The model seeks to minimize two objectives. The first is to minimize the sum of errors committed by each of the discriminators between classes (of the form $xw = \gamma$). The second objective considers the function that 1 will give if the feature is other than 0 in any discriminator, and it should give 0, if it is 0 in all the discriminators.

Thus the optimization model takes the following form:

$$Min \sum_{i=1}^{N-1} \sum_{j=i+1}^{N} \left[\frac{1}{card(S_i)} ||max\{0, -S_i w_{ij} + I\gamma_{ij} + I\}||_1 + \frac{1}{card(S_j)} ||max\{0, S_j w_{ij} - I\gamma_{ij} + I\}||_1 \right]$$

$$Min \; I^t(I - exp(-\alpha \sum_{i=1}^{N-1} \sum_{j=i+1}^{N} |w_{ij}|))$$

where S_i and $card(S_i)$, $i = 1,..,N$ are the N classes and the number of examples of each class, $w_{ij} \in R^n$, $\gamma_{ij} \in R$ are the parameters which define the discriminator between classes S_i and S_j, α is a control parameter of the model, I is a vector of 1 and $||z||_1 = \sum_i^n |z_i|$, $z \in R^M$.

As can be observed, the model is constraint free biobjective and the discriminators obtained are those which commit the smallest error possible when separating the classes with the least number of features.

Fuzzy clustering. In this section we are going to carry out a fuzzy clustering upon a data set. The same objective function that FCM uses could have been used here, but we have used a different one. The fuzziness index characteristic of the function (which is a very small intuitive parameter) has been removed and a parameter called proximity distance has been introduced.

The proximity distance extends the idea of the classic k-mean algorithm. In this algorithm the all that is added is the distance of the example from the nearest centre.

The idea that we propose is that every example must be assigned to a center if the distance of this centre from the example is smaller than the distance of the example from the nearest centre added to the proximity distance. These distances have a value of membership, which describes the degree of membership of the example to the cluster. These degrees of membership are normalized.

The fuzzy mathematical programming model, [2], that solves this problem is:

$$Min \sum_{i=1}^{M} \sum_{j=1}^{k} d_{ij}^2 \frac{p_{ij}}{p_i}$$

s.t. :

$$\left. \begin{array}{l} p_i = \sum_{j=1}^{k} p_{ij} \\ p_{ij} = max(0, (min_i - d_{ij}^2) + t_{ij}) \\ t_{ij} \gtrsim 0 \end{array} \right\} \forall\ i, j$$

with $min_i = \min_{j=1,..,k} \{d_{ij}^2\}$, M the number of examples and k the number of the cluster we wish to find. Each cluster, $Clus_h$, will be defined by its centre C_h and the examples which belong to it, along with their degree of membership.

d_{ij} is the distance from the i-th example to the j-th center, and p_{ij} is a value of membership which describes the degree of membership (still not normalized) of the i-th example to the j-th cluster. In order to normalize the degrees of membership, these are divided by p_i. The fuzzy constraint controls the proximity distance value (t_{ij}). When this distance is 0 then we have the case of the classic k-mean problem. When this distance is greater than 0, then we have the case of fuzzy clusters.

The fuzzy clusters are generated using the solution to the problem. The clusters are defined by a mean and by a membership function of the examples of the clusters. These functions can be defined in several ways, some of which are:

- By using as membership functions the same functional expression obtained when obtaining fuzzy clusters. Thus we have k- fuzzy clusters with membership functions given by;

$$\mu_{ij} = \frac{max(0, (min_i - d_{ij}^2 + t_{ij})}{\sum_{l=1}^{k} max(0, (min_i - d_{il}^2 + t_{il})}, \quad i = 1, .., M, \ \ j = 1, .., k$$

- Another kind of membership functions are those of an exponential type. Hence the membership function of the h-th cluster is given by

$$\mu_h(x) = exp\left[-\frac{||x - C_h||_{Cov_h}^2}{2}\right]$$

where C_h is the centre and Cov_h the fuzzy covariance matrix for the h-th cluster. This Cov_h matrix is defined as follows:

$$Cov_h^{ij} = \frac{\sum_{x \in X} \mu_{xh}(x^i - C_h^i)(x^j - C_h^j)}{\sum_{x \in X} \mu_{xh}} \quad i,j = 1,..,n, \quad h = 1,..,k$$

where X is the set of examples, x^i is the i-th coordinate of the examples vector, and C_h^j is the j-th coordinate of the centre vector of cluster h.

The rules are generated using the fuzzy clusters. The rules are of the form: "If antecedent then conclusion", where the antecedent is the fuzzy cluster and the conclusion is the class with which the cluster is labelled. The class which is labelled as conclusion will be $Clase_i$, that which predominates in the fuzzy cluster $Clus_h$. For each class we will add the degrees of membership of all the examples to the $Clus_h$ cluster, and we select the class giving the highest value.

$$Class_j = \max_{i=1,..,N} \{\sum_{x \in X} min(\mu_h(x), class_i(x))\}, \quad h = 1,..,k$$

where $class_i(x) = \begin{cases} 1 & if\ x\ is\ of\ class\ i \\ 0 & in\ the\ other\ case \end{cases}$

Optimization rules. In this section we focus on a series of problems aimed at optimizing the set of rules learned, [2]. We study two problems in particular; the selection of features from rules, and the selection of rules. Regarding the former, we propose a mathematical programming model capable of selecting features that the selection method from examples is not capable of so doing. As for the selection of rules, the aim is to eliminate unnecessary rules from the classification process, since the task they perform can be done by the remaining rules.

- Feature selection from rules
  In this section we are going to select features from the rules generated in the previous section. The rules that we have used are Gaussian, defined by a mean (m) and a covariance matrix. However, the idea that we propose can be generalized for every type of rule. We select features by looking for a linear discriminant between each pair of rules which have a different conclusion, that is, are labelled with a different class, and which use as few of the features as possible. Finally, the features that use the discriminant will be the features that we will select. The error that a discriminant commits when separating two rules is the membership degree of the discriminant's point which has the highest degree of membership to any of the two rules. In order to determine this, we have to transform the Gaussian to Gaussian with mean equal to 0 and variance equal to 1. This transforms the points from original space to the destiny space in the

following way, $y = \Lambda^{1/2} V^t (x - m) = \theta(x - m)$, where x represents the points of the original space and y the points of the destiny space. Λ is the eigenvalues matrix of the covariance matrix of the Gaussian, and V is the eigenvectors matrix. By calling C_i to the center of the fuzzy set A_i of the rule R_i, w_{ij} and γ_{ij} to the discriminant between the sets A_i and A_j, and θ_i to the matrix θ of the fuzzy set A_i, which can be computed before beginning the optimization process, then the mathematical programming model can be formulated in the following way:

$$Min \sum_{i=1}^{N-1} \sum_{\substack{j=i+1 \\ class(j) \neq class(i)}}^{N} exp(-min\left(\frac{\gamma 1_{ij}^2}{||w1_{ij}||^2}, \frac{\gamma 2_{ij}^2}{||w2_{ij}||^2}\right)/2)$$

$$Min\ I^t(I - exp(-\alpha \sum_{i=1}^{N-1} \sum_{j=i+1}^{N} |w_{ij}|))$$

s.t. :

$$\left.\begin{array}{l} \gamma 1_{ij} = \gamma_{ij} - w_{ij} C_i, \quad w1_{ij} = w_{ij} \theta_i^{-1} \\ \gamma 2_{ij} = \gamma_{ij} - w_{ij} C_j, \quad w2_{ij} = w_{ij} \theta_j^{-1} \\ -w_{ij} C_i + \gamma_{ij} + 1 \leq 0 \\ w_{ij} C_i - \gamma_{ij} + 1 \leq 0 \end{array}\right\} \begin{array}{l} \forall\ i,j, \quad j > i \\ class(j) \neq class(i) \end{array}$$

where class(i) is the class that represents the conclusion of the rule R_i, I is the identity matrix.

- Selection of rules
 We select rules by looking for a reduced set of fuzzy rules that have as much, or almost as much, cover, as the original set of fuzzy rules. The fuzzy mathematical programming model that solves this is:

$$Min\ I^t(I - e^{\alpha v})$$
s.t. :
$$maxcover - \sum_{i=1}^{n} \max_{j=1..k}\{(1 - e^{\alpha v_j})\mu_{ij}\} \lesssim 0$$
$$v \geq 0$$

The first constraint is the fuzzy constraint. Maxcover represents the cover of the whole set of rules, and $\sum_{i=1}^{N} \max_{j=1,...,k}\{(1 - e^{-\alpha v_j})\mu_{ij}\}$ represents the cover of the selected rules, where μ_{ij} is the degree of membership of the i-th example to the j-th cluster and $\alpha > 0$ (a good value is α=5).

2.3 Results obtained with IRIS data

Features selection. The model which represents this problem is a multi-objective one. We can use an EA, [7], to solve the problem directly, or we can use the MATLAB environment to solve the model which has been transformed into a uniobjective one.

By applying the EA we obtain a set of non-dominated solutions (for $\alpha = 5$). We select the solution below as satisfactory:

Table 1. Solution of the biobjective problem

f_1	f_2	CLASSES	w	γ	errors
0.437	1.47	setosa-versicolour	(0,0,0,-4.929)	-3.929	0
		setosa-virginica	(0,0,0,-3.235)	-3.3160	0
		versicolour-virginica	(0,0,-0.126,-3.588)	-6.385	6

By applying the MATLAB environment with $\lambda = 0.5$ as a parameter for weighting the two functions, we obtain the following solution:

Table 2. Solution of the weighted objective

$(1-\lambda)f_1 + \lambda f_2$	CLASSES	w	γ	errors
1.709	setosa-versicolour	(0,0,-0.856,0)	-2.316	0
	setosa-virginica	(0,0,-0.561,0)	-1.847	0
	versicolour-virginica	(0,0,0,-5.408)	-8.904	6

Thus the features selected are the third and the fourth.

Fuzzy clustering and generation of rules. The features selection process selected the third and fourth features. Let us look at the results with these features for clusters 3, 5 and 7 (To solve this model, and the following ones, we will use the method provided by the MATLAB environment).

Numbers of clusters (k)	errors
3	5
5	4
7	4

The "errors" column shows the number of errors obtained when we classify the learning examples with the model learned.

We therefore select the set of 5 clusters as the set of rules defined by the following rules.

Table 3. The set of 5 rules

μ_{zw}	S	class
(1.464,0.244)	$\begin{pmatrix} 0,029504 & 1,517904 \\ 1,517904 & 0,011264 \end{pmatrix}$	setosa
(3.781,1.155)	$\begin{pmatrix} 0,103436 & 7,049124 \\ 7,049124 & 0,019734 \end{pmatrix}$	versicolour
(4.525,1.427)	$\begin{pmatrix} 0,044009 & 9,742719 \\ 9,742719 & 0,020948 \end{pmatrix}$	versicolour
(5.155,1.914)	$\begin{pmatrix} 0,05935 & 10,694579 \\ 10,694579 & 0,067335 \end{pmatrix}$	virginica
(6.027,2.140)	$\begin{pmatrix} 0,152875 & 15,309235 \\ 15,309235 & 0,059998 \end{pmatrix}$	virginica

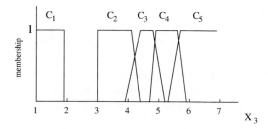

Fig. 2. Projected Clusters on the x_3 feature

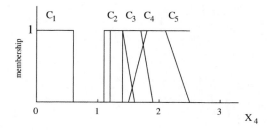

Fig. 3. Projected Clusters on the x_4 feature

Optimization of rules. Features 3 and 4 have been selected already from the examples. Now, new features are not selected from the rules.

We can not apply the rule selection on 3 or 5 rules, since these rules are necessary.

If we apply the rule selection to the set with 7 rules (clusters), we perform the following reasoning: Classes Iris setosa and Iris virginica are defined by rules 1 and 2 rules, respectively. These are necessary for defining them. For the Iris versicolour class (with four rules), we apply the rule selection model, obtaining the following result:

	Solution		
Rule	v	cover	errors
versicolour	(0,0.658,4.359,2.397)	23.843	4

In this way, we can remove the first rule that defines the versicolour class, and so, the set with 7 clusters is reduced to 6 clusters, producing the same number of errors, 4.

3 Fuzzy rule learning in MFGN framework

3.1 Introduction

As we have observed above, we are going to tackle the problem of obtaining a set of rules which describe a system. The objective that we set ourselves in generating this set of rules was that it should be a small (reduced) set of simple rules that best describe the system under study.

However, this is not possible within the framework of the MFGN technique since, although the ultimate aim is to describe the system in the best way possible, the underlying heuristic technique in MFGN is the maximization of the likelihood of the examples available, and it does not take into account that the necessary set of rules be the simplest.

In other words, there is no process defined in MFGN to make a features selection in such a way that the set of rules obtained can be simplified. Hence, in order to perform this process, it is necessary to resort to a another technique which, from a set of rules, will allow features to be eliminated. We have, therefore, to resort to a hybrid method.

The aim is to find, within the MFGN framework, a system description based in a joint density function which includes the dependences of the features of interest. The real density function which describes the system under study is only approximate, finding a series of components which capture objects with independent features. Each component can, therefore, be considered as a cluster and, given that the component is described through normal densities, we can consider the cluster is fuzzy and that the degree of membership of an example to a cluster is given by these normals.

The density function parameters are obtained through the extended EM algorithm, [10], which repeatedly modifies the parameters of the mixture. It begins with a random initialization of the set of parameters, and the E and M steps are repeated until the likelihood of the examples ceases to improve.

Hence the problem set out on page 5 is not solved within the framework of the MFGN technique. In this problem the aim is to generate clusters which correspond to natural groupings of the examples. In MFGN, though, the clusters groups together examples of independent features. Thus, we make an interpretation of the structure of the density function found in this technique in order to generate the set of rules which describes the system.

We are now, therefore, going to describe how to perform the fuzzy clustering with this technique, and then we will apply it to the Iris data base.

3.2 Fuzzy clustering

This section describes how we can, within the framework of the MFGN technique, solve the problem of obtaining a set of rules that describe the behaviour of a system.

Within the framework of the MFGN technique, the system is described through the joint density in the form of a finite mixture with l of the factorized components $C_1, C_2, ...C_l$, [10]:

$$p(z) = p(z^1, z^2, ..., z^n) = \sum_{i=1}^{l} P\{C_i\} p(z/C_i) = \sum_{i=1}^{l} P\{C_i\} \prod_{j=1}^{n} p(z^j/C_i)$$

where:

- The numbers of components, l, is an input parameter.
- $P\{C_i\}$ is the weight of the i-th component.
- n is the number of features with which an object is described.
- For a continuous z^j feature, $p(z^j/C_i)$ is a Normal function with mean μ_i^j and deviation σ_i^j, $N(z^j, \mu_i^j, \sigma_i^j)$.
- For a discrete or nominal z^j feature, $p(z^j/C_i) = \sum_\omega t_{i\omega}^j N(z^j, \omega)$ where $t_{i\omega} = P(z^j = \omega/C_i)$ and $N(z^j, \omega)$ is a Dirac delta.

The learning stage is performed by maximizing the likelihood of the examples available as representatives of the system. In other words, if we have a data base S with M examples, the likelihood of the examples is given by:

$$p(S; \theta) = \prod_{k=1}^{M} p(S^{(k)}; \theta)$$

where θ is an approximation of the parameters vector which maximizes the likelihood of the data of S.

The learning phase in which the θ vector is obtained is performed using the extended EM algorithm, which is based on the EM (Expectation-Maximization) algorithm, but which allows uncertain and imprecise data to appear in the data base, [6,10]. This is one of the main characteristics of the system description within the framework of the MFGN technique, which allows the introduction of data expressed with uncertainty (both objective and subjective) and with imprecision (both fuzzy and crisp). It furthermore allows us to work with discrete or nominal features and with continuous ones.

Using the expression of the above mentioned technique, we can infer the unknown value of any feature of an object, given the known value of the remaining objects. This is another of the main characteristics of the technique since, once the joint density $p(z)$, is obtained, it captures the joint dependency between the features, thus making it unnecessary to perform a particular learning for each partial dependence.

We can say that within the framework of the MFGN technique, the system is described by a set of l clusters, where each cluster groups together those examples with independence in their features. Thus, the set of clusters obtained is expressed by a set of l rules of the following structure:

$$P\{C_i\} \prod_{j=1}^{n} p(z^j/C_i)$$

Since l is an a priori unknown input parameter, then once the learning for a value of a given l has been performed, those clusters whose weight, $P\{C_i\}$ is below a certain, previously established threshold can be eliminated.

With this structure of rules we can obtain the class of an object from the rest of the features which describe it. In this case, if we suppose that z^n is the class feature to be estimated for an object, given the rest of the features, then the expression of the classifier from the set of rules obtained will the following:

$$z^n(z^1, z^2, ..., z^{n-1}) = argmax_\omega q_\omega$$

where:

$$q_\omega = \sum_i \alpha_i I(z^n = \omega)$$

The function I will take the value 1 in those rules with the value $z^n = \omega$. The expression for α_i is the following:

$$\alpha_i = \frac{P\{C_i\}\beta_i}{\sum_h P\{C_h\}\beta_h}$$

where β is factor which represents the degree of membership of the example to the C_i cluster. Its form is obtained in a different way for each type of information ([6,10]).

3.3 Results of the MFGN technique with IRIS data

When carrying out the MFGN technique learning on the Iris data base, and for several values of parameter l, we obtained the following results:

Numbers of clusters (l)	errors
7	4
5	4
3	6

The "errors" column of the above table indicates the number of errors committed on classifying the whole set of data with the system description learning from the set of data itself.

Thus we select the set of 5 clusters, defined by the rules below, as the system description:

Table 4. The set of 5 rules

$P\{C_i\}$	μ_x	σ_x	μ_y	σ_y	μ_z	σ_z	μ_w	σ_w	$t_{Cl,1}$	$t_{Cl,2}$	$t_{Cl,3}$
0.2	6.92	0.5	3.11	0.29	5.84	0.46	2.15	0.23	0	0	1
0.07	5.41	0.36	2.4	0.19	3.61	0.33	1.03	0.05	0	1	0
0.15	6.06	0.38	2.79	0.24	5.06	0.26	1.81	0.21	0	0.14	0.86
9 0.33	5	0.35	3.42	0.38	1.46	0.17	0.24	0.11	1	0	0
0.25	6.05	0.46	2.85	0.26	4.38	0.31	1.38	0.12	0	1	0

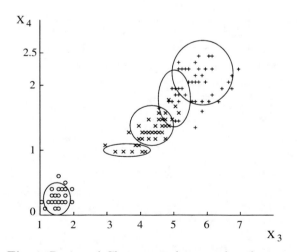

Fig. 4. Projected Clusters on the x_3 and x_4 features

4 Analysis of results

A table that summarizes the results obtained is:

Table 5. Iris data base

Method	Choiced Features	Rules	Choiced Rules	Errors
FMP	3, 4	3	none	5
	3, 4	5	none	4
	3, 4	7	6	4
MFGN	all	3	none	6
	all	5	none	4
	all	7	none	4
FCM	all	4	none	11
	all	11	none	4
GC-ANFIS	all	4	none	3
FM[a]	all	3	none	6
	all	4	none	6

[a] Method proposed in [4]

Methods like GC-Anfis, obtain better results due to their carrying out a supervised adjust process, something that the others methods do not do.

We can observe that the results obtained with the methods (FMP, MFGN) dealt with and other methods are good and similar. But, when we compare the results obtained it may happen that the comparison is not performed in the same framework, due to the fact that the procedures considered different aspects of problem. Because of this, sometimes we can select a rule set which obtain more errors than others.

5 Conclusions

In this paper we have described several methods that can be used in the learning context. These methods solve three types of problems in this area: feature selection, rule selection and clustering. In order to find solutions to these problems, we have used fuzzy mathematical programming and a heuristic based factoring method. The solutions described show that, on the one hand, we can solve problems within the cognitive learning context in an effective way, using the formalism that fuzzy mathematical programming gives us. With this formalism we only have to be concerned about defining the optimization function, since the solutions searching process is implemented by the problem resolution technique that we are using.

On the other hand, we can solve problems, directly, using a resolution technique. The specific technique that we use imposes its restrictions when it solves the problem.

Acknowledgements

The authors thank the Comisión Interministerial de Ciencia y Tecnología (CICyT) for the support given to this work under the projects TIC2000-0062-P4-03.

References

1. Cadenas, J.M., Verdegay, J.L. (1999) Modelos de Optimización con datos Imprecisos. Servicio de Publicaciones. Universidad de Murcia
2. Cadenas, J.M., Garrido, M.C. et al. (2001) Fuzzy Modelling using Fuzzy Mathematical Programming. International Conference in Fuzzy Logic and Technology, 10-13
3. Delgado, M., Gómez-Skarmeta, A.F. et al. (1995) Un enfoque aproximativo para la generación de reglas mediante análisis cluster. V Congreso Español sobre Tecnologías y lógica fuzzy, 43-48
4. Flores, A., Cadenas, J.M. et al. (1998) A local geometrical propierties application to fuzzy clustering. Int. Jour. Fuzzy Sets and Systems **100**, 245-256
5. Flores, A., Cadenas, J.M. et al. (1999) Membership Function in the Fuzzy C-Means algorithm. Int. Jour. Fuzzy Sets and Systems **101** (1), 49-58
6. Garrido, M.C. (1999) Un método de aprendizaje e inferencia a partir de información imperfecta. Tesis Doctoral. Universidad de Murcia
7. Jimenez, F., Gómez-Skarmeta, A.F. et al. (2001). Un algoritmo evolutivo para optimización multi-objetivo con restricciones. Procs. de la IX Conferencia de la Asociación Española para la Inteligencia Artificial, 335-344, Gijón, España
8. Liu, H., Motoda, H. (1998) Feature selection for knowledge discovery and data mining. Kluwer Academic Publishers
9. Mangasarian, O.L. (1997) Mathematical Programming in Data Mining. Data Mining and Knowledge Discovery 1, 183-201. Kluwer Academic Publishers.
10. Ruiz, A., Lopez de Teruel, P.E. et al. (1998) Probabilistic Inference from Arbitrary Uncertainty using Mixtures of Factorized Generalized Gaussians. Journal of Artificial Intelligent Research, **9**, 167-217

Optimization with linguistic variables

Christer Carlsson[1] and Robert Fullér[1,2]

[1] Institute for Advanced Management Systems Research, Åbo Akademi University, DataCity B6734, FIN-20520 Åbo, Finland
[2] Department of Operations Research, Eötvös Loránd University, Pázmány Péter sétány 1C, P. O. Box 120, H-1518 Budapest, Hungary

Abstract. We consider fuzzy mathematical programming problems (FMP) in which the functional relationship between the decision variables and the objective function is not completely known. Our knowledge-base is supposed to consists of a block of fuzzy if-then rules, where the antecedent part of the rules contains some linguistic values of the decision variables, and the consequence part is either a linguistic value of the objective function or a linear combination of the crisp values of the decision variables. In this paper we suggest the use of an adequate fuzzy reasoning method to determine the crisp functional relationship between the objective function and the decision variables, and to solve the resulting (usually nonlinear) programming problem to find a fair optimal solution to the original fuzzy problem. Furthermore, we illustrate how the optimal solution may change if we are able to refine the rule base by introducing some non-monotonicity (dependency) rules.

1 Introduction

After Bellman and Zadeh [1], and then Zimmermann[14] introduced fuzzy sets into methods for handling optimization problems, they cleared the way for a new family of methods to deal with problems which had been inaccessible to and not solvable with standard mathematical programming (MP) techniques. These problems include a number of limitations and simplifications, which have become apparent, as the use of MP models became widespread and common.

The parameters used in MP models need to be valid and accurate, and the estimates used to define them should be built from reliable data sets, which are extensive enough to allow repeated verification and validation. This is not often the case as data may be spotty and incomplete, the data sets are limited and quite often ad hoc, and the validation procedures need to be simplified out of necessity. Nevertheless, MP models are used with the simplifying assumption, that the parameter estimates are good enough for the actual problem - or for the time being (which includes a promise of later adjustments and modifications). The end-result appears to be that optimal solutions found with MP models do not fare too well in the long run, nor do they survive for very long. We need methods to deal with the problems of poor, incomplete data head-on.

The traditional use of one-objective MP models is in many cases an oversimplification as the intentions formulated in decision problems cannot be adequately dealt with without using multiple objectives. It is true that MP models with multiple objectives become complex and difficult to handle and work with, which is why they are avoided. There is even more to multiple objective models: in the standard case we assume that the objectives are independent, which has given us a set of standard tools for finding optimal solutions in cases with multiple objectives. In practical cases, we normally have interdependence among the objectives: some of the objectives may be conflicting, other may be supportive; in cases with greater numbers of objectives, we may have complex combinations of interdependencies [2,7,3]. Model builders shy away from these more complex issues as the methods for finding identifiable optimal solutions have been rather few and, in some cases, have not yet been developed. Thus, we need methods for handling interdependence issues in multiple objective MP models.

In practical problem solving work, and in handling decision problems in the business world, it has become evident for us that probably a majority of the problem solving and decision processes is not of the type, which can be resolved with the precise MP-type models and methods. These practical situations have actors, who are vague about objectives and constraints in the beginning of a process, and then they become more focused and precise as the process is progressing towards some results. This appears to be more of a search-learning process than a systematic process of analysis progressing in well-defined steps towards a predefined goal. Then, for the early stages of the search-learning process we need tools, models and methods, which allow for knowledge-rich imprecision but which have an inner core of methodological structure, such that we in later stages of the process can become analytically more precise as objectives and constraints become more focused. Nevertheless, we want to keep the knowledge-rich substance of the models and methods as we progress to more MP-like modelling. This has long been considered a methodological contradiction, but it has become apparent in recent years [5] that this standard objection is not necessarily true. A combination of linguistic variables and fuzzy logic is emerging as a good approach to have knowledge-rich imprecision, a systematic and firm methodological structure, and effective and fast analytical MP-algorithms.

In this paper, we explore the conditions for optimisation with linguistic variables.

Fuzzy sets were introduced by Zadeh[12] as a means of representing and manipulating data that was not precise, but rather fuzzy. Let X be a nonempty set. A fuzzy set A in X is characterized by its membership function

$$\mu_A \colon X \to [0, 1]$$

and $\mu_A(x)$ is interpreted as the degree of membership of element x in fuzzy set A for each $x \in X$. Frequently we will write simply $A(x)$ instead of $\mu_A(x)$. The family of all fuzzy (sub)sets in X is denoted by $\mathcal{F}(X)$. A fuzzy set A in X is called a fuzzy point if there exists a $u \in X$ such that $A(t) = 1$ if $t = u$ and $A(t) = 0$ otherwise. We will use the notation $A = \bar{u}$.

The use of fuzzy sets provides a basis for a systematic way for the manipulation of vague and imprecise concepts. In particular, we can employ fuzzy sets to represent linguistic variables. A linguistic variable [13] can be regarded either as a variable whose value is a fuzzy number or as a variable whose values are defined in linguistic terms. If x is a linguistic variable in the universe of discourse X and $u \in X$ then we simple write "$x = u$" or "x is $\bar{u}$" to indicate that u is a crisp value of x.

Fuzzy optimization problems can be stated in many different ways [8]. Authors usually consider fuzzy optimization problems of the form

$$\max/\min f(x); \text{ subject to } x \in X,$$

where f or/and X are defind in fuzzy terms. Then they are searching for a crisp x^* which (in a certain) sense maximizes f on X. For example, fuzzy linear programming (FLP) problems can be stated as

$$\max/\min f(x) = \tilde{c}x; \text{ subject to } \tilde{A}x \lesssim \tilde{b}, \qquad (1)$$

where the fuzzy terms are denoted by tilde.

Unlike in (1) the fuzzy value of the objective function $f(x)$ may not be known for any $x \in \mathbb{R}^n$. More often than not we are only able to describe the partial causal link between x and $f(x)$ linguistically using some fuzzy if-then rules. In [6] we have considered constrained fuzzy optimization problems of the form

$$\max/\min f(x); \text{ subject to } \{\Re_1(x), \ldots, \Re_m(x) \mid x \in X \subset \mathbb{R}^n\}, \qquad (2)$$

with

$$\Re_i(x) : \text{if } x_1 \text{ is } A_{i1} \text{ and } \ldots \text{and } x_n \text{ is } A_{in} \text{ then } f(x) \text{ is } C_i, \qquad (3)$$

where A_{ij} and C_i are fuzzy numbers with strictly monotone membership functions; and we have suggested the use of the Tsukamoto fuzzy reasoning method [11] to determine the crisp values of f. In [4] we assumed that the knowledge base is given in the form

$$\Re_i(x) : \text{if } x_1 \text{ is } A_{i1} \text{ and } \ldots \text{and } x_n \text{ is } A_{in} \text{ then } f(x) = \sum_{j=1}^n a_{ij}x_j + b_i \qquad (4)$$

where A_{ij} is a fuzzy number, and a_{ij} and b_i are real numbers. Crisp values of f have been determined by the Takagi and Sugeno[10] fuzzy reasoning method. In both cases the firing levels of the rules have been computed by the product t-norm [9], $T(a,b) = ab$, (to have a smooth output function), and then a solution to the original fuzzy problem (2) has been defined as a solution to the resulting deterministic (usually nonlinear) optimization problem

$$\max/\min f(u), \text{ subject to } u \in X.$$

where the crisp value of the objective function f at $u \in \mathbb{R}^n$, denoted also by $f(u)$, has been determined by Zadeh's compositional rule of inference.

2 Optimization with linguistic variables

To find a fair solution to the fuzzy optimization problem

$$\max/\min f(x); \text{ subject to } \{\Re_1(x), \dots, \Re_m(x) \mid x \in X\}, \qquad (5)$$

with fuzzy if-then rules of form (3) or (4) we first determine the crisp value of the objective function f at $u \in \mathbb{R}^n$, denoted also by $f(u)$, by the compositional rule of inference

$$f(u) := (x \text{ is } \bar{u}) \circ \{\Re_1(x), \cdots, \Re_m(x)\}.$$

That is, in case (3) we apply the Tsukamoto fuzzy reasoning method as

$$f(u) := \frac{\alpha_1 C_1^{-1}(\alpha_1) + \cdots + \alpha_m C_m^{-1}(\alpha_m)}{\alpha_1 + \cdots + \alpha_m}$$

where the firing levels are computed according to

$$\alpha_i = \prod_{j=1}^n A_{ij}(u_j). \qquad (6)$$

Furthermore, in case (4) we apply the Takagi and Sugeno fuzzy reasoning method as

$$f(u) := \frac{\alpha_1 z_1(u) + \cdots + \alpha_m z_m(u)}{\alpha_1 + \cdots + \alpha_m}.$$

where the firing levels of the rules are computed by (6) and the individual rule outputs, denoted by z_i, are derived from the relationships

$$z_i(u) = \sum_{j=1}^n a_{ij} u_j + b_i.$$

In this manner our constrained optimization problem (5) turns into the following crisp (usually nonlinear) mathematical programmimg problem

$$\max/\min f(u); \text{ subject to } u \in X.$$

If X is a fuzzy set with membership function μ_X (e.g. given by soft constraints as in [14]) then following Bellman and Zadeh[1] we define the fuzzy solution to problem (5) as

$$D = \mu_X \cap \mu_f, \qquad (7)$$

where μ_f is an appropriate transformation of the values of f to the unit interval, and an optimal solution to (5) is defined to be as any maximizing element of D.

3 Examples

Example 1 *Consider the optimization problem*

$$\min f(x); \text{ subject to } \{x_1 + x_2 = 1/2,\ 0 \le x_1, x_2 \le 1\}, \tag{8}$$

where

$$\Re_1(x) : \text{if } x_1 \text{ is small and } x_2 \text{ is small then } f(x) = x_1 + x_2,$$
$$\Re_2(x) : \text{if } x_1 \text{ is small and } x_2 \text{ is big then } \quad f(x) = -x_1 + x_2.$$

Let $u = (u_1, u_2)$ be an input to the fuzzy system. Then the firing levels of the rules are

$$\alpha_1 = (1 - u_1)(1 - u_2), \quad \alpha_2 = (1 - u_1)u_2,$$

It is clear that if $u_1 = 1$ then no rule applies because $\alpha_1 = \alpha_2 = 0$. So we can exclude the value $u_1 = 1$ from the set of feasible solutions. The individual rule outputs are computed by

$$z_1 = u_1 + u_2, \quad z_2 = -u_1 + u_2.$$

and, therefore, the overall system output, interpreted as the crisp value of f at u is

$$f(u) = \frac{(1 - u_1)(1 - u_2)(u_1 + u_2) + (1 - u_1)u_2(-u_1 + u_2)}{(1 - u_1)(1 - u_2) + (1 - u_1)u_2} =$$

$$u_1 + u_2 - 2u_1 u_2.$$

Thus our original fuzzy problem turns into the following crisp nonlinear mathematical programming problem

$$\min (u_1 + u_2 - 2u_1 u_2)$$

$$\text{subject to } \{u_1 + u_2 = 1/2,\ 0 \le u_1 < 1,\ 0 \le u_2 \le 1\}.$$

which has the optimal solution $u_1^* = u_2^* = 1/4$ and its optimal value is $f(u^*) = 3/8$. Even though the individual rule outputs are linear functions of u_1 and u_2, the computed input/output function $f(u) = u_1 + u_2 - 2u_1 u_2$ is a nonlinear one.

Example 2 *Consider the problem*

$$\max_X f \tag{9}$$

where X is a fuzzy susbset of the unit interval with membership function

$$\mu_X(u) = 1 - (1/2 - u)^2,$$

for $u \in [0,1]$, and the fuzzy rules are

$$\Re_1(x) : \text{if } x \text{ is small then } f(x) = 1 - x,$$
$$\Re_2(x) : \text{if } x \text{ is big then} \quad f(x) = x.$$

Let $u \in [0,1]$ be an input to the fuzzy system $\{\Re_1(x), \Re_2(x)\}$. Then the firing levels of the rules are $\alpha_1 = 1 - u, \alpha_2 = u$. The individual rule outputs are $z_1 = (1-u)(1-u)$, $z_2 = u^2$ and, therefore, the overall system output is

$$f(u) = (1-u)^2 + u^2 = 2u^2 + 2u + 1.$$

Then according to (7) our original fuzzy problem (9) turns into the following crisp biobjective mathematical programming problem

$$\max \min\{2u^2 + 2u + 1, 1 - (1/2 - u)^2\}; \text{ subject to } u \in [0,1],$$

which has the optimal value of 0.8333 and two optimal solutions

$$\{0.09, 0.91\}.$$

The rules represent our knowledge-base for the fuzzy optimization problem. The fuzzy partitions for lingusitic variables will not ususally satisfy ε-completeness, normality and convexity. In many cases we have only a few (and contradictory) rules. Therefore, we can not make any preselection procedure to remove the rules which *do not play any role* in the optimization problem. All rules should be considered when we derive the crisp values of the objective function. We have chosen the Takagi and Sugeno and the Tsukamoto fuzzy reasoning scheme, because the individual rule outputs are crisp functions, and therefore, the functional relationship between the input vector u and the system output $f(u)$ can be easily identified.

4 Extensions

Assume that besides $\{\Re_1, \ldots, \Re_m\}$ we are able to justify some monotonicity properties in the functional link between x and $f(x)$, for example, "if x_i is very A_{ij} then $f(x)$ is very C_i", where "*very A_{ij}*" and "*very C_i*" are new values of linguistic variables x_i and C_i, respectively. Consider the following very simple optimization problem

$$\max f(x); \text{ subject to } \{\Re_1(x), \Re_2(x) \mid x \in X = [0,1]\}, \tag{10}$$

where

$$\Re_1(x) : \text{if } x \text{ is } small \text{ then } f(x) \text{ is } small$$
$$\Re_2(x) : \text{if } x \text{ is } big \text{ then} \quad f(x) \text{ is } big$$

Let small$(x) = 1 - x$ and big$(x) = x$, and let u be an input to the rule base $\Re = \{\Re_1, \Re_2\}$ then the firing levels of the rules are computed by

$$\alpha_1 = 1 - u, \quad \alpha_2 = u.$$

Then we get

$$f(u) = (1 - u)u + u \times u = u.$$

Thus our original fuzzy problem turns into the following trivial crisp problem

$$\max u; \text{ subject to } u \in [0, 1]. \tag{11}$$

which has the optimal solution $u^* = 1$.

Assume that besides $\{\Re_1, \Re_2\}$ we are able to justify the following monotonicity properties between x and $f(x)$, "if x is very small then $f(x)$ is very small" and " if x is very big then $f(x)$ is very big". Assume further that these monotonicity rules are implemented by

$\Re_3(x)$: if x is *very small* then $f(x)$ is *very small*
$\Re_4(x)$: if x is *very big* then $\quad f(x)$ is *very big*

where

$$(\text{very small})(u) = \begin{cases} 1 - 2u & \text{if } 0 \leq u \leq 1/2, \\ 0 & \text{otherwise,} \end{cases}$$

and

$$(\text{very big})(u) = \begin{cases} 2u - 1 & \text{if } 1/2 \leq u \leq 1, \\ 0 & \text{otherwise,} \end{cases}$$

Then the fuzzy problem

$$\max f(x); \text{ subject to } \{\Re_1(x), \Re_2(x), \Re_3(x), \Re_4(x) \mid x \in X = [0, 1]\},$$

where

$\Re_1(x)$: if x is *small* then $\quad f(x)$ is *small*
$\Re_2(x)$: if x is *big* then $\quad\quad f(x)$ is *big*
$\Re_3(x)$: if x is *very small* then $f(x)$ is *very small*
$\Re_4(x)$: if x is *very big* then $\quad f(x)$ is *very big*

turns into the same crisp optimization problem (11), and therefore, the optimal solution remains the same, $u^* = 1$. If, however, $\Re_2$ does not entail $\Re_4$, but we have

if x is *very big* then $f(x)$ is not *very big*

instead, then the optimal solution changes. Really, the problem

$$\max f(x); \text{ subject to } \{\Re_1(x), \Re_2(x), \Re_3(x), \Re_4(x) \mid x \in X = [0,1]\},$$

where

$\Re_1(x)$: if x is *small* then $\quad f(x)$ is *small*
$\Re_2(x)$: if x is *big* then $\quad f(x)$ is *big*
$\Re_3(x)$: if x is *very small* then $f(x)$ is *very small*
$\Re_4(x)$: if x is *very big* then $\quad f(x)$ is *not very big*

has the following crisp objective function

$$f(u) = \begin{cases} u & \text{if } 0 \leq u \leq 1/2, \\ \frac{5u - 2u^2 - 3/2}{2u} & \text{otherwise,} \end{cases}$$

and the solution to the resulting crisp optimization problem

$$\max f(u); \text{ subject to } u \in [0,1].$$

is $u^* = 0.865$ and its optimal value is $f(u^*) = 0.766$.

5 Summary

We have addressed FMP problems where the functional relationship between the decision variables and the objective function is known linguistically. We have suggested the use of an appropriate fuzzy reasoning method to determine the crisp functional relationship between the decision variables and the objective function and solve the resulting (usually nonlinear) programming problem to find a fair optimal solution to the original fuzzy problem.

We have shown that the refinement of the intial fuzzy rule base (by introducing some non-monotonicity rules) can result in a substantial change of the solution.

References

1. R. E. Bellman and L.A.Zadeh, Decision-making in a fuzzy environment, *Management Sciences*, Ser. B 17(1970) 141-164.
2. C. Carlsson and R. Fullér, Interdependence in fuzzy multiple objective programming, *Fuzzy Sets and Systems*, 65(1994) 19-29.
3. C. Carlsson and R. Fullér, Multiple Criteria Decision Making: The Case for Interdependence, *Computers & Operations Research*, 22(1995) 251-260.
4. C. Carlsson, R. Fullér and S. Giove, Optimization under fuzzy rule constraints, *The Belgian Journal of Operations Research, Statistics and Computer Science* 38(1998) 17-24.

5. C. Carlsson and R. Fullér, Multiobjective linguistic optimization, *Fuzzy Sets and Systems*, 115(2000) 5-10.
6. C. Carlsson and R. Fullér, Optimization under fuzzy if-then rules, *Fuzzy Sets and Systems*, 119(2001) 111-120.
7. R.Felix, Relationships between goals in multiple attribute decision making, *Fuzzy sets and Systems*, 67(1994) 47-52.
8. M.Inuiguchi, H.Ichihashi and H. Tanaka, Fuzzy Programming: A Survey of Recent Developments, in: Slowinski and Teghem eds., *Stochastic versus Fuzzy Approaches to Multiobjective Mathematical Programming under Uncertainty*, Kluwer Academic Publishers, Dordrecht 1990 45-68.
9. B.Schweizer and A.Sklar, Associative functions and abstract semigroups, *Publ. Math. Debrecen*, 10(1963) 69-81.
10. T.Takagi and M.Sugeno, Fuzzy identification of systems and its applications to modeling and control, *IEEE Trans. Syst. Man Cybernet.*, 1985, 116-132.
11. Y. Tsukamoto, An approach to fuzzy reasoning method, in: M.M. Gupta, R.K. Ragade and R.R. Yager eds., *Advances in Fuzzy Set Theory and Applications* (North-Holland, New-York, 1979).
12. L.A.Zadeh, Fuzzy Sets, *Information and Control*, 8(1965) 338-353.
13. L.A.Zadeh, The concept of linguistic variable and its applications to approximate reasoning, Parts I,II,III, *Information Sciences*, 8(1975) 199-251; 8(1975) 301-357; 9(1975) 43-80.
14. H.-J. Zimmermann, Description and optimization of fuzzy systems, *Internat. J. General Systems* 2(1975) 209-215.

Interactive Algorithms Using Fuzzy Concepts for Solving Mathematical Models of Real Life Optimization Problems

C. Mohan[1] and S.K. Verma[2]

[1] Prof. & Dean Amity School of Computer Sciences, Sector 44, Noida-201303, U.P Formerly. Prof. & Head Mathematics , Roorkee University, Roorkee-247667, India
[2] Department of Mathematics, Govt. Engineering College, Sagar,M.P., India

Abstract. In this paper we first briefly survey our work on computational algorithms developed by us for solving mathematical models or real life optimization problems and then present in brief an interactive type computational algorithm which has been developed by us for solving mathematical models of real life optimization problems using fuzzy concepts. The working of this algorithm has been also demonstrated on some test problems taken from literature.

1 Introduction

Optimization problems are encountered in almost every sphere of human activity, such as engineering design, business, management, agriculture etc., whenever a single objective or simultaneously several objectives are to be optimized (maximized or minimized) subject to certain constraints. To efficiently solve such real life optimization problems, the following three points have to be kept in view. First, the mathematical formulation of real life optimization problems must be as close to the realistic situation as possible. The resulting optimization models can be single-objective as well as multiobjective, linear as well as nonlinear, wherein decision variables can have any real value or some or all are restricted to have integer values only. These mathematical models may turn out to be crisp, fuzzy, stochastic, or mixed fuzzy-stochastic depending on the nature of the realistic problem under investigation. Second, the methods developed for solving such models (particularly methods for solving multiobjective optimization problems in interactive manner) must efficiently support the decision making process. In other words, they must enable the DM to get a reasonable insight of the problem structure and support him in continuously updating and revising his decisions during his search for a satisfying solution. Third, the computational techniques used for obtaining numerical solutions must be efficient and reliable. It would also be desirable that these techniques are robust in terms of mathematical properties of the objective and constraint functions so that they can be used to solve a large variety of real life optimization problems.

Classical mathematical programming is insufficient to handle many real-world situations such as those arising in long term planning problems and programming of development strategies. The nature of these problems requires taking into account multiple objectives on the one hand, and various kinds of uncertainties, on the other hand. The classical multiobjective programming is also inadequate to solve such problems as it fails to account for various kinds of uncertainties. These uncertainties (vagueness or ambiguity) can be treated with the help of fuzzy set theory in which there is no sharp transition from non-membership to membership. A multi-objective programming problem modeled in fuzzy environment is called multiobjective fuzzy programming problem (MOFPP). One of the most effective approaches to solve MOFPP is to use interactive approach (Goicoechea (1982), Zimmermann [1985], [1987], Slowinski [1986], Kacprzyk and Orlovski (eds) [1987], Sakawa and Yano [1989], [1994],[1996], Buckley[1990a],[1990b], Rommefanger [1989], [1990], Fedrizzi et al. (eds) [1991], Czyzak [1991], Dutta [1992], Czyzak [1993], Lai and Hwang [1994], Delgado [1994], Carlsson et al. [1996], Thanh [1996], Inuiguchi et al. [1997]. Several of these methods have been successfully applied to realistic problems. However, a majority of these methods deal with multi-objective liner fuzzy programming problems (MOFLPP) and only few deal with multi-objective non-liner fuzzy programming problems (MOFNLPP), particularly those in which some or all variables are restricted to have integer values only (MOFMIPP/MOFIPP). Some of the desirable features which methods for solving MOFPPS should have are: (i) the method should be simple in structure and reliable and efficient in supporting decision making process (ii) it should provide adequate facilities for treatment of fuzziness (iii) the method should preferably be interactive in nature enabling the decision maker to update his / her preferences from time to time during the search for a satisficing solution, and (iv) it should also be supported by a reliable computational algorithm for solving resultant deterministic optimization problem and finally (v) the solution provided should possess some kind of Pareto optimality.

We have also been trying to develop user friendly, efficient and reliable computational algorithms for solving real life optimization problems. RST2 algorithm (Mohan and Shanker (1994)) and RST1AN and RST2 ANU algorithms (Mohan and Nguyen (1999) have been designed to provide global solutions of nonlinear optimization problems with or without integer restrictions on values of decision variables. These algorithms are to an extent heuristic in nature. Whereas RST2 algorithm utilizes controlled random approach to search for global optimal solution, RST2AN and RST2ANU algorithms merge annealing concept with controlled random search to locate for the global optimal solution. These algorithms have proved quite efficient and reliable in solution of a large variety of realistic optimization problems. More recently Deep and Pant have been trying to develop computational algorithms which merge genetic algorithmic approach with controlled random search.

Interactive type computational algorithms named PRELIM and RDIM have also been developed by us for solving multiobjective linear as well as a class of nonlinear optimization problems in crisp, fuzzy, stochastic and mixed fuzzy stochastic environments (Mohan and Nguyen (1997, 1998, 2001).

In view of the vast variety of real life situations wherein optimal solutions as per users specification are to be obtained, there is still need for developing more efficient and reliable interactive type computational algorithms for solving multiobjective optimization problems in crisp, fuzzy and / or stochastic environments.

In this paper we propose a new method for solving multiobjective fuzzy programming problems. The method is interactive in nature and is applicable to linear as well as a class of nonlinear multiobjective problems. In the proposed method multiobjective fuzzy, problems are first converted into deterministic equivalent with the help of Max-min approach and then solved by interactive style of "loosing constraints with the help of safety parameters and gaining objective preference". The proposed method can be used to solve MOFLPP's as well as a class of MOFNLPP's in which some or all decision variables can have real values or restricted to integer values. This method is an intermix of FULPAL method of Rommelfanger [1990] and comparison principle of Slowinski [1986]. It has several of the desirable features mentioned above. It has four components: (i) interpretation and treatment of fuzziness and aggregations of fuzzy goals derived, (ii) a method for interpretation and treatment of fuzziness in constraints which is independent of the style in which fuzziness is treated in the objectives, (iii) an interactive style supporting the DM to understand basic problem structure and his preferences so as to help him in updating his decisions in each interactive phase in the search for a final satisficing solution (which possesses some kind of Pareto optimality) and (iv) use of RSANSDC/ RSANSDCU algorithms for obtaining the numerical solution of the resultant deterministic max-min problem to be solved in each interactive phase.

The remaining sections of this paper are organized as follows. Mathematical formulation of MOFPP's which can be solved by the proposed algorithm and its interpretation in fuzzy environment is given in section 2. The concept of PL-Pareto optimal solutions of MOFPP's is discussed in Section 3. The calculation of aspiration level is next considered in section 4. The proposed interactive algorithm for solving MOFPP's is next presented in section 5. The Working of the proposed algorithm is illustrated by solving some test examples taken from literature in section 6. Comparison of the results with the results available in literature is also given in this section. Certain concluding observations are finally made in section 7. RSANSDC / RSANSDCU computational algorithms which can be used for obtaining global optimal solution of equivalent nonlinear optimization problems are briefly described in an appendix to the paper.

2 Non-Symmetric Treatment of Linear and a Class of Non-Linear Multiobjective Fuzzy Programming Problems

In many of the realistic situations MOFPP can be expressed in the following format:

$$\text{Max } \tilde{c}_{j1}y_1(X) \oplus \tilde{c}_{j2}y_2(X) \oplus \ldots \oplus \tilde{c}_{jn}y_n(X) \tag{1i}$$

$$X = (x_1, x_2, \ldots, x_p), j = 1, 2, \ldots, k \tag{1}$$

s.t.

$$\tilde{a}_{j'1}y_1(X) \oplus \tilde{a}_{j'2}y_2(X) \oplus \ldots \oplus \tilde{a}_{j'n}y_n(X) \leq \tilde{b}_{j'}; j' = 1, 2, \ldots, m \tag{1ii}$$

$$a_i \leq x_i \leq b_i, i = 1, 2, \ldots, p \tag{1iii}$$

where $y_k(X), \forall \kappa$ are linear or non-linear functions of $x_1, x_2, x_3, \ldots, x_p$ in crisp environment. Superscript $\sim$ stands for fuzziness and $\oplus$ stands for extended addition in fuzzy environment (in case no misunderstanding occurs instead of $\oplus$ symbol + may be used). For the sake of convenience in notations, integer restrictions, if any imposed on x_i for some or all indices are not mentioned separately.

In problem (1) each function $y_k(X), \forall \kappa$ is a crisp function defined in the domain $S = [a_1, b_1] \times [a_2, b_2] \times \ldots \times [a_p, b_p] \subset R^p$. Writing $Y = (y_1(X), y_2(X), \ldots, y_n(X))$, $Y = Y(X)$ is a vector function defined in S. Problem (1) which is in general a non linear multiobjective optimization problem with fuzzy parameters becomes a multiobjective linear programming problem with fuzzy parameters in case $p = n$ and $y_k(X) = x_k, \forall k$. Also for the sake of notational convenience, we assume:

$$y_k : S \to R^+ U\{0\} \text{ for } \forall k \tag{2}$$

Problem (1) in compact form may be written as:

$$\text{Max } \sum_{i=1}^{n} \tilde{c}_{ji} y_i(X), j = 1, 2, \ldots, k \tag{3i}$$

$$\tag{3}$$

s.t.

$$\sum_{i=1}^{n} \tilde{a}_{j'i} y_i(X) \leq \tilde{b} \tag{3ii}$$

$$a_i \leq x_i \leq b_i \quad i = 1, 2, \ldots, p \tag{3iii}$$

Each fuzzy parameter may be modeled as a fuzzy interval or fuzzy number which is written as a quadruple of its left and right referent points and left and right spreads.

$$\tilde{c}_{ji} = (c_{ij}^L, c_{ji}^U, \underline{c}_{ji}, \bar{c}_{ji})_{LR'} \quad j = 1, 2, \ldots, k \; ; \; i = 1, 2, \ldots, n$$
$$\tilde{a}_{j'i} = (a_{j'i}^L, a_{j'i}^U, \underline{a}_{ji}, \bar{a}_{j'i})_{LR} \quad j' = 1, 2, \ldots, k \; ; \; i = 1, 2, \ldots, n$$
$$\tilde{b}_{j'} = (b_{j'}^L, \underline{b}_{j'}^U, \underline{b}_{j'}, \bar{b}_{j'})_{LR} \quad j' = 1, 2, \ldots, k$$

For the sake of simplicity we shall assume that the reference functions L and R are linear, although other types of reference functions can also be considered. Usually LHS fuzzy parameters $\tilde{b}_{j'}$ in j'-th constraint (for each $j' \in \{1, 2, \ldots, m\}$) is modelled as a triangular fuzzy number $\tilde{b}_{j'} = (b_{j'}, \underline{b}_{j'}, \bar{b}_{j'})_{LR'}$. In case the left spread of $\tilde{b}_{j'}$ is of no consequence to the DM, we shall be writing $\tilde{b}'_j = (b'_j, \bar{b}'_j)_{RR}$.

Using the extension principle, (3) can be written as

$$\text{Max } \tilde{C}_j(Y) \quad j = 1, 2, \ldots, k$$

s.t.
$$\tilde{A}_{j'}(Y) \lesssim \tilde{b}_{j'} \text{ for } j' = 1, 2, \ldots, m \quad (4)$$
$$a_i \leq x_i \leq b_i \text{ for } i = 1, 2, \ldots, p$$

where $\tilde{C}_j(Y) = \left(C_j^L(Y), C_j^U(Y), \underline{C}_j(Y), \overline{C}_j(Y)\right)_{LR}$ (5)

wherein

$$C_j^L(Y) = \sum_{i=1}^n c_{ji}^L y_i \; ; \; C_{ji}^U(Y) = \sum_{i=1}^n c_{ji}^U y_i,$$
$$\underline{C}_j(Y) = \sum_{i=1}^n \underline{c}_{ji} y_i \; ; \; \overline{C}_j(Y) = \sum_{i=1}^n \bar{c}_{ji} y_i \quad j = 1, 2, \ldots, k$$

and

$$\tilde{A}_{j'}(Y) = \left(A_{j'}^L(Y), A_{j'}^U(Y), \underline{A}_{j'}(Y), \overline{A}_{j'}(Y)\right)_{LR} \quad (6)$$

where in

$$A_{j'}^L(Y) = \sum_{i=1}^n a_{j'i}^L y_i; \; A_{j'}^U(Y) = \sum_{i=1}^n a_{j'i}^U y_i;$$
$$\underline{A}'_j(Y) = \sum_{i=1}^n \underline{a}_{ji} y_i; \quad \overline{A}_{j'}(Y) = \sum_{i=1}^n \bar{a}_{j2} y_2 \quad j' = 1, 2, \ldots, m$$

Fuzzy objective (1i) is treated based on the fuzzy inequality concept of FULPAL method of Rommelfanger (1990). Let $\tilde{d}_j$, the fuzzy aspiration level of $j - th$ fuzzy objective be written as $\tilde{d}_j = (d_j, \underline{d}_j)_{LL}$ if the right spread is of no consequence to the DM. Then the fuzzy inequality relations

$$\tilde{C}_j(Y) \gtrsim \tilde{d}_{j'} \; j = 1, 2, \ldots, k \quad (7)$$

may be converted into following system of conditions

$$\sum_{i=1}^n \left(c_{ji}^L - \underline{c}_{ji}(1 - \varepsilon)\right) y_i \geq d_j - \underline{d}_j(l - \varepsilon) \quad (8)$$

$$\mu_{\tilde{G}j}(\sum_{i=1}^n c_{j'}^L y_i) \to \text{Max}$$

The membership functions $\mu_{\tilde{G}j}(.)$ (which may be interpreted as the subjective evaluation of $\sum_{i=1}^{n} c_{ji}^L y_i$ with regard to the RHS $\tilde{d}_j$), can be defined as a two piecewise linear function,

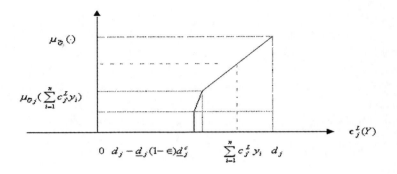

$$\mu_{\tilde{G}j}(.) = \begin{cases} 0 & \text{if } \sum_{i=1}^{n} c_{ji}^L y_i < d_j - \underline{d}_j(1-\varepsilon) \\ \varepsilon + \frac{\sum_{i=1}^{n} c_{ji}^L - (d_j - \underline{d}_j(1-\varepsilon))}{\underline{d}_j^c - (d_j - \underline{d}_j(1-\varepsilon))}(\lambda^* - \varepsilon) & \text{if } d_j - \underline{d}_j(1-\varepsilon) \leq \\ & \qquad \sum_{i=1}^{n} c_{ji}^L y_i < \underline{d}_j^c \\ \lambda^* + \frac{\sum_{i=1}^{n} c_{ji}^L y_i - \underline{d}_j^c}{d_j - \underline{d}_j^c}(1 - \lambda^*) & \text{if } \underline{d}_j^c \leq \sum_{i=1}^{n} c_{ji}^L y_i < d \\ 1 & \text{if } \sum_{i=1}^{n} c_{ji}^L y_i \geq d_j \end{cases} \quad (9)$$

where $\underline{d}_j^c$ corresponds to λ^* and can be changed during the interactive process. Clearly $\underline{d}_j^c$ is a sort of preference level for $\sum_{i=1}^{n} c_{ji}^L y_i$ corresponding to the critical membership level λ^*.

Fuzzy inequality constraints (1ii) may be treated based on comparison principle of fuzzy numbers described in Flip method of Slowanski and is converted into system of conditions:

$$A_{\tilde{j}'}^L(Y) - b_{\tilde{j}'}^U \leq L^{-1}(\tau_{\tilde{j}'})(A_{j'}(Y) + b_{\tilde{j}'}) \tag{10}$$
$$b_{\tilde{j}'}^U + b_{\tilde{j}'} L^{-1}(\eta_{\tilde{j}'}) - A_{\tilde{j}'}^U(Y) - \tilde{A}_{\tilde{j}'}(Y) + R^{-1}(\eta_{\tilde{j}'}) \geq \theta_{\tilde{j}'} j = 1, 2, \ldots, m \tag{11}$$

In consequence, we arrive at the following deterministic mathematical programming problem equivalent to problem(1):

$$\text{Max } \mu_{\tilde{G}_{j'}}\left(\sum_{i=1}^{n} c_{ji}^{L} y_i\right), j = 1.2, \ldots, k \qquad (12)$$
subject to (8), (10), (11) and $a_i \leq x_i \leq b_i$, $i = 1, 2\ldots, p$.

Using Bellmann- Zadeh's min operator as aggregation operator, (12) can be turned into the following max-min type deterministic single objective optimization problem:

$$\text{Max. Min } \mu_{\hat{G}_j}(.), j = 1, 2, \ldots, k \qquad (13)$$
subject to (8), (10), (12) and $a_i \leq x_i \leq b_i$, $i = 1, 2, \ldots, p$.
(it is assumed that the optimal objective function value λ of (13) exists.)

3 PL-Pareto Optimal Solution

Definition: A solution X of (12) is called weak PL-Pareto optimal (PL stands for preference level) for the MOFPP (1) at the specified credibility level ε if there does not exist another solution X' of (12) at which all the values with highest realization (left reference points of the fuzzy objectives $\sum_{i=1}^{n} c_{ji}^{L} y_i$), are respectively better than the corresponding values attained at X.

Definition: A solution X of (12) is called PL-Pareto optimal for the MOFPP (1) at the specified credibility level ε, if there does not exist another solution X' of (12) at which all the values with highest realization (left reference points of the fuzzy objectives) are respectively not worse than the corresponding values attained at X, and at least one of these values is better than the corresponding value attained at X.

With the above definitions of PL-Pareto optimality we can establish the following result.

Theorem

(i) If X is a global optimal solution of (13) with $0 < \lambda < 1$ then X is a weak PL-Pareto optimal solution of the original problem (1) at the chosen credibility level.
(ii) If X is the unique global optimal solution of (13) then X is a PL-Pareto optimal solution of the original problem (1) at the chosen credibility level.

Proof: Suppose X is not a weak PL-Pareto optimal solution of (1). Then there exists another feasible solution X' of (1) such that

$$\sum_{i=1}^{n} c_{j'i}^{L} y_i(X') > \sum_{i=1}^{n} c_{j'i}^{L} y_i(X) \ \forall j, \ j = 1.2, \ldots, k \qquad (14)$$

and X' satisfies constraints of optimization problem (1):
$$\sum_{i=1}^{n} \tilde{a}_{j'i}\, y_i(X') \le \tilde{b}_{j'},\, j' = 1.2,\ldots,m.$$

By comparison principle of Slowinski (1ii) is equivalent to system of conditions (10) and (11) i.e.

$$\sum_{i=1}^{n} a_{j'i}^{L} y_i(X') - b_j'^{U} \le L^{-1}(\tau_{j'}) \sum_{i=1}^{n} a_{j'i} y_i(X\prime) + b_{j'} \tag{15}$$

$$b_{j'}^{U} + b_{j'} L^{-1}(\eta_{j'i}) - \sum_{i=1}^{n} \underline{a}_{j'i} \underline{y}_i(X\prime) - \sum_{i=1}^{n} \bar{a}_{j'i} \underline{y}_i(X\prime) R^{-1}(\eta_{j'}) \ge \theta_{j'}$$
$$j' = 1,2,\ldots,m$$

Also (14) $\to \sum_{i=n}^{n} [c_{ji}^{L} - \underline{c}_{ji}(1-\varepsilon)]\, y_i(X') \ge d_j - \underline{d}_j(1-\varepsilon),\, j = 1,2,\ldots,k.$
This together with (10) and (11) implies that X' is a feasible solution of (13).

Now without loss of generality suppose $\lambda = \min_{x}\{\mu_{\tilde{G}_j}(.)\forall j\}$ is uniquely achieved at $\mu_{\tilde{G}_1}(.)$. Since the membership functions $\mu_{\tilde{G}_j}(\sum_{i=1}^{n} c_{ji}^{L} y_i(X)), \forall j$ are increasing functions, (14) implies
$\mu_{\tilde{G}_j}(\sum_{i=1}^{n} c\,_{ji}^{L}\, y_i(X')) \ge \mu_{\tilde{G}_j}(\sum_{i=1}^{n} c\,_{ji}^{L}\, y_i(X))$ with $j = 1,2,\ldots,k.$

In particular

$\mu_{\tilde{G}_j}(\sum_{i=1}^{n} c\,_{ji}^{L} y_i(X')) > \mu_{\tilde{G}_j}(\sum_{i=1}^{n} c\,_{ji}^{L} y_i(X))$

(since $\mu_{\tilde{G}_j}(\sum_{i=1}^{n} c_{ji}^{L} y_i(X')) = \lambda \in (0,1)$, $\mu_{\tilde{G}_j}(\sum_{i=1}^{n} c_{ji}^{L} y_i(X))$ is strictly increasing).

Hence
$$\min_{x}\{\mu_{\tilde{G}_j}(\sum_{i=1}^{n} c_{ji}^{L} y_i)\,\forall j\} \text{ is greater than } \mu_{\tilde{G}_1}(.) = \lambda = \min_{x}\{\mu_{\tilde{G}_j}(\sum_{i=1}^{n} c_{ji}^{L} y_i)\,\forall j\}$$

This together with the fact that X' is a feasible solution of (13) implies that X in not a global solution of (13) This contradicts our assumption that X is a global optimal solution of (13). The proof of the first of the theorem is thus completed. The second part of the theorem can be proved similarly.

Since problem (13) is in general a nonlinear optimization problem, any suitable global optimization technique can be used for obtaining a global optimal solution to (13) which according to the above theorem will be a PL-Pareto optimal solution of the original MOFPP (1). In our proposed algorithm we recommend use of RSANSDC/RSANSDCU algorithms for this purpose, Situations in which (2) does not hold can also be handled. Since the recommended global optimization computational techniques depend purely on functional evaluations for their working, restriction (2) can also be removed and replaced by:

$$Y'_K S \to R, \qquad \forall k \qquad (17)$$

However, since the extended multiplication in fuzzy environment follows the rule: $(a^L, a^R, \underline{a}, \bar{a})_{LR}\, y = (a_Y^L, a_y^R, \underline{a}y, \bar{a}y)_{L\hat{R}}$ in case $y \geq 0$ and $(a^L, a^R, \underline{a}, \bar{a})_{LR}\, y = (a_Y^R, a_y^L, \underline{a}y, \bar{a}y)_{LR}$ in case $y < 0$, suitable changes have to be made while calculating $C_j(Y)^\varepsilon$. Namely:

$$C_j(Y)^\varepsilon = \sum_{i=1}^{n}(c_{j\bar{i}} - \tilde{\underline{c}}_{ji}(i-\varepsilon))\, y_i \qquad (18)$$

where

$$\tilde{\underline{c}}_{ji} = \begin{cases} \underline{c}_{ji} & \text{if } y_i \geq 0 \\ -\underline{c}_{ji} & \text{if } y_i < 0 \end{cases} \quad \text{and} \quad c_{ji} = \begin{cases} c_{ji}^L & \text{if } y_i \geq 0 \\ c_{ji}^R & \text{if } y_i < 0 \end{cases}$$

and (8) becomes

$$\sum_{i=1}^{n}(c_{ji}^L - \tilde{\underline{c}}_{ji}(1-\varepsilon))\, y_i > d_j - \underline{d}_j(1-\varepsilon) \qquad (19)$$

Similar changes will have also to be made when calculating $\underline{d}_j$. These changes however, do not effect the working of RSANSDC/RSANSDCU algorithms as these algorithms depend purely on function evaluations at randomly generated feasible points $X \in S$.

The disadvantage of the min operator used for aggregating the fuzzy goals of the objective and the limited information at the disposal of the DM can be compensated by an interactive procedure which helps him/her to gradually learn more and more about the problem structure and his/her preferences, so as to suitably revise them at each interactive phase while moving gradually towards a satisfying solution.

4 Calculation of Fuzzy Aspiration Levels

The fuzzy aspiration level $\tilde{d}_j = (d_j, \underline{d}_j)_{LL}$ may be determined as:

$$d_j = \sum_{i=1}^{n} c_{ji}^L y_i^*$$

where $Y^* = (y_1^*, y_2^*, \ldots, y_n^*) = Y(X^*)$ with $X^* = (x_1^*, x_2^*, \ldots, x_p^*)$, being the optimal solution of the following optimization problem:

$$\text{Max} \sum_{i=1}^{n} c_{ji}^L y_i$$

s.t. $\qquad\qquad\qquad\qquad\qquad\qquad\qquad\qquad\qquad\qquad$ (19i)

$$\sum_{i=1}^{n}(a_{j'i}^R + \bar{a}_{j'i}(1-\varepsilon))y_i \leq b_{j'i} + \bar{b}_j(1-\varepsilon)$$

and $a_i \leq x_i \leq b_i, \qquad i = 1, 2, \ldots, p$

$\underline{d}_j$ is determined by:

$$d_j - \underline{d}_j(1-\varepsilon) = Min\left[\sum_{i=1}^{n}(c_{ji}^L - \underline{c}_{ji}(1-\varepsilon))y_i^*\right] \qquad (20)$$

where minimization is taken on the set $y_s^* = Y(X_s^*)$, $s = 1, 2, \ldots, 2k$, wherein $X_S^*, s = 1, 2, \ldots, k$ are optimal solutions of the optimization problems:

$$\text{Max} \sum_{i=1}^{n} c_{\ell i}^L y_i$$

s.t

(19) and $a_{j'} \leq x_i \leq b_i$; $\quad i = 1, 2, \ldots, p$ for $\ell = 1, 2, \ldots, k$

Also X_S^*, $s = k+1, k+2, \ldots, 2k$, are the optimal solutions of the optimization problems:

$$\text{Max} \sum_{i=1}^{n} \left(c_{\ell i}^L - \underline{c}_{\ell i}(1-\varepsilon)\right) y_i \qquad (21)$$

s.t

(19) and $\sum_{i=1}^{n} a_{j'i}^R y_i \leq b_{j'}$ $j' = 1, 2, \ldots, m$ and $\ell = 1, 2, \ldots, k$.

The determination of fuzzy aspiration levels $\tilde{d}_j \forall j$ as shown above is, in fact, based on the pay off type information inherent in the problem structure (for more details see Rommelfanger [1990]). However, if desired, the DM may specify these fuzzy aspiration levels on his own incorporating the above pay off information.

5 The Proposed IMMOFP Interactive Algorithm

The computational steps of the proposed algorithm may be summarized as follows:

Initialization.

Specify[1] ε, λ^*. Obtain d_j, $\underline{\tilde{d}}_j$ and specify $\underline{d}_j^C \in \left[d_j^L, d_j^L - \underline{d}_j(1-\varepsilon)\right]$ for $j = 1, 2, \ldots, k$. Tighten all the constraints with safety parameters $\tau_{j'}, \eta_{j'}$, i.e. $\tau_{j'}=1$, $\eta_{j'} = 1$ for $j' = 1, 2, \ldots, m$. Set the counter for the number of iterations $\ell = 1$.

Iterations:

Step 1: Solve the deterministic optimization problem (13) using RSANSDC / RSANSDCU algorithms and provide the DM with:

---

[1] The word 'specify' is referred to the DM's actions.

(I) The compromising solution λ_ℓ, $X_\ell = (x_1^\ell, x_2^\ell, \ldots, x_p^\ell)$.

(II) The value (the left reference point) of fuzzy objectives
$$\sum_{i=1}^{n} c_{ji}^L y_i, \quad j = 1, 2, \ldots, k.$$

(III) The values (the right reference point) of LHS' of the fuzzy constraints
$$\sum_{i=1}^{n} a_{j'i}^R y_i \quad j' = 1, 2, \ldots, m.$$

Step 2:

(I) if $\lambda_l < \lambda^*$ then X_ℓ is not a satisfactory solution as the values of some of the membership function $\mu_{\tilde{G}_j}(.)$ are less than λ^* (implying thereby that the corresponding preference levels have not been achieved). The DM may like to loosen some of the constraints with parameters τ_j', $\eta_{j'}$ and /or increase at least one of $\underline{d}_j^c$, $j = 1, 2, \ldots, k$ and go to (III).

(II) if $\lambda > \lambda^*$ then all the preference levels have been achieved, in case DM is satisfied with X_ℓ or if $|\lambda - \lambda^*| < \delta$ (where δ is a user prescribed small positive value) the interactive procedure may be terminated with the satisficing solution X_ℓ. However if the DM wishes to continue the exploration further then he may loosen the constraints with the help of safety parameters and / or increase at least one of $\underline{d}_j^c, j = 1, 2, \ldots, k$.

(III) Set $\ell = \ell + 1$ and go to step 1.

In the proposed algorithm DM's preferences with regard to the objectives are progressively articulated through the critical membership level enabling the DM to efficiently compromise his/her preferences.[2]

6 Illustrative Examples

The working of the IMMOFP algorithm has been tested on several examples taken from literature. In each of these problems the solution available in literature was assumed as the target solution of that problem which DM wants to achieve as closely as possible. The results show that IMMOFP method can be used to solve single and multiobjective linear as well as a class of nonlinear programming problems of the type (1). In this section we demonstrate the working of the algorithm on industrial water pollution management in an artificial river basin.

---

[2] Note: Here we have considered only MOFPP's with objectives of two types: minimization and maximization. In realistic situations there may be a need of considering a fuzzy equal goal of the type $\tilde{z} \cong \tilde{d}$, i.e. the objective $\tilde{z}$ should be in vicinity of $\tilde{d}$. By replacing each of such equal goals by two fuzzy goals $\tilde{z} \leq \tilde{d}$ and $\tilde{z} \geq \tilde{d}$ such type of MOFPP's can also be solved using IMMOFP method.

Example: Industrial Water Pollution Management Problem The problem is of industrial water pollution management in an artificial river basin. The mathematical formulation of the water quality and economic objectives and the constraints as given in Monarchi et. al. [1975] for this problem is:

$$\text{Max } f_1(X) = 4.75 + (5.68 \times 10^{-5})(4.00 \times 10^4)(x_1 - 0.3)$$

$$\begin{aligned}\text{Max } f_2(X) = &\, 2.00 + (1.31 \times 10^{-5})(4.00 \times 10^4)(x_1 - 0.3) + \\ & (2.18 \times 10^{-5})(1.28 \times 10^5)(x_2 - 0.3) + \\ & (3.15 \times 10^{-5})(2.80 \times 10^4)(w_1 - 0.3) + (5.53 \times 10^{-5}) \\ & (4.80 \times 10^4)(w_2 - 0.3).\end{aligned}$$

$$\begin{aligned}\text{Max } f_3(X) = &\, 5.10 + (0.422 \times 10^{-5})(4.00 \times 10^4)(x_1 - 0.3) + \\ & (0.764 \times 10^{-5})(1.28 \times 10^5)(x_2 - 0.3) + \\ & (0.771 \times 10^{-5})(2.80 \times 10^4)(w_1 - 0.3) + \\ & (1.600 \times 10^{-5})(4.80 \times 10^4)(w_2 - 0.3).\end{aligned}$$

$$\text{Max } f_4(X) = 0.2 \times 10^{-4}[3.75 \times 10^{-5} - 0.6(59/(1.09 - x_1^2) - 59) \times 10^3]$$

$$\text{Max } f_5(X) = 2.4 \times 10^{-3}(0.75)(532/(1.09 - x_2^2) - 532).$$

$$\text{Max } f_6(X) = (3.33 \times 10^{-3}(0.75)(450/(1.09 - x_3^2) - 450).$$

subject to:

$$\begin{aligned}1.00 & + (8.30 \times 10^{-7})(4.00 \times 10^4)(x_1 - 0.3) \\ & + (1.45 \times 10^{-7})(1.28 \times 10^5)(x_2 - 0.3) \\ & + (3.49 \times 10^{-7})(9.57 \times 10^4)(x_3 - 0.3) \\ & + (7.30 \times 10^{-7})(2.80 \times 10^4)(w_1 - 0.3) \\ & + (1.62 \times 10^{-6})(4.80 \times 10^4)(w_2 - 0.3) \\ & + (7.33 \times 10^{-5})(3.57 \times 10^4)(w_3 - 0.3) \geq 3.5,\end{aligned}$$

and $0.3 \leq x_i \leq 1.0$, $i = 1, 2, 3$ where $w_i = 0.39/(1.39 - x_i^2), i = 1, 2, 3$.

Several solutions to the model are available in literature. The solution as obtained by the sequential multiobjective problem solving (SEMOFS) method reported in Goicocehea et al. [1982] is $X^1 = (0.807, 0.837, 0.817)$ with $f_1 = 5.9, f_2 = 4.67, f_3 = 5.97, f_4 = 6.6, f_5 = 1.5, f_6 = 1.54$. The solution as obtained by the interactive group compromise method (IGCM) reported in Sharma [1991] is $X^2 = (0.86, 0.90, 0.83)$ with $f_1 = 6.02, f_2 = 5.22, f_3 = 6.14, f_4 = 6.18, f_5 = 2.46, f_6 = 1.67$. Assuming that the DM takes X^2 as the target solution to be achieved, the problem has been solved by using IMMOFP method. For this the fifth and sixth objective functions were first converted into the format of IMMOFP method as:

$$\text{Max } f'_5(X) = -f_5(X)$$
$$\text{Max } f'_6(X) = -f_6(X)$$

Initialization : Set $\varepsilon = 0.1, \lambda^* = 0.5$ and obtain

$$d_1 = 6.340400 \quad , d_1 - \underline{d}_1(1-\varepsilon) = 4.750000$$
$$d_2 = 6.795560 \quad , d_2 - \underline{d}_2(1-\varepsilon) = 2.304601$$
$$d_3 = 6.597020 \quad , d_3 - \underline{d}_3(1-\varepsilon) = 5.192968$$
$$d_4 = 7.500000 \quad , d_4 - \underline{d}_4(1-\varepsilon) = 0.341333$$
$$d_5 = 0,000000 \quad , d_5 - \underline{d}_5(1-\varepsilon) = -9.682400$$
$$d_6 = -1.605783 \, , d_6 - \underline{d}_6(1-\varepsilon) = -4.653551$$

The DM initializes the interactive solution process with loose preference levels keeping in his mind the target solution to be achieved and sets $\underline{d}_1^c = 5, \underline{d}_2^c = 3.5, \underline{d}_3^c = 5.2, \underline{d}_4^c = 3.5, \underline{d}_5^c = -5, \underline{d}_6^c = -2.5$

Iterations.

Iter No.	λ_ℓ	Fuzzy Obj.1	Fuzzy Obj.2	Fuzzy Obj.3	Fuzzy Obj.4	Fuzzy Obj.5	Fuzzy Obj.6
1.	0.759610	6.039910	5.209169	6.132337	6.107301	-2.393520	-1.693790

$$X_1 = (0.868405, 0.897289, 0.830534)$$

The DM is not satisfied with the obtained solution. He wants to increase the fourth objective function value. For this, he sets: $\underline{d}_1^c = 5.5, \underline{d}_2^c = 4.5$,$\underline{d}_3^c = 5.8, \underline{d}_4^c = 6.0, \underline{d}_5^c = -3, \underline{d}_6^c = -2$ to obtain:

2.	0.626389	5.965535	5.080268	6.094704	6.405389	-2.224631	-1.894065

$$X_2 = (0.835007, 0.888301, 0.847114)$$

The DM is still not satisfied with the obtained solution. He wants to increase the first three objective function values. For this he sets: $\underline{d}_1^c = 6.01, \underline{d}_2^c = 5.22$,$\underline{d}_3^c = 6.13, \underline{d}_4^c = 6.18, \underline{d}_5^c = -2.5, \underline{d}_6^c = 1.67$.

Now he obtains:

3.	0.498585	6.025337	5.247311	6.144662	6.174183	-2.527290	-1.673912

$$X_3 = (0.859957, 0.902810, 0.830102)$$

The DM is satisfied with the obtained solution, since it is reasonably close to the target solution. The results of this problem are reported in Table 1. For comparison the results earlier obtained in literature are also reported in the Table.

Table 1. Solutions of Industrial water pollution management in an artificial river basin problem.

Method	Fuzzy Obj.1	Fuzzy Obj.2	Fuzzy Obj.3	Fuzzy Obj.4	Fuzzy Obj.5	Fuzzy Obj.6
PRELI	6.022	5.229	6.139	6.188	2.486	1.668
(Mohan et.al., 1999)	\multicolumn{6}{c}{$X = (0.86, 0.901, 0.829)$}					
IMMOFP	6.025	5.247	6.145	6.174	2.527	1.674
	\multicolumn{6}{c}{$X = (0.86, 0.903, 0.830)$}					
SEMOPS	5.9	4.67	5.97	6.6	1.5	1.54
(Goicoechea et.al.,1982)	\multicolumn{6}{c}{$X = (0.807, 0.837, 0.817)$}					
IGCM	6.02	5.22	6.14	6.18	2.46	1.67
Sharma,1991	\multicolumn{6}{c}{$X = (0.86, 0.90, 0.83)$}					

7 Concluding Observations

In this paper an interactive method has been proposed for solving realistic liner as well as a class of non-linear multi-objective programming problems. This interactive method (named IMMOFP) is applicable to non-symmetric models in which fuzzy objectives are treated as fuzzy goals and constraints are treated by comparison principle of Slowinski [1986]. The decision variables in the problems can have real or integer values as desired by the DM. Some other methods are also available in literature for solving MOFPP's. However, most of these methods are applicable only to symmetric models of linear problems in which the objectives as well as the constraints are treated equally as fuzzy goals. In many real life situations it may not be possible to formulate the mathematical models of the problems which are linear. Moreover it may not be always desirable to treat the fuzzy objectives and fuzzy constraints equally as fuzzy goals. Unlike symmetric methods (for example PRELIM method of Mohan *et. al.* IMMOFP method does not treat fuzzy objectives and fuzzy constraints equally. Fuzzy goals are set for fuzzy objectives and safety parameters are defined for fuzzy constraints. These parameters are responsible for the safety of crisp constraints.

The method provides the DM with coherent and systematic way for solving the multi-objective programming models where the parameters involved in the objectives and constraints are ambiguously known and fuzzy objectives are understood in vague manner. Whereas FULPAL method of Rommelfanger

[1990] and FLIP method of Slowinski [1986] solve the multi-objective linear programming problems, IMMOFP method can solve linear as well as a class of non-linear multi-objective fuzzy programming problems wherein integer restrictions for some or all decision variables can also be present.

The proposed method however, assumes that there is only a single decision maker or a group of decision makers whose consensus decision is known at each stage and is to be implemented. It is possible to suitably modify this method to solve problems in stochastic and fuzzy / stochastic environments.

References

1. Buckley J. J., *Stochastic versus possibilistic programming*, Fuzzy Sets and Systems, 34, 173-177, I 990(b).
2. Carlsson C. and Fuller R., *Fuzzy multiple criteria decision making: Recent developments*, Fuzzy Sets and Systems, 78, 139-153, 1996
3. (i) Czyzak P. and Slowinski R., *FLIP: Multiobjective fuzzy linear programming software with graphical facilities*, in Fedrizzi M., Kacprzyk J. and Roubens M. (eds), *Interactive fuzzy optimization*, Lecture Notes in Economics and Mathematical Systems, 368, 168-187, 1991(a).
   (ii) Czyzak P. and Slowinski R., *A fuzzy MOLP method with graphical display of fuzziness*, Annales Univ., Sci., Budapest Compo 12, 59-67, 1991(b).
   (iii) Czyzak P. and Slowinski R., *A visual interactive method for MOLP problems with fuzzy coefficients in* : R. Lawen and M. Roubens (eds.), *Fuzzy Logic: State of Art*, Kluwer, Dordrecht, 321-332, 1993.
4. Delgado M., Kacprzyk J., Verdegay J. L. and Vila M. A. (eds), *Fuzzy optimization: Recent advances*, Physica Verlag, 1994.
5. Deep K . and Pant M. Genetic random search techniques for global optimization, submitted to International Journal of Computer Mathematics.
6. Dutta D., Tiwari R.N. and Rao J.R., Multiple objective linear fractional programming-a fuzzy set theoretic approach, Fuzzy Sets and Systems, 52, 39-45, 1992
7. Fedrizzi M., Kacprzyk J. and Roubens M. (eds), Interactive fuzzy optimization, Lecture Notes in Economics and Mathematical Systems, 368, 1991.
8. Goicoechea A., Hansen D.R. and Duckstein L., Multiobjective decision making with engineering and business applications, John Wiley & Sons, NY, 1982.
9. Inuiguchi M. and Sakawa M. and Abboud N., *A fuzzy programming approach to multiobjective 0-1 Knapsack problems*, Fuzzy sets and systems, 86, 1-14, 1997.
10. Kacprzyk J. and Orlovski S. (eds), *Optimization models using fuzzy sets and possibility theory*, D. Reidal, Dordrecht, 1987.
11. (i) Lai Y.J. and Hwang C.L., Fuzzy mathematical programming: methods and applications, Springer-Verlag, Heidelberg, 1992.
    (ii) Lai Y.J. and Hwang C.L., Interactive fuzzy multiple objective decision makin, in Delgalo M., Kacprzyk J., Verdegay J.A. and Vila M.A. (eds), Fuzzy optimization: Recent advances, Physica-Verlag, Germany, 178-198, 1994.
12. (i) Mohan C. and Shanker K.. *Computational algorithms based on random search for solving global optimization problems*, International Journal of Computer Mathematics, 33. 115-126, 1990.

(ii) Mohan C. and Shanker K., *A controlled random search technique for global optimization using quadratic appro.timation,* Asia-Pacific Journal of Operational Research, 11, 93-101, 1994.

(iii) Mohan C and Nguyen, H.T., A fuzzyfying approach to stochastic programming, Opsearch, 34, 73-96, 1997.

(iv) Mohan C and Nguyen, H.T., Reference direction interactive method for solving multi-objective fuzzy programming problems, European Journal of Operation Research, 107, 599-613, 1998.

(v) Mohan C and Nguyen, H.T., A controlled random search technique incorporating the simulated annealing concept for solving integer and mixed integer global optimization problems, Computational Optimization and Applications, 14, 105-132, 1999.

(vi) Mohan C and Nyuyen, H.T. Preference label interactive method for solving multi objective fuzzy programming problems, Asia Pacific Journal of Operational Research, 16, 63-86, 1999.

(vii) Mohan C and Nguyen, H T. An interview satisficing method for solving multiobjective mixed fuzzy stochastic programming problems., Fuzzy Sets and System, 117, 61-79,2001.

13. Monarchi D. E., Kisiel C.C. and Duckstein Lk, Interactive multiobjective programming in water resources: A case study, Water Resource Research, 9, 837-880, 1975.

14. (i) Rommelfanger H., *Interactive decision making in fuzzy linear optimization problems*, European Journal of Operational Research, 41, 210-217, 1989.

(ii) Rommelfanger H., *FULPAL-An interactive method for solving multiobjective fuzzy linear programming problems,* in Slowinski R. and Teghem J. (eds), *Stochastic versus fuzzy approaches to multiobjective mathematical programming under uncertainty,* Kluwer Academic Publisher, Dordrecht, 279-299, 1990.

(iii) Rommelfanger H., *FULP-A PC supported procedure for solving multicriteria linear programming problem.\ " with fuzzy data,* in Fedrizzi, Kacprzyk J. and Roubens M., (eds), *Interactive fuzzy optimization,* Lecture Notes in Economics and Mathematical Systems, 360, 154-167, 1991.

15. (i) Sakawa M. and Yano H., *An interactive fuzzy satisficing method for multiobjective nonlinear programming problems with fuzzy parameters,* Fuzzy Sets and Systems, 30, 221-238, 1989.

(ii) Sakawa M. and Yano H., *Feasibility and Pareto optimality for multiobjective nonlinear programming problems with fuzzy parameters,* Fuzzy Sets and Systems, 43, 1-15, 1991.

(iii) Sakawa M. and Yano H., *A three level optimization method for fuzzY large scale multiobjective non-linear programming problems,* Fuzzy Sets and Systems, 81, 141-155, 1996.

16. (i) Slowinski R., *A multicriteria fuzzy linear programming method for water supply system development planning,* Fuzzy Sets and Systems, 19, 217-237, 1986.

(ii) Slowinski R., *Interactive multiobjective optimization based on ordinal regression,* in Lewandowski A. and Volkovich V. (eds), *Multiobjective problems of mathematical programming,* Springer Verlag, Heidelberg, 93-99, 1991.

17. Thanh N. H., *Some global optimization techniques and their use in solving optimization problems in crisp and fuzzy environments,* Ph.D. Thesis, Department of Mathematics, University of Roorkee, Roorkee, 1996.

18. Verma S.K. Solution of optimization problems in crisp, fuzzy and stochastic environments, Ph.D. Thesis, 1997, University of Roorkee, Roorkee, India.
19. (i) Zimmermann H. J., *Fuzzy set theory and its applications*, 2nd edition, Kluwer Academic Publisher, Dordrecht, 1985.
    (ii) Zimmermann H. J., *Fuzzy set, decision making and expert systems*, Kluwer Academic Publisher, Dordrecllt, 1987.

Appendix

The *RSANSDC* algorithm combines steepest descent concept with the annealing concept. The proposed algorithm first generates an initial array of N (say 2(n+1)) random feasible points and stores these in an array A. These points are then arranged in ascending order of their objective function values. Next the point with the highest function value (in case of a minimization problem) is replaced with a feasible point in the direction of the line joining this point to the point with the lowest objective function value (called steepest descent direction) with probability determined in the manner of annealing algorithm. In cases where the algorithm decides not to replace the point with highest function value in array A with such a newly generated point, it is replaced by a new randomly generated point. The algorithm terminates when array has suitably clustered around a single point.

The main computational steps of the proposed RSANSDC algorithm can be summarized as below:

Step I: Choose randomly N (say $N = 2(n+1)$ where n is number of unknown independent variables) feasible points and evaluate the objective function at each of these points. Store these points and their function values in an N by $(n+1)$ array A.

Step II: Out of these N points find M and L the points with the greatest and the smallest objective function values $f(M)$ and $f(L)$ respectively, and arrange these points in ascending order of their objective function values.

Step III: Calculate *check*
$check = |if(M-1) - f(L)|/DN$ where $DN = 1$ if $|f(M-1)| < 1$ and $|f(M-1)|$ otherwise.
If $check < \varepsilon$ go to step XI:
else if $ITER > Itl$ go to step XI;
else $ITER = ITER + 1$;
if $ITER = Itin$;
then $ix = ix + ixinc$ and $Itin = Itin + Itinc$;
else go to step IV.

Step IV: Calculate the probability p_r
$p_r = exp(-\beta(f(M) - f(M - 1))/(f(M) - f(L)))$
where β is the controlling parameter.
Now compare p_r with a newly generated random number $r_1 \in [0..1]$.
If $p_r > r_1$ go to step VII;
Else find $P = 2L - M$
If P is not feasible go to step V.
Else find $f(P)$ and compare with $f(M - 1)$.
If $f(P) < f(M - 1)$ replace M by P and go to step VIII.

Step V: Find $P^* = 1.5L - 0.5M$
If $f(P^*)$ is not feasible go to VI
Else find $f(P^*)$ and compare with $f(M - 1)$.
If $f(P^*) < f(M - 1)$, replace M by P^* and go to step VIII,
Else go to step VI.

Step VI: Find the centroid C of all the N points in array A and check the feasibility of C.
If C is infeasible go to step VII,
Else find $f(C)$ and compare with $f(M - 1)$.
If $f(C) < f(M - 1)$, replace M by C and go to step VIII,
Else go to step IX.

Step VII: Find a new random feasible point P^{**} and compute $f(P^{**})$ and compare with $f(M - 1)$.
If $f(P^{**}) < f(M - 1)$ then replace M by P^{**} and go to step VIII,
Else go to Step X.

Step VIII: Rearrange the points in the array A and go to step III.

Step IX: Replace M by C, and go to step IV.

Step X: Replace M by P^{**}, and go to step IV.

Step XI: Print L and $f(L)$ and stop.

A modified version of this algorithm which can be used to solve integer and mixed integer optimization problems has been also developed and named RSANSDCU algorithm. More details about these algorithms may be found in Verma (1997).

Fuzzy Data Envelopment Analysis: A Credibility Approach*

Saowanee Lertworasirikul[1,3,][**], Shu-Cherng Fang[1], Jeffrey A. Joines[2], and Henry L. W. Nuttle[1]

[1] Department of Industrial Engineering, North Carolina State University, Raleigh NC 27695-7906, USA
[2] Department of Textile Engineering, North Carolina State University, Raleigh, NC 27695-8301, USA
[3] Department of Product Development, Faculty of Agro-Industry, Kasetsart University, 10900 Thailand

Abstract. While the traditional data envelopment analysis (DEA) requires precise input and output data, available data is usually imprecise and vague. "Fuzzy DEA" integrates the concept of fuzzy set theory with the traditional DEA by representing imprecise and vague data with fuzzy sets. In this paper, a credibility approach is proposed as a way to solve the fuzzy DEA model. The approach transforms a fuzzy DEA model into a well-defined credibility programming model, in which fuzzy variables are replaced by "expected credits" in terms of credibility measures. It is shown that when the membership functions of fuzzy data are trapezoidal, the credibility programming model becomes a linear programming model. Numerical examples are given to illustrate the proposed approach and results are compared with those obtained with alternative approaches.

1 Introduction

Over the past two decades, Data Envelopment Analysis (DEA) has emerged as a useful tool for business entities and organizations to evaluate their activities and to find opportunities for improvement. DEA evaluates the efficiencies of a set of homogenous decision making units (DMUs) with multiple performance criteria. Data used in evaluating the efficiency usually consists of multiple inputs and multiple outputs which represent possibly conflicting criteria. DEA evaluates the relative efficiency of DMUs by using a ratio of the weighted sum of outputs to the weighted sum of inputs with the weights being variable. Unlike statistical approaches such as regression analysis, DEA is a non-parametric technique, i.e., no specific functional form is required. DEA identifies the source and the amount of inefficiency in each input relative to

---

* This research was supported, in part, by the National Textile Center of the United States of America (Grant Number: I01-S01).
** Corresponding author.
  Fax: +1-919-515-5281; E-mail address: slertwo@unity.ncsu.edu.

each output for a target decision making unit based on peer-group comparisons. DEA has been applied in a number of contexts including education systems, health care units, agricultural productions, and military logistics [1,5,6].

The most frequently used DEA model is the CCR model, named after Charnes, Cooper, and Rhodes [7]. For the CCR model, a DMU is inefficient if it is possible to reduce any input without increasing any other inputs and achieve the same levels of output. A relative efficiency, which is the ratio of a weighted sum of outputs to a weighted sum of inputs, falls in the range of (0, 1]. If the efficiency of a DMU is equal to 1, the DMU is weakly efficient (technically efficient). A DMU is Pareto-Koopmans efficient [4] if its efficiency value is equal to 1 and its input excesses and output shortfalls are zero. The interpretation of the CCR model can be found in many references [5,6,13]. In addition to the CCR model, other well-known DEA models include the "BCC model," named after Banker, Charnes, and Cooper [2], the "Additive model," the "Free Disposal Hull" (FDH) model, and the "Slacks-Based Measure of Efficiency" (SBM) model. More details on other DEA models and their applications can be found in [5,6,19,21].

The traditional DEA requires accurate and precise performance data since it is a methodology which focuses on frontiers or boundaries. Small changes in data can change efficient frontiers significantly. However, in some situations such as in a manufacturing system, a production process or a service system, inputs and outputs are volatile and complex. In addition, the only data available for efficiency analysis will often be in the form of qualitative, linguistic data, e.g., "old" equipment and "high" inventory.

Most of the previous studies that deal with inaccurate and imprecise data in DEA models have simply used simulation techniques like the one in Banker et al. [3]. However, these methods have shortcomings as mentioned by Cooper et al. [6]. Fuzzy set theory, established by Zadeh [22], has been proven to be useful as a way to quantify imprecise and vague data in DEA models. The DEA models with fuzzy data ("fuzzy DEA" models) can represent real-world problems more realistically than the conventional DEA models. Fuzzy DEA models usually take the form of fuzzy linear programming models [24]. However, most fuzzy linear programming models (FLP) are not well defined due to ambiguity which arises in the ranking of fuzzy sets.

Nevertheless, as a class of classical mathematical models, fuzzy linear programming models should be interpreted and understood in a consistent manner. In recent papers [14,15], a possibility approach has been proposed for solving fuzzy DEA models. This approach deals with the uncertainty in fuzzy objectives and fuzzy constraints through the use of possibility measures. It transforms fuzzy DEA models into well-defined possibility DEA models. In this paper, we propose an alternative way to solve fuzzy DEA models using the credibility approach. Similar to the expected value approach to stochastic programming where random variables are replaced by their expected values, in the credibility programming-DEA (CP-DEA) model fuzzy variables are replaced by "expected credits," which are derived by using credibility measures.

The rest of this paper is organized as follows. Section 2 presents the fuzzy DEA model and existing approaches for solving fuzzy DEA models. Section 3 introduces the basic concepts of possibility, necessity, and credibility measures, as well as the expected credit operator for fuzzy variables. In this section, the derivation of the expected credits of normal, convex fuzzy variables is given. It is also shown that if such a fuzzy variable has a trapezoidal membership function, then an explicit expression for the expected credit of the fuzzy variable can be obtained. Section 4 presents the CP-DEA model using the credibility approach. When the membership functions of fuzzy data are of trapezoidal or triangular types, the CP-DEA model becomes a linear programming model. Section 5 provides numerical examples to illustrate the credibility approach to determining relative efficiencies for both the case of symmetrical triangular fuzzy parameters and the case of asymmetrical triangular fuzzy parameters. Finally, Section 6 concludes the paper, and discusses some future research directions.

2 Fuzzy DEA Model

This paper focuses on the CCR model because it is the original DEA model. All other models are extensions of the CCR model obtained by either modifying the production possibility set of the CCR model or adding slack variables in the objective function. Hence, the approach developed for solving the CCR model can be adapted for other DEA models.

Suppose that there are n DMUs, each of which has m inputs and r outputs of the same type. All inputs and outputs are assumed to be nonnegative, but at least one input and one output are positive. The following notation will be used throughout this paper.

Indices:
$i = 1, 2, ..., n$, $j = 1, 2, ..., m$, $k = 1, 2, ...r$.

Notation:
DMU_i is the ith DMU, DMU_o is the target DMU,
$\widetilde{x}_{ji}$ is the (fuzzy) amount of input j consumed by DMU_i,
$\widetilde{\mathbf{x}}_i = (\widetilde{x}_{ji})_{m \times 1}$ is a column vector of m fuzzy inputs consumed by DMU_i
$\widetilde{\mathbf{x}}_o = (\widetilde{x}_{jo})_{m \times 1}$ is a column vector of m fuzzy inputs consumed by DMU_o,
$\widetilde{\mathbf{X}} = [\widetilde{x}_{ji}]_{m \times n}$ is a matrix of m fuzzy inputs of n DMUs,
$\widetilde{y}_{ki}$ is the (fuzzy) amount of output k produced by DMU_i,
$\widetilde{\mathbf{y}}_i = (\widetilde{y}_{ki})_{r \times 1}$ is a column vector of r fuzzy outputs produced by DMU_i
$\widetilde{\mathbf{y}}_o = (\widetilde{y}_{ko})_{r \times 1}$ is a column vector of r fuzzy outputs produced by DMU_o,
$\widetilde{\mathbf{Y}} = [\widetilde{y}_{ki}]_{r \times n}$ is a matrix of r fuzzy outputs of n DMUs,
$\mathbf{u} \in R^{m \times 1}$ is a column vector of input weights, and
$\mathbf{v} \in R^{r \times 1}$ is a column vector of output weights.

The CCR model with fuzzy coefficients is as follows:

$$\text{(FCCR)} \max_{\mathbf{u},\mathbf{v}} \quad \mathbf{v}^T \widetilde{\mathbf{y}}_o$$
$$\mathbf{u}^T \widetilde{\mathbf{x}}_o = 1$$
$$-\mathbf{u}^T \widetilde{\mathbf{X}} + \mathbf{v}^T \widetilde{\mathbf{Y}} \leq \mathbf{0}$$
$$\mathbf{u}, \mathbf{v} \geq \mathbf{0}.$$

Even though several approaches have been developed to solve fuzzy DEA models, each has shortcomings in the way it treats uncertain data in DEA models as mentioned in [14]. The key point is that fuzzy DEA models take the form of fuzzy linear programming models, which are not well defined since ranking of fuzzy sets, such as $\mathbf{u}^T \widetilde{\mathbf{x}}_o = 1$ and $-\mathbf{u}^T \widetilde{\mathbf{X}} + \mathbf{v}^T \widetilde{\mathbf{Y}} \leq 0$ are ambiguous. Also, since the input/output data are fuzzy sets, an efficiency value of the fuzzy DEA model $\left(\frac{\mathbf{v}^T \widetilde{\mathbf{y}}_o}{\mathbf{u}^T \widetilde{\mathbf{x}}_o}\right)$ may take a value higher than 1 with certain degree of membership. Therefore, there is no universally accepted approach for solving fuzzy DEA models.

The few papers that have been published on solving fuzzy DEA problems can be categorized into four distinct approaches: tolerance approach, defuzzification approach, α-level based approach, and fuzzy ranking approach. The tolerance approach [11,20] incorporates uncertainty into the DEA models by defining tolerance levels on constraint violations. In the defuzzification approach [13], fuzzy inputs and fuzzy outputs are first defuzzified into crisp values. Using these crisp values, the resulting crisp model can be solved by an LP solver. In the α-level based approach, the fuzzy DEA model is solved by parametric programming using α cuts. Solving the model at a given α level produces a corresponding interval efficiency for the target DMU. A number of such intervals can be used to construct the corresponding fuzzy efficiency. More details on this approach can be found in Maeda, Entani and Tanaka [18], Kao and Liu [12], and Lertworasirikul [13]. The fuzzy ranking approach was developed by Guo and Tanaka [10]. Both fuzzy inequalities and fuzzy equalities in the fuzzy CCR model are defined by ranking methods so that the resulting model is a bi-level linear programming model. For the case of symmetrical triangular fuzzy inputs and outputs, Guo and Tanaka construct non-symmetrical triangular fuzzy efficiencies. They define a nondominated set for a given α-level. A DMU is said to be α-possibilistic nondominated if the maximum value of the fuzzy efficiency at that α level is greater than or equal to 1.

Recently, a possibility approach has been proposed by Lertworasirikul et al. [14]. The approach uses the concepts of possibility measures and chance-constrained programming to transform a fuzzy DEA model into a well-defined possibility model.

In this paper, a credibility approach is proposed as an alternative way for solving fuzzy DEA models. This approach uses the "expected credits" of fuzzy variables to deal with the uncertainty in fuzzy objectives and fuzzy constraints. The expected credits of fuzzy variables are derived by using credibility measures. With this approach, the problem of ranking fuzzy sets in fuzzy DEA models is handled in a meaningful way.

3 Possibility, Necessity, and Credibility Measures

Zadeh [23] proposed possibility theory in the context of the fuzzy set theory as a mathematical framework for modeling and characterizing situations involving uncertainty. A good reference on possibility theory is Dubois and Prade [8]. Zadeh also introduced the concept of "fuzzy variable," which is associated with a possibility distribution in the same manner that a random variable is associated with a probability distribution. Liu [16] introduced an "expected value" operator for fuzzy variables which is obtained by using credibility measures, which, in turn, are derived from possibility and necessity measures. In this paper, the term "expected credit" instead of "expected value" is used for a fuzzy variable to distinguish it from the expected value of a random variable.

3.1 Possibility Measure

Let $(\Theta_i, \mathcal{P}(\Theta_i), \pi_i)$, for each $i = 1, 2, ..., n$, be a possibility space with Θ_i being the nonempty set of interest, $\mathcal{P}(\Theta_i)$ the collection of all subsets of Θ_i, and π_i the possibility measure from $\mathcal{P}(\Theta_i)$ to $[0, 1]$.

Given a possibility space $(\Theta_i, \mathcal{P}(\Theta_i), \pi)$ with

(i) $\pi(\phi) = 0$, $\pi(\Theta_i) = 1$, and
(ii) $\pi(\bigcup_i A_i) = \sup_i \{\pi(A_i)\}$ with each $A_i \in \mathcal{P}(\Theta_i)$,

Zadeh defined a fuzzy variable, $\widetilde{\xi}$, as a real-valued function defined over Θ_i with the membership function:

$$\mu_{\widetilde{\xi}}(s) = \pi(\{\theta_i \in \Theta_i | \widetilde{\xi}(\theta_i) = s\}) \qquad (1)$$
$$= \sup_{\theta_i \in \Theta_i} \{\pi(\{\theta_i\}) | \widetilde{\xi}(\theta_i) = s\}, \forall s \in R.$$

Suppose $\widetilde{a}$ and $\widetilde{b}$ are two fuzzy variables on the possibility spaces $(\Theta_1, \mathcal{P}(\Theta_1), \pi_1)$ and $(\Theta_2, \mathcal{P}(\Theta_2), \pi_2)$, respectively. Then $\widetilde{a} \leq \widetilde{b}$ is a fuzzy event defined on the product possibility space $(\Theta = \Theta_1 \times \Theta_2, \mathcal{P}(\Theta), \pi)$, with

$$\pi(\widetilde{a} \leq \widetilde{b}) = \sup_{\substack{\theta_1 \in \Theta_1 \\ \theta_2 \in \Theta_2}} \{\pi\{(\theta_1, \theta_2) | \widetilde{a}(\theta_1) \leq \widetilde{b}(\theta_2)\}\}$$

Furthermore, from the definition of fuzzy variables (1), we have

$$\pi(\tilde{a} \leq \tilde{b}) = \sup_{s,t \in R} \left\{ \min(\mu_{\tilde{a}}(s), \mu_{\tilde{b}}(t)) \mid s \leq t \right\}.$$

Similarly, possibilities of the fuzzy events $\tilde{a} < \tilde{b}$, and $\tilde{a} = \tilde{b}$ defined on the product possibility space $(\Theta, \mathcal{P}(\Theta), \pi)$ are given as

$$\pi(\tilde{a} < \tilde{b}) = \sup_{s,t \in R} \left\{ \min(\mu_{\tilde{a}}(s), \mu_{\tilde{b}}(t)) \mid s < t \right\},$$

$$\pi(\tilde{a} = \tilde{b}) = \sup_{s,t \in R} \left\{ \min(\mu_{\tilde{a}}(s), \mu_{\tilde{b}}(t)) \mid s = t \right\}, \text{respectively.}$$

3.2 Necessity Measure

Associated with each possibility measure is a necessity measure $\mathcal{N}$. Necessity measure and possibility measure are mutually dual. In particular, given a possibility space $(\Theta, \mathcal{P}(\Theta), \pi)$, if A and $\overline{A}$ are two opposite events ($\overline{A}$ is the complement of A in Θ), then

$$\mathcal{N}(A) = 1 - \pi(\overline{A}).$$

The necessity measures of a fuzzy event is defined as the impossibility of the opposite event. Thus, given fuzzy variables $\tilde{a}_1, \tilde{a}_2, ..., \tilde{a}_n$ and $f_j : R^n \to R$, $j = 1, ..., m$, the necessity of the fuzzy events "$f_j(\tilde{a}_1, \tilde{a}_2, ..., \tilde{a}_n) \leq 0$", "$f_j(\tilde{a}_1, \tilde{a}_2, ..., \tilde{a}_n) \geq 0$", and "$f_j(\tilde{a}_1, \tilde{a}_2, ..., \tilde{a}_n) = 0$", $j = 1, ..., m$ are defined respectively by

$$\mathcal{N}(f_j(\tilde{a}_1, \tilde{a}_2, ..., \tilde{a}_n) \leq 0, \ j = 1, ..., m)$$
$$= 1 - \sup_{s_1,...,s_n \in R} \left\{ \min_{1 \leq i \leq n} \{\mu_{\tilde{a}_i}(s_i)\} \mid \exists j \in \{1, ..., m\}, \text{ s.t. } f_j(s_1, ..., s_n) > 0 \right\}.$$

$$\mathcal{N}(f_j(\tilde{a}_1, \tilde{a}_2, ..., \tilde{a}_n) \geq 0, \ j = 1, ..., m)$$
$$= 1 - \sup_{s_1,...,s_n \in R} \left\{ \min_{1 \leq i \leq n} \{\mu_{\tilde{a}_i}(s_i)\} \mid \exists j \in \{1, ..., m\}, \text{ s.t. } f_j(s_1, ..., s_n) < 0 \right\},$$

$$\mathcal{N}(f_j(\tilde{a}_1, \tilde{a}_2, ..., \tilde{a}_n) = 0, \ j = 1, ..., m)$$
$$= 1 - \sup_{s_1,...,s_n \in R} \left\{ \min_{1 \leq i \leq n} \{\mu_{\tilde{a}_i}(s_i)\} \mid \exists j \in \{1, ..., m\}, \text{ s.t. } f_j(s_1, ..., s_n) \neq 0 \right\}.$$

3.3 Credibility Measure

Liu [16,17] defines a credibility measure (Cr) of a fuzzy event as the average of its possibility and necessity measures, i.e.,

$$Cr(\cdot) = \frac{\pi(\cdot) + \mathcal{N}(\cdot)}{2}.$$

The relationship among possibility, necessity and credibility measures is:

$$\pi(\cdot) \geq Cr(\cdot) \geq \mathcal{N}(\cdot).$$

Similar to the expected value operator for a random variable in probability theory, an expected credit operator of a fuzzy variable $\tilde{\xi}$ on a possibility space $(\Theta, \mathcal{P}(\Theta), \pi)$ is defined by Liu [17] as

$$E(\tilde{\xi}) = \int_0^{+\infty} Cr(\tilde{\xi} \geq t) dt - \int_{-\infty}^0 Cr(\tilde{\xi} \leq t) dt. \qquad (2)$$

Note that if the fuzzy variable $\tilde{\xi}$ and the credibility measure $Cr(\cdot)$ in (2) are replaced with a random variable and a probability measure, respectively, then (2) becomes the expected value of the random variable. In addition, $E(p\tilde{\xi} + q\tilde{\varphi}) = pE(\tilde{\xi}) + qE(\tilde{\varphi})$ for any real numbers p and q, which is similar to the case of the expected value of random variables.

In what follows, the expected credit of a normal, convex fuzzy variable is derived. Normal, convex fuzzy variables are suitable for representing fuzzy parameters in fuzzy DEA models because fuzzy inputs and outputs usually have this type of membership functions.

Expected credits of normal, convex fuzzy variables The definitions of normal and convex fuzzy variables are given below. Consider a fuzzy variable $\tilde{a}$ on a possibility space $(\Theta, \mathcal{P}(\Theta), \pi)$.

Definition 1. (Normal fuzzy variables) *The fuzzy variable $\tilde{a}$ is normal if*

$$\sup_{s \in R} \{\mu_{\tilde{a}}(s)\} = 1.$$

Definition 2. (α-level sets) *The α-level set of the fuzzy variable $\tilde{a}$ is defined by the set of elements that belong to the fuzzy variable $\tilde{a}$ with membership of at least α, i.e.,*

$$\tilde{a}_\alpha = \{s \in R | \mu_{\tilde{a}}(s) \geq \alpha\}.$$

Definition 3. (Convex fuzzy variables) *The fuzzy variable $\tilde{a}$ is convex if*

$$\mu_{\tilde{a}}(\lambda s_1 + (1-\lambda)s_2) \geq \min\{\mu_{\tilde{a}}(s_1), \mu_{\tilde{a}}(s_2)\} \text{ for all } s_1, s_2 \in R \text{ and } \lambda \in [0,1].$$

Alternatively, the fuzzy variable $\tilde{a}$ is convex if all α-level sets are convex.

Let $\tilde{\xi}_i$, $i = 1, ..., n$, be a normal, convex fuzzy variable on a possibility space $(\Theta_i, \mathcal{P}(\Theta_i), \pi_i)$, and $\tilde{\xi} = \tilde{\xi}_1 + ... + \tilde{\xi}_n$. It follows that $\tilde{\xi}$ is also a normal, convex fuzzy variable on the product possibility space $(\Theta, \mathcal{P}(\Theta), \pi)$. Using the relationship among possibility, necessity, and credibility measures, the

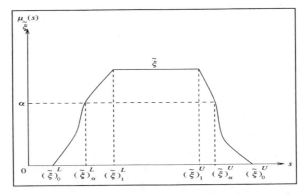

Fig. 1. The lower and upper bounds of the α-level of $\mu_{\tilde{a}}$.

expected credit of $\tilde{\xi}$ can be derived as follows.

$$\begin{aligned}
E(\tilde{\xi}) &= \lim_{M \to \infty} \left(\int_0^{+M} Cr(\tilde{\xi} \geq t) dt - \int_{-M}^0 Cr(\tilde{\xi} \leq t) dt \right) \\
&= \tfrac{1}{2} \lim_{M \to \infty} \left(\int_0^{+M} \left[\pi(\tilde{\xi} \geq t) + \mathcal{N}(\tilde{\xi} \geq t) \right] dt - \int_{-M}^0 \left[\pi(\tilde{\xi} \leq t) + \mathcal{N}(\tilde{\xi} \leq t) \right] dt \right) \\
&= \tfrac{1}{2} \lim_{M \to \infty} \left(\int_0^{+M} 1 dt - \int_{-M}^0 1 dt + \int_0^{+M} \pi(\tilde{\xi} \geq t) dt + \int_{-M}^0 \pi(\tilde{\xi} > t) dt \right. \\
&\qquad \left. - \int_0^{+M} \pi(\tilde{\xi} < t) dt - \int_{-M}^0 \pi(\tilde{\xi} \leq t) dt \right)
\end{aligned}$$

For a normal fuzzy variable $\tilde{\xi}$, the values of $\pi(\tilde{\xi} < t)$ and $\pi(\tilde{\xi} > t)$ are very close to $\pi(\tilde{\xi} \leq t)$ and $\pi(\tilde{\xi} \geq t)$, respectively. This leads to the following lemma.

Lemma 1. *Let $\tilde{\xi}$ be a normal, convex fuzzy variable. Let $(\cdot)_\alpha^L$ and $(\cdot)_\alpha^U$ denote the lower and upper bounds of the α-level set of $\mu_{\tilde{\xi}}$, the membership function of $\tilde{\xi}$ (see Figure 1). If $\pi(\tilde{\xi} < t)$ and $\pi(\tilde{\xi} > t)$ are approximated by $\pi(\tilde{\xi} \leq t)$ and $\pi(\tilde{\xi} \geq t)$, respectively, then*

$$E(\tilde{\xi}) = \tfrac{1}{2} \left((\tilde{\xi})_1^U + (\tilde{\xi})_1^L + \int_{(\tilde{\xi})_1^U}^{(\tilde{\xi})_0^U} \pi(\tilde{\xi} \geq t) dt - \int_{(\tilde{\xi})_0^L}^{(\tilde{\xi})_1^L} \pi(\tilde{\xi} \leq t) dt \right).$$

Proof. If $\pi(\widetilde{\xi} < t)$ and $\pi(\widetilde{\xi} > t)$ are approximated by $\pi(\widetilde{\xi} \leq t)$ and $\pi(\widetilde{\xi} \geq t)$, respectively, then

$$E(\widetilde{\xi}) = \frac{1}{2} \lim_{M \to \infty} \left(\int_0^{+M} 1 dt - \int_{-M}^0 1 dt + \int_0^{+M} \pi(\widetilde{\xi} \geq t) dt \right.$$
$$\left. + \int_{-M}^0 \pi(\widetilde{\xi} \geq t) dt - \int_0^{+M} \pi(\widetilde{\xi} \leq t) dt - \int_{-M}^0 \pi(\widetilde{\xi} \leq t) dt \right)$$
$$= \frac{1}{2} \lim_{M \to \infty} \left(\int_0^{+M} 1 dt - \int_{-M}^0 1 dt + \int_{-M}^{+M} \pi(\widetilde{\xi} \geq t) dt - \int_{-M}^{+M} \pi(\widetilde{\xi} \leq t) dt \right).$$

To determine the terms $\lim_{M \to \infty} \left(\int_{-M}^{+M} \pi(\widetilde{\xi} \geq t) dt \right)$ and $\lim_{M \to \infty} \left(\int_{-M}^{+M} \pi(\widetilde{\xi} \leq t) dt \right)$, there are three possible cases to be considered.

Case 1: $(\widetilde{\xi})_1^L \geq 0$

For $-\infty \leq t \leq (\widetilde{\xi})_1^U$, $\pi(\widetilde{\xi} \geq t) = \sup_{s \in R} \{\mu_{\widetilde{\xi}}(s) | s \geq t\} = 1$. Because $(\widetilde{\xi})_1^L \geq 0$, it follows that $(\widetilde{\xi})_1^U \geq 0$. Thus, we have

$$\lim_{M \to \infty} \left(\int_{-M}^{+M} \pi(\widetilde{\xi} \geq t) dt \right) = \lim_{M \to \infty} \left(\int_{-M}^0 1 dt \right) + \int_0^{(\widetilde{\xi})_1^U} 1 dt + \int_{(\widetilde{\xi})_1^U}^{(\widetilde{\xi})_0^U} \pi(\widetilde{\xi} \geq t) dt.$$

For $(\widetilde{\xi})_1^L \leq t \leq +\infty$, $\pi(\widetilde{\xi} \leq t) = \sup_{s \in R} \{\mu_{\widetilde{\xi}}(s) | s \leq t\} = 1$.

Also, $\lim_{M \to \infty} \left(\int_{(\widetilde{\xi})_1^L}^{+M} \pi(\widetilde{\xi} \leq t) dt \right)$ can be written as $\lim_{M \to \infty} \left(\int_0^{+M} \pi(\widetilde{\xi} \leq t) dt \right) - \int_0^{(\widetilde{\xi})_1^L} \pi(\widetilde{\xi} \leq t) dt = \lim_{M \to \infty} \left(\int_0^{+M} 1 dt \right) - \int_0^{(\widetilde{\xi})_1^L} 1 dt$. Thus, we have

$$\lim_{M \to \infty} \left(\int_{-M}^{+M} \pi(\widetilde{\xi} \leq t) dt \right) = \int_{(\widetilde{\xi})_0^L}^{(\widetilde{\xi})_1^L} \pi(\widetilde{\xi} \leq t) dt + \lim_{M \to \infty} \left(\int_0^{+M} 1 dt \right) - \int_0^{(\widetilde{\xi})_1^L} 1 dt.$$

Substituting into the expression for $E(\widetilde{\xi})$, we obtain

$$E(\widetilde{\xi}) = \frac{1}{2} \left((\widetilde{\xi})_1^U + (\widetilde{\xi})_1^L + \int_{(\widetilde{\xi})_1^U}^{(\widetilde{\xi})_0^U} \pi(\widetilde{\xi} \geq t) dt - \int_{(\widetilde{\xi})_0^L}^{(\widetilde{\xi})_1^L} \pi(\widetilde{\xi} \leq t) dt \right).$$

Case 2: $(\widetilde{\xi})_1^L \leq 0$ and $(\widetilde{\xi})_1^U \geq 0$

Similar to the derivation in Case 1, for this case we have

$$\lim_{M\to\infty}\left(\int_{-M}^{+M}\pi(\widetilde{\xi}\geq t)dt\right) = \lim_{M\to\infty}\left(\int_{-M}^{0}1dt\right) + \int_{0}^{(\widetilde{\xi})_1^U}1dt + \int_{(\widetilde{\xi})_1^U}^{(\widetilde{\xi})_0^U}\pi(\widetilde{\xi}\geq t)dt.$$

$$\lim_{M\to\infty}\left(\int_{-M}^{+M}\pi(\widetilde{\xi}\leq t)dt\right) = \int_{(\widetilde{\xi})_0^L}^{(\widetilde{\xi})_1^L}\pi(\widetilde{\xi}\leq t)dt + \lim_{M\to\infty}\left(\int_{0}^{+M}1dt\right) + \int_{(\widetilde{\xi})_1^L}^{0}1dt.$$

Substituting these terms into the expression for $E(\widetilde{\xi})$, we obtain

$$E(\widetilde{\xi}) = \tfrac{1}{2}\left[(\widetilde{\xi})_1^U + (\widetilde{\xi})_1^L + \int_{(\widetilde{\xi})_1^U}^{(\widetilde{\xi})_0^U}\pi(\widetilde{\xi}\geq t)dt - \int_{(\widetilde{\xi})_0^L}^{(\widetilde{\xi})_1^L}\pi(\widetilde{\xi}\leq t)dt\right].$$

Case 3: $(\widetilde{\xi})_1^U \leq 0$

If $(\widetilde{\xi})_1^U \leq 0$, then it follows that $(\widetilde{\xi})_1^L \leq 0$ and we have

$$\lim_{M\to\infty}\left(\int_{-M}^{+M}\pi(\widetilde{\xi}\geq t)dt\right) = \lim_{M\to\infty}\left(\int_{-M}^{0}1dt\right) - \int_{(\widetilde{\xi})_1^U}^{0}1dt + \int_{(\widetilde{\xi})_1^U}^{(\widetilde{\xi})_0^U}\pi(\widetilde{\xi}\geq t)dt.$$

$$\lim_{M\to\infty}\left(\int_{-M}^{+M}\pi(\widetilde{\xi}\leq t)dt\right) = \int_{(\widetilde{\xi})_0^L}^{(\widetilde{\xi})_1^L}\pi(\widetilde{\xi}\leq t)dt + \lim_{M\to\infty}\left(\int_{0}^{+M}1dt\right) + \int_{(\widetilde{\xi})_1^L}^{0}1dt.$$

Therefore, by substitution we obtain

$$E(\widetilde{\xi}) = \tfrac{1}{2}\left((\widetilde{\xi})_1^U + (\widetilde{\xi})_1^L + \int_{(\widetilde{\xi})_1^U}^{(\widetilde{\xi})_0^U}\pi(\widetilde{\xi}\geq t)dt - \int_{(\widetilde{\xi})_0^L}^{(\widetilde{\xi})_1^L}\pi(\widetilde{\xi}\leq t)dt\right).$$

Thus, Lemma 1 is proved. ∎

From Lemma 1, in order to obtain an explicit form of $E(\widetilde{\xi})$, the terms $\int_{(\widetilde{\xi})_1^U}^{(\widetilde{\xi})_0^U}\pi(\widetilde{\xi}\geq t)dt$ and $\int_{(\widetilde{\xi})_0^L}^{(\widetilde{\xi})_1^L}\pi(\widetilde{\xi}\leq t)dt$ have to be evaluated.

Lemma 2. *For a given* $\widetilde{\xi} = \widetilde{\xi}_1 + ... + \widetilde{\xi}_n$ *with* $\widetilde{\xi}_i, i = 1,...,n$ *being normal, convex fuzzy variables and* $\alpha \in [0,1]$, $\pi(\widetilde{\xi}\geq t)$ *and* $\pi(\widetilde{\xi}\leq t)$ *can be evaluated by solving respectively the following two problems,*

$$\max_{\alpha} \alpha \qquad (3)$$
$$s.t.\ (\widetilde{\xi}_1)_\alpha^U + (\widetilde{\xi}_2)_\alpha^U + ... + (\widetilde{\xi}_n)_\alpha^U \geq t.$$

$$\max_{\alpha} \alpha \qquad (4)$$
$$s.t.\ (\widetilde{\xi}_1)_\alpha^L + (\widetilde{\xi}_2)_\alpha^L + ... + (\widetilde{\xi}_n)_\alpha^L \leq t.$$

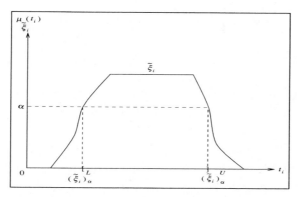

Fig. 2. α-level of $\mu_{\widetilde{\xi}_i}(t_i)$.

Proof. From Section 3.1, we know that

$$\pi(\widetilde{\xi} \geq t) = \sup_{t_1,t_2,...,t_n \in R} \left\{ \min \left\{ \mu_{\widetilde{\xi}_1}(t_1), \mu_{\widetilde{\xi}_2}(t_2), ..., \mu_{\widetilde{\xi}_n}(t_n) \right\} \mid t_1 + t_2 + ... + t_n \geq t \right\}.$$

For a given value of t, finding $\pi(\widetilde{\xi} \geq t)$ is equivalent to solving the following mathematical programming model:

$$\max_{t_1,t_2,...,t_n \in R, \alpha \in [0,1]} \alpha$$
$$\text{s.t.} \quad \mu_{\widetilde{\xi}_1}(t_1) \geq \alpha$$
$$\vdots \qquad (5)$$
$$\mu_{\widetilde{\xi}_n}(t_n) \geq \alpha$$
$$t_1 + t_2 + ... + t_n \geq t.$$

Considering the α-level of $\mu_{\widetilde{\xi}_i}(t_i)$, $i = 1, ..., n$, as depicted in Figure 2, it follows that $\mu_{\widetilde{\xi}_1}(t_1) \geq \alpha$ is equivalent to $(\widetilde{\xi}_i)^L_\alpha \leq t_i \leq (\widetilde{\xi}_i)^U_\alpha$. Using this result, (5) can be written as follows:

$$\max_{t_1,t_2,...,t_n \in R, \alpha \in [0,1]} \alpha$$
$$\text{s.t.} \quad (\widetilde{\xi}_1)^L_\alpha \leq t_1 \leq (\widetilde{\xi}_1)^U_\alpha$$
$$\vdots \qquad (6)$$
$$(\widetilde{\xi}_n)^L_\alpha \leq t_n \leq (\widetilde{\xi}_n)^U_\alpha$$
$$t_1 + t_2 + ... + t_n \geq t.$$

To maximize α subject to $t_1 + t_2 + ... + t_n \geq t$, $t_1, t_2, ..., t_n$ should take the highest possible value from their ranges. Consequently, (6) reduces to:

$$\max \alpha$$
$$\text{s.t.} \quad (\widetilde{\xi}_1)^U_\alpha + (\widetilde{\xi}_2)^U_\alpha + ... + (\widetilde{\xi}_n)^U_\alpha \geq t. \qquad (7)$$

Analogously (4) can be derived by using

$$\pi(\widetilde{\xi} \leq t) = \sup_{t_1, t_2, \ldots, t_n \in R} \left\{ \min \left\{ \mu_{\widetilde{\xi}_1}(t_1), \mu_{\widetilde{\xi}_2}(t_2), \ldots, \mu_{\widetilde{\xi}_n}(t_n) \right\} \mid t_1 + t_2 + \ldots + t_n \leq t \right\}$$

Note that the results in this section also apply for the fuzzy variables of the LR types [24].

A special case: Expected credits of trapezoidal fuzzy variables It can be shown that an explicit expression for $E(\widetilde{\xi})$ can be obtained when $\widetilde{\xi}$ is a normal, convex trapezoidal fuzzy variable.

Lemma 3. *Given that* $\widetilde{\xi} = \widetilde{\xi}_1 + \ldots + \widetilde{\xi}_n$ *with* $\widetilde{\xi}_i, i = 1, \ldots, n$, *being normal, convex trapezoidal fuzzy variables,*

$$\int_{(\widetilde{\xi})_1^U}^{(\widetilde{\xi})_0^U} \pi(\widetilde{\xi} \geq t) dt = \frac{(\widetilde{\xi})_0^U - (\widetilde{\xi})_1^U}{2}$$

$$\int_{(\widetilde{\xi})_0^L}^{(\widetilde{\xi})_1^L} \pi(\widetilde{\xi} \leq t) dt = \frac{(\widetilde{\xi})_1^L - (\widetilde{\xi})_0^L}{2}.$$

Proof. For the case that $\widetilde{\xi}_i, i = 1, \ldots, n$, is a normal, convex trapezoidal fuzzy variable, we have

$$(\widetilde{\xi}_i)_\alpha^U = (1 - \alpha)(\widetilde{\xi}_i)_0^U + \alpha(\widetilde{\xi}_i)_1^U.$$

Thus the constraint, $(\widetilde{\xi}_1)_\alpha^U + (\widetilde{\xi}_2)_\alpha^U + \ldots + (\widetilde{\xi}_n)_\alpha^U \geq t$, in (3) becomes

$$(1 - \alpha)(\widetilde{\xi}_1)_0^U + \alpha(\widetilde{\xi}_1)_1^U + \ldots + (1 - \alpha)(\widetilde{\xi}_n)_0^U + \alpha(\widetilde{\xi}_n)_1^U \geq t.$$

By rearranging this equation, we obtain

$$(\widetilde{\xi}_1)_0^U + \ldots + (\widetilde{\xi}_n)_0^U - \alpha \left[(\widetilde{\xi}_1)_0^U + \ldots + (\widetilde{\xi}_n)_0^U - (\widetilde{\xi}_1)_1^U - \ldots - (\widetilde{\xi}_n)_1^U \right] \geq t.$$

Since $(\widetilde{\xi}_1)_0^U + \ldots + (\widetilde{\xi}_n)_0^U \geq (\widetilde{\xi}_1)_1^U + \ldots + (\widetilde{\xi}_n)_1^U$,

$$\alpha \leq \frac{(\widetilde{\xi}_1)_0^U + \ldots + (\widetilde{\xi}_n)_0^U - t}{(\widetilde{\xi}_1)_0^U + \ldots + (\widetilde{\xi}_n)_0^U - (\widetilde{\xi}_1)_1^U - \ldots - (\widetilde{\xi}_n)_1^U}.$$

Since in (3) we wish to maximize α subject to $(\widetilde{\xi}_1)_\alpha^U + (\widetilde{\xi}_2)_\alpha^U + \ldots + (\widetilde{\xi}_n)_\alpha^U \geq t$, it follows that

$$\alpha = \frac{(\widetilde{\xi}_1)_0^U + \ldots + (\widetilde{\xi}_n)_0^U - t}{(\widetilde{\xi}_1)_0^U + \ldots + (\widetilde{\xi}_n)_0^U - (\widetilde{\xi}_1)_1^U - \ldots - (\widetilde{\xi}_n)_1^U} = \pi(\widetilde{\xi} \geq t). \quad (8)$$

Since $\tilde{\xi} = \tilde{\xi}_1 + ... + \tilde{\xi}_n$,

$$\int_{(\tilde{\xi})_1^U}^{(\tilde{\xi})_0^U} \pi(\tilde{\xi} \geq t)dt = \frac{[(\tilde{\xi})_0^U][(\tilde{\xi})_0^U - (\tilde{\xi})_1^U]}{(\tilde{\xi})_0^U - (\tilde{\xi})_1^U} - \frac{[((\tilde{\xi})_0^U)^2 - ((\tilde{\xi})_1^U)^2]}{2[(\tilde{\xi})_0^U - (\tilde{\xi})_1^U]}$$

$$= \frac{(\tilde{\xi})_0^U - (\tilde{\xi})_1^U}{2}.$$

In a similar manner, the solution of (4) for the case of normal, convex trapezoidal fuzzy variables can be derived by substituting

$$(\tilde{\xi}_i)_\alpha^L = (1 - \alpha)(\tilde{\xi}_i)_0^L + \alpha(\tilde{\xi}_i)_1^L$$

into the constraint $(\tilde{\xi}_1)_\alpha^L + (\tilde{\xi}_2)_\alpha^L + ... + (\tilde{\xi}_n)_\alpha^L \leq t$. The result after doing some calculations is that

$$\alpha \leq \frac{(\tilde{\xi}_1)_0^L + ... + (\tilde{\xi}_n)_0^L - t}{(\tilde{\xi}_1)_0^L + ... + (\tilde{\xi}_n)_0^L - (\tilde{\xi}_1)_1^L - ... - (\tilde{\xi}_n)_1^L}.$$

Since in (4) we wish to maximize α subject to $(\tilde{\xi}_1)_\alpha^L + (\tilde{\xi}_2)_\alpha^L + ... + (\tilde{\xi}_n)_\alpha^L \leq t$, it follows that

$$\alpha = \frac{(\tilde{\xi}_1)_0^L + ... + (\tilde{\xi}_n)_0^L - t}{(\tilde{\xi}_1)_0^L + ... + (\tilde{\xi}_n)_0^L - (\tilde{\xi}_1)_1^L - ... - (\tilde{\xi}_n)_1^L} = \pi(\tilde{\xi} \leq t). \quad (9)$$

Using (9) and doing some calculations, we obtain

$$\int_{(\tilde{\xi})_0^L}^{(\tilde{\xi})_1^L} \pi(\tilde{\xi} \leq t)dt = \frac{(\tilde{\xi})_1^L - (\tilde{\xi})_0^L}{2}.$$

This proves Lemma 3. ∎

From Lemma 3, the expected credit of $\tilde{\xi}$ becomes

$$E(\tilde{\xi}) = \frac{1}{2}\left[(\tilde{\xi})_1^U + (\tilde{\xi})_1^L + \int_{(\tilde{\xi})_1^U}^{(\tilde{\xi})_0^U} \pi(\tilde{\xi} \geq t)dt - \int_{(\tilde{\xi})_0^L}^{(\tilde{\xi})_1^L} \pi(\tilde{\xi} \leq t)dt\right]$$

$$= \frac{1}{4}\left[(\tilde{\xi})_1^U + (\tilde{\xi})_1^L + (\tilde{\xi})_0^U + (\tilde{\xi})_0^L\right].$$

Thus, for the case that $\tilde{\xi}$ is a normal, convex trapezoidal fuzzy variable, an explicit expression of the expected credit for the fuzzy variable can be derived. Also notice that when the membership functions are symmetrical, the expected credit of normal, convex trapezoidal fuzzy variables are located at the central point of their membership functions.

4 CP-DEA model

In this section, the credibility approach to solving fuzzy DEA models is presented. This approach treats the uncertainty in fuzzy objectives and fuzzy constraints by replacing fuzzy variables with their expected credits. In this way, the fuzzy DEA model (FCCR) is transformed into the following credibility programming-DEA model (CP-DEA):

$$\text{(CP-DEA)} \max_{\mathbf{u},\mathbf{v}} \quad E(\mathbf{v}^T \widetilde{\mathbf{y}}_o)$$
$$s.t. \quad E(\mathbf{u}^T \widetilde{\mathbf{x}}_o) = 1$$
$$E(-\mathbf{u}^T \widetilde{\mathbf{X}} + \mathbf{v}^T \widetilde{\mathbf{Y}}) \leq 0$$
$$\mathbf{u}, \mathbf{v} \geq \mathbf{0}.$$

Note that no matter what types of fuzzy variable $\widetilde{\xi}$ and functional form, g, we have, for a given decision u, $g(u, \widetilde{\xi})$ is also a fuzzy variable. Thus, the fuzzy constraints can be treated by taking expected credit operators. In this model we wish to select the decision vectors, $\mathbf{u}$ and $\mathbf{v}$ to maximize the expected return, $E(\mathbf{v}^T \widetilde{\mathbf{y}}_o)$, while satisfying constraints on the expected credits of $\mathbf{u}^T \widetilde{\mathbf{x}}_o$ and $-\mathbf{u}^T \widetilde{\mathbf{X}} + \mathbf{v}^T \widetilde{\mathbf{Y}}$.

Similar to the traditional CCR model, the second and third constraints in the CP-DEA model are used for normalization of the value of $E(\mathbf{v}^T \widetilde{\mathbf{y}}_o)$. Therefore, an expected efficiency value $E(\mathbf{v}^T \widetilde{\mathbf{y}}_o)$ of the target DMU falls in the range of (0,1]. In the traditional CCR model, the $\left(\frac{\mathbf{v}^T \mathbf{y}_o}{\mathbf{u}^T \mathbf{x}_o}\right)$ value of one signifies that the DMU_o is technically efficient. Following this concept, we use $\mathbf{v}^T \widetilde{\mathbf{y}}_o$ to determine if a DMU is technically efficient for the FCCR model. Correspondingly, the $E(\mathbf{v}^T \widetilde{\mathbf{y}}_o)$ in the CP-DEA model is used to determine if a target DMU is technically efficient (in the credibilistic sense).

Definition 4. A DMU is a credibilistically efficient DMU if its $E(\mathbf{v}^T \widetilde{\mathbf{y}}_o)$ value is equal to one, otherwise, it is a credibilistically inefficient DMU.

From the derivation of the expected credits of fuzzy variables in Section 3, it follows that the CP-DEA model is either a nonlinear or a linear programming model depending upon the form of the membership functions (of fuzzy inputs and outputs). When these membership functions are normal, convex trapezoidal, the CP-DEA model is transformed into the following linear programming model, which can be solved by a standard LP solver [9].

$$\max_{\mathbf{u},\mathbf{v}} \quad \tfrac{1}{4} \left[(\mathbf{v}^T \widetilde{\mathbf{y}}_o)_1^U + (\mathbf{v}^T \widetilde{\mathbf{y}}_o)_1^L + (\mathbf{v}^T \widetilde{\mathbf{y}}_o)_0^U + (\mathbf{v}^T \widetilde{\mathbf{y}}_o)_0^L \right]$$
$$s.t. \quad \tfrac{1}{4} \left[(\mathbf{u}^T \widetilde{\mathbf{x}}_o)_1^U + (\mathbf{u}^T \widetilde{\mathbf{x}}_o)_1^L + (\mathbf{u}^T \widetilde{\mathbf{x}}_o)_0^U + (\mathbf{u}^T \widetilde{\mathbf{x}}_o)_0^L \right] = 1$$
$$\tfrac{1}{4} \left[\left(-\mathbf{u}^T \widetilde{\mathbf{X}} + \mathbf{v}^T \widetilde{\mathbf{Y}}\right)_1^U + \left(-\mathbf{u}^T \widetilde{\mathbf{X}} + \mathbf{v}^T \widetilde{\mathbf{Y}}\right)_1^L \right.$$
$$\left. + \left(-\mathbf{u}^T \widetilde{\mathbf{X}} + \mathbf{v}^T \widetilde{\mathbf{Y}}\right)_0^U + \left(-\mathbf{u}^T \widetilde{\mathbf{X}} + \mathbf{v}^T \widetilde{\mathbf{Y}}\right)_0^L \right] \leq 0$$
$$\mathbf{u}, \mathbf{v} \geq \mathbf{0}.$$

5 Numerical Examples

In this section, some numerical examples are presented to illustrate the credibility approach. The first example is taken from Guo and Tanaka [10] so that the results from the credibility approach can be compared with those from other approaches. Table 1 provides the data for the example, which has two fuzzy inputs and two fuzzy outputs. These fuzzy inputs and fuzzy outputs have symmetrical triangular membership functions, a special case of trapezoidal membership functions. The membership functions are denoted by (c, d) where c is the center, and d is the spread. The efficiency values $E(\mathbf{v}^T \widetilde{\mathbf{y}}_o)$ obtained from the credibility approach are also given at the bottom of Table 1. Table 2 provides the results from the possibility approach for five different possibility levels (0, 0.25, 0.5, 0.75, 1).

For this example with symmetrical triangular membership functions, the results from the credibility approach are the same as those from the ranking approach [10], the α-level based approach [14] and the possibility approach [14] at the confidence level = 1. The reason is that in this case the expected credits of fuzzy variables coincide with the points where their membership functions are equal to 1. With the credibility approach, DMUs 2, 4 and 5 are credibilistically efficient, while DMUs 1 and 3 are credibilistically inefficient.

The second example illustrates the credibility approach for a fuzzy DEA model with asymmetric triangular fuzzy inputs and outputs. Table 3 provides the fuzzy data and the efficiency values from the credibility approach. The membership functions in Table 3 are denoted by (c, b, d) where c is the point at which the membership function is 1, b is the left spread and d is the right spread. Table 4 provides the results from the possibility approach for five different possibility levels (0, 0.25, 0.5, 0.75, 1).

For this example, with the credibility approach the efficiency values of DMUs 2, 3, 4 and 5 are equal to 1, while DMU 1 is 0.857. Thus, DMUs 2, 3, 4 and 5 are credibilistically efficient, while DMU 1 is credibilistically inefficient. The results from the credibility approach are different from the case of symmetrical triangular membership functions, even though the points, c, at which the membership functions equal 1 for the fuzzy inputs and outputs are the same. This shows that the credibility approach takes into account the spreads of the membership functions. With the asymmetric triangular fuzzy parameters, DMU 3 becomes credibilistically efficient. The results from the possibility approach (Table 4) are different from the first example except at the possibility (confidence) level $\alpha_o = ... = \alpha_n = \beta = 1$ due to the asymmetrical triangular fuzzy data. Table 4 also shows that at some possibility levels DMUs 1 and 3 which are inefficient for the case of symmetrical triangular membership functions become possibilistically efficient for the case of asymmetrical triangular membership functions. Thus, efficiency values from the possibility approach depend on the shapes of the membership functions.

Table 1. DMUs with two symmetrical triangular fuzzy inputs and outputs.

DMU (i)	1	2	3	4	5
input 1	(4.0,0.5)	(2.9,0.0)	(4.9,0.5)	(4.1,0.7)	(6.5,0.6)
input 2	(2.1,0.2)	(1.5,0.1)	(2.6,0.4)	(2.3,0.1)	(4.1,0.5)
output 1	(2.6,0.2)	(2.2,0.0)	(3.2,0.5)	(2.9,0.4)	(5.1,0.7)
output 2	(4.1,0.3)	(3.5,0.2)	(5.1,0.8)	(5.7,0.2)	(7.4,0.9)
$E(\mathbf{v}^T \widetilde{\mathbf{y}}_o)$	0.855	1.000	0.861	1.000	1.000

Table 2. Efficiency values at 5 possibility levels obtained by possibility approach for symmetrical triangular fuzzy data.

Possibility levels	$\overline{f}$				
$\alpha_o = ... = \alpha_n = \beta$	DMU1	DMU2	DMU3	DMU4	DMU5
0.00	1.107	1.238	1.276	1.52	1.296
0.25	1.032	1.173	1.149	1.386	1.226
0.50	0.963	1.112	1.035	1.258	1.159
0.75	0.904	1.055	0.932	1.131	1.095
1.00	0.855	1.000	0.861	1.000	1.000

Table 3. DMUs with two asymmetrical triangular fuzzy inputs and outputs.

DMU (i)	1	2	3	4	5
input 1	(4.0,0.3,0.9)	(2.9,0.2,0.5)	(4.9,1.2,0.4)	(4.1,0.8,0.9)	(6.5,0.5,0.5)
input 2	(2.1,0.2,0.4)	(1.5,0.7,0.8)	(2.6,0.8,0.5)	(2.3,1.0,0.7)	(4.1,0.7,1.2)
output 1	(2.6,0.3,0.5)	(2.2,0.2,0.2)	(3.2,0.2,1.2)	(2.9,0.7,0.4)	(5.1,0.3,0.9)
output 2	(4.1,0.9,0.6)	(3.5,0.6,0.8)	(5.1,0.0,1.0)	(5.7,0.9,0.3)	(7.4,1.2,0.3)
$E(\mathbf{v}^T \widetilde{\mathbf{y}}_o)$	0.857	1.000	1.000	1.000	1.000

Table 4. Efficiency values at 5 possibility levels obtained by possibility approach for asymmetrical triangular fuzzy data.

Possibility levels	$\overline{f}$				
$\alpha_o = ... = \alpha_n = \beta$	DMU1	DMU2	DMU3	DMU4	DMU5
0.00	1.414	2.740	2.119	2.353	1.603
0.25	1.301	2.144	1.748	1.949	1.460
0.50	1.195	1.726	1.449	1.646	1.316
0.75	1.052	1.381	1.153	1.297	1.144
1.00	0.855	1.000	0.861	1.000	1.000

6 Concluding Remarks

This paper has presented a credibility approach as a way to solve fuzzy DEA models. The approach transforms a fuzzy DEA model into a well-defined CP-

DEA model similar to the expected value models in stochastic programming. It uses the expected credits of fuzzy variables to deal with the uncertainty in fuzzy objectives and fuzzy constraints. The expected credits of fuzzy variables are derived by using credibility measures, which are the average of possibility and necessity measures. For the case of normal, convex trapezoidal fuzzy inputs/outputs, the CP-DEA model is in the form of a linear programming model, which can be easily solved by a standard LP solver. Two numerical examples are given to illustrate the implementation of the approach.

With the credibility approach, the problem of ranking fuzzy sets in fuzzy DEA models is handled in a meaningful way. The credibility approach combines the information from both the degrees of possibility and the dual degrees of necessity of fuzzy events. It provides an efficiency value for each DMU as a representative of its possible range in such a way that decision makers do not have to specify any parameters (confidence levels) as in the possibility approach. Comparing with other approaches, this approach can be implemented more easily and requires less interactive data from decision makers than tolerance, α-level based, fuzzy ranking, and possibility approaches, while it deals with uncertain data by using the expected credit instead of defuzzifying uncertain data into crisp values like the defuzzification approach. Also, this approach can be interpreted in a similar way as that of the expected value model in stochastic programming.

An interesting topic for further research is to extend this approach to other types of fuzzy DEA models and study the sensitivity of the efficiency to fuzzy data. It would also be interesting to study applications to the real-world systems such as supply chain and production systems.

References

1. Arnade, C. A. (1994): Using data envelopment analysis to measure international agricultural efficiency and productivity, United States Department of Agriculture, Economic Research Service, Technical Bulletin, 1831, 1-30.
2. Banker, R. D., Charnes, A. and Cooper, W. W. (1984) Some models for estimating technical and scale inefficiency in data envelopment analysis, Management Science, 30, 1078-1092.
3. Banker, R. D., Chang, H. and Cooper, W. W. (1996) Simulation studies of efficiency, returns to scale and misspecification with nonlinear functions in DEA, Annals of Operations Research, 66, 233-253.
4. Charnes, A., Cooper, W. W., Golany, B. and Seiford, L. M. (1985) Foundation of data envelopment analysis for Pareto-Koopmans efficient empirical production functions, Journal of Econometrics, 30, 91-107.
5. Charnes, A., Cooper, W. W., Lewin, A. Y. and Seiford, L. M. (1994) Data Envelopment Analysis: Theory, Methodology, and Application, Kluwer Academic Publishers, London.
6. Cooper, W. W., Seiford, L. M. and Tone, K. (2000) Data Envelopment Analysis: A Comprehensive Text with Models, Applications, References and DEA-Solver Software, Kluwer Academic Publishers, London.

7. Charnes, A., Cooper, W. W. and Rhodes, E. (1978) Measuring the efficiency of decision-making units, European Journal of Operational Research, 2, 429-444.
8. Dubois, D. and Prade, H. (1988) Possibility Theory: An Approach to Computerized Processing of Uncertainty, Plenum Press, New York.
9. Fang, S. -C. and Puthenpura, S. (1993) Linear Optimization and Extensions: Theory and Algorithms, Prentice Hall, Englewood Cliffs, New Jersey.
10. Guo, P. and Tanaka, H. (2001) Fuzzy DEA: A perceptual evaluation method, Fuzzy Sets and Systems, 119, 149-160.
11. Kahraman, C. and Tolga, E. (1998) Data envelopment analysis using fuzzy concept, Proceedings of the 28th International Symposium on Multiple-Valued Logic, 338-343.
12. Kao, C. and Liu, S. -T. (2000) Fuzzy efficiency measures in data envelopment analysis, Fuzzy Sets and Systems, 113, 427-437.
13. Lertworasirikul, S. (2001) Fuzzy Data Envelopment Analysis for Supply Chain Modeling and Analysis, Dissertation Proposal in Industrial Engineering, North Carolina State University.
14. Lertworasirikul, S., Fang, S.-C., Joines, J. A. and Nuttle, H. L. W. (2001) Fuzzy data envelopment analysis (DEA): A possibility approach, submitted to Fuzzy Sets and Systems.
15. Lertworasirikul, S., Fang, S.-C., Joines, J. A. and Nuttle, H. L. W. (2002) A possibility approach to fuzzy data envelopment analysis, to appear in the Proceedings of 8th International Conference on Fuzzy Theory and Technology.
16. Liu, B. (2002) Toward fuzzy optimization without mathematical ambiguity, to appear in Fuzzy Optimization and Decision Making, 1-1.
17. Liu, B. (2001a) Uncertain programming: A unifying optimization theory in various uncertain environments, Applied Mathematics and Computation, 120, 227-234.
18. Meada, Y., Entani, T. and Tanaka, H. (1998) Fuzzy DEA with interval efficiency, Proceedings of the 6th European Congress on Intelligent Techniques and Soft Computing, 2, 1067-1071.
19. Seiford, L. M. and Thrall, R. M. (1990) Recent development in DEA: The mathematical programming approach to frontier analysis, Journal of Econometrics, 46, 7-38.
20. Sengupta, J. K. (1992) A fuzzy systems approach in data envelopment analysis, Computers and Mathematics with Applications, 24, 259-266.
21. Sengupta, J. K. (1995) Dynamics of Data Envelopment Analysis: Theory of Systems Efficiency, Kluwer Academic Publishers, London.
22. Zadeh, L. A. (1965) Fuzzy sets, Information and Control 8, 338-353.
23. Zadeh, L. A. (1978) Fuzzy sets as a basis for a theory of possibility, Fuzzy Sets and Systems, 1, 3-28.
24. Zimmermann, H. J. (1996) Fuzzy Set Theory and Its Application, Kluwer Academic Publishers, London.

Fuzzy Optimization using Simulated Annealing:An Example Set

Rita Almeida Ribeiro[1] and Leonilde Rocha Varela[2]

[1] Universidade Lusiada
  Dept. Economics & Man. Sciences
  Rua da Junqueira 190
  1349-001 Lisbon, Portugal
[2] Universidade do Minho
  Escola de Engenharia
  Dept. Produo e Sistemas
  4800-058 Guimares, Portugal

Abstract. In this paper we tested a set of examples, presented in the literature, with a general fuzzy optimization approach using Simulated Annealing (SA). For linear fuzzy problems, we selected a set of examples from well-known authors. For non-linear fuzzy optimization problems we selected two crisp problems and only two fuzzy examples because there are not many fuzzy non-linear examples in the literature. The comparison of the results, obtained with our approach and the ones shown in the literature, allow us to highlight the flexibility, generality and performance of this fuzzy approach to solve either linear or non-linear fuzzy optimization problems.

1 Introduction

The main objective of this paper is to assess the generality, flexibility and performance of a general fuzzification model solved with the simulated annealing algorithm (SA) [1], [2], [3]. The fuzzification model allows fuzzy coefficients, fuzzy resources, fuzzy goals and combinations. This general fuzzy optimization model is able to handle linear and non-linear problems and even unfeasible problems. However, when we have to deal with fuzzy coefficients there is a price to pay because the fuzzification model transforms the problem into a larger dimension one (more constraints and variables) and which also is non-linear.

Since the formulation of the model is non-linear, when we have fuzzy coefficients, we should use an algorithm independent of the formulation, such as SA, to solve the problem. Our rational in this work is that instead of discussing and comparing details of the fuzzy approach used, with other models proposed in the literature, we use their examples to compare their results with the ones obtained with our approach. The set of examples tested can act as a test-bed for any fuzzy approach proposed in the literature.

Specifically, for linear problems we selected a set of examples from representative authors and methods, such as: Carlsson and Korhonen [4]; [5]; Lai

& Hwang [6]; Sakawa [7]; Tanaka, Ichihashi and Asai [8]; Delgado, Verdegay and Vila [9, 10]; and Zimmermann [11]. In addition, we solved a linear unfeasible solution example [12] to show that the approach formulation, can also provide solutions for unsolvable crisp problems. For non-linear problems we selected four examples, two fuzzy non-linear problem by Sakawa [7] and two crisp problems [13]. The latter selection was based on the scarcity of non-linear fuzzy examples (exceptions can be seen in [14] [7] [15]).

The selected set of examples are, first, formulated with the fuzzification model of Moura-Pires and Ribeiro [1] [2] and then solved with an SA algorithm implementation proposed by the authors [3]. With this comparison we can show how general and flexible both the fuzzification model and the SA algorithm are to solve different types of fuzzy optimization problems, from linear to non-linear, as well as unfeasible problems.

This paper is organized as follows. This first section introduces the objectives of this work and gives a brief introduction of fuzzy optimization concepts. Section 2 describes the main characteristics of the fuzzification model, as well as of the solution algorithm used to solve all the examples (different methods). Section 3 introduces the set of linear examples from different authors, solves them and discusses the results obtained by these authors and with our approach. Section 4 follows the same logic as section 3, but introduces the set of non-linear examples tested. Section 5 presents the conclusions.

2 Basics on the approach used for formulating and solving the example set

In this section we introduce the main characteristics of the fuzzy approach used in this paper [1] [2] [3], to formulate and solve a set of examples in fuzzy optimization. To make this section more easy to read we divided it in four subsections: the first briefly introduces fuzzy optimization main concepts; the second subsection presents the fuzzification model used in this work to formalize the example set; in the third subsection we describe the implemented solution process and the reasons to use the simulated annealing algorithm.

2.1 Introduction to fuzzy optimization

The main objective of a fuzzy optimization method is to find the "best" solution (decision alternative) in the presence of incomplete information, i.e., imprecise information and/or in the presence of "vague" limits in the information. There exist many forms of imprecision in fuzzy optimization problems, as for example, variable coefficients that we do not know precisely (for example, "processing times of about one hour for assembling a piece") and constraint satisfaction levels with imprecise limits (for example, "the total processing time available is around 100 hours").

A classical linear optimization problem consists on maximizing or minimizing a certain objective function subject to a set of constraints, which express, for example, the resource limitations. Formally,

$$\begin{aligned} \max/\min Z &= Cx \\ Ax\{\geq, \leq, =\}&B \\ x &\geq 0 \end{aligned} \quad (1)$$

The fuzzy version of this problem is generally formalized as,

$$\begin{aligned} \max/\min \tilde{Z} &= \tilde{C}x \\ \tilde{A}x\{\geq, \leq, =\}&\tilde{B} \\ x &\geq 0 \end{aligned} \quad (2)$$

where $\tilde{Z}$ represents a fuzzy goal, $\tilde{C}$ is the vector of fuzzy costs, $\tilde{A}$ is the matrix that contains the fuzzy coefficients of the objective(s) and of the constraints and $\tilde{B}$ is the corresponding vector of the limits of the resources. The "tilde" on top of the parameters means that they are defined by fuzzy sets. We opted for using the "tilde" on top of the right hand side parameter and not in the constraint sign has a uniform concept of fuzzy parameter for resource limits, fuzzy coefficients and fuzzy goals. In section 2.2 this point is discussed further.

The first fuzzy extension of the classical optimization problem to a fuzzy environment is due to Bellman and Zadeh [16]. Based on the similarity model of the latter authors, Zimmerman was the first author to propose a method to solve fuzzy linear programming problems with fuzzy resources (constraints) and fuzzy goals [17] [11]. Nowadays there are many fuzzification methods proposed in the literature, for resources fuzzification, goal fuzzification as well as for coefficients fuzzification (see good overviews in [18] [19] [6] [14, 20]).

2.2 Flexible fuzzification model used

The set of examples that are tested in this paper were formulated with a fuzzy model proposed by Ribeiro and Moura-Pires [1] [2]. This model allows using the following types of fuzzifications in the optimization problem: (a) fuzzy coefficients in the objective function; (b) fuzzy coefficients in the left-hand side of the constraints; (c) fuzzy resource limits of the constraints; (d) fuzzy goals; (e) any combination of the previous.

Assuming the general formulation for fuzzy optimization problems defined in (2) the fuzzy model used in this work is,

$$\max Z = \sum_{k=1}^{K} \tilde{c}_k \cdot x_k$$

$$\sum_{k=1}^{K} \tilde{a}_{ik} \cdot x_k \{\geq, \leq, =\} \tilde{b}_i, i = 1, \ldots, N$$

$$x \geq 0 \qquad (3)$$

The optimization model (3) is then transformed into the following system of non-linear fuzzy constraints:

$$\begin{cases} \max Z = \sum_{k=1}^{K} w_k \cdot x_k \\ \begin{cases} \max M = \min(\mu_{a_{ik}}, \mu_{b_i}, \mu_{c_k}) \\ \sum_{k=1}^{K} y_{ik} \cdot x_k \{\leq, =, \geq\} \tilde{b}_i \\ y_{ik} = \tilde{a}_{ik} \\ w_k = \tilde{c}_k \end{cases} \\ k = 1, \ldots, K \quad i = 1, \ldots, N \quad x, y, z \geq 0 \end{cases} \qquad (4)$$

This mathematical transformation implies the addition of as many new fuzzy constraints and as many new variables as the fuzzy coefficients of the problem (e.g. for 2 fuzzy coefficients we add two new variables and two new constraints to the formulation). Any new added constraint is represented as an equality constraint with a fuzzy resource limit; this process allows the handling of all constraints in a similar fashion. The fuzzification model includes two objectives, the initial one and another objective (M) to obtain the best of the worst violations of the fuzzy parameters, in the sense of the maxmin model [11]. In general, the objectives of this fuzzification model are two fold:

1) Find the best values for x, y and w that maximize the minimum aggregated membership values (denoted by M), considering a threshold value for a minimum acceptable violation level of the constraints.

2) Find the optimal value of Z that satisfies all the constraints as well as the first step.

The aggregation described in 1) represents the intersection of all the membership values of the fuzzy parameters considered (i.e. coefficients, and/or resource limits, and/or goals), to indicate that all the constraints/coefficients have to be satisfied with a certain level (min). In addition, a threshold value was included in our implementation to allow the decision maker to specify if he/she wants an optimistic scenario (high satisfaction level), average scenario (average satisfaction level) or a pessimistic scenario (low satisfaction level). In point 2) it must be pointed that we are describing just a single objective function, but the fuzzification model can handle multi-objective problems [21].

To clarify the non-linearity of the model lets consider that $\tilde{c}_k$ is a fuzzy set that indicates how acceptable are the values around c_k. Let $w = (w_1, w_2, \ldots, w_K)$ be a set of objective function costs such that $\mu_{\tilde{c}_k}(w_k) > 0$. In fact, for each different combination of values w_k a different function to maximize over x is obtained and each set has a satisfaction level defined by: $\mu_{\tilde{C}}(w) = \min_k \mu_{\tilde{c}_k}(w_k)$ and $W_\alpha = \{w : \mu_{\tilde{C}}(w) > \alpha\}$. The same logic applies to constraint coefficients a_{ik}. For more details about the fuzzification model, see [1]).

One of the drawbacks of the fuzzification model is the addition of more constrains and variables, whenever there are fuzzy coefficients. However, it provides the advantage of handling all fuzzy coefficients in the same manner of fuzzy constraints, hence, we can know how much each constraint (being a resource or coefficient) is being violated. Another important drawback is the non-linearity of the model. The compensation for this disadvantage is the generality of the model since it can handle either linear or non-linear fuzzy optimization problems

In conclusion we can say that this fuzzification model provides tradeoffs between constraint satisfaction (objective M) and the original problem objective (Z). Both fuzzy coefficients $\tilde{c}_{ik}$ and $\tilde{a}_{ik}$ and the resources $\tilde{b}_i$ are all handled in a similar fashion, i.e. as fuzzy constraints. Further, this method allows manipulating either linear or non-linear fuzzy optimization problems, as well as unfeasible crisp problems [1] [21] [2].

2.3 Solution algorithm: Simulated Annealing (SA)

Before discussing the main characteristics of the SA algorithm and of our implementation, we need to clarify why we selected the SA algorithm for solving the general fuzzification model (4), used in this work. First, and most important, since our fuzzification model is non-linear we need an algorithm that was independent of the problem to be solved, i.e. an algorithm that could solve either linear or non-linear optimisation problems. Second, it is to easy to understand because the parameters have an analogy with the annealing process of a solid [22] and the algorithm does not have too many parameters to handle. Third, since it is a guided-random search algorithm, it allows us to control the search for the ""optimum" (e.g. with parameter temperature and the stopping criteria) [22]. Fourth, the algorithm is quite simple to implement and the computational time to achieve "good" results is quite acceptable.

In 1983, Kirkpatrick and others originally proposed the Simulated Annealing algorithm using, as mentioned, an analogy with the annealing process of a solid [23] [22]. The objective of an algorithm of this nature is to find the best solution among a finite number of possible solutions, however, it does not guarantee that the solution found is indeed the global optimum. This last characteristic restricts its use to the cases where "good" local optima are acceptable. The SA technique is also interesting because it allows finding near-optimal solutions within a reasonable computational time frame. The worst drawback of the SA algorithm is the need to provide "good" initial points to run the algorithm – without good initial points the simulation can get easily stuck in a local minima [2].

The SA algorithm requires the definition of a neighbourhood structure, as well as the parameters for the cooling process [22]. A temperature parameter allows distinguishing among deep or slight alterations in the objective function. Drastic alterations occur at high temperatures and small or slight modifications at low temperatures. The four basic requirements for using the SA algorithm in combinatorial optimization problems are: a concise problem description; a random generation of the alterations from one configuration to another; an objective function that includes the utility function of the trade-offs; and a definition of the initial state, of the number of iterations to be executed for each temperature and its annealing process.

In general we can say that we have to specify: (a) how to generate a state y, neighbour of x; (b) which aggregation function (M) to use; (c) the selection of number of neighbours to generate; (d) the temperature decrease function; (e) and finally the algorithm stopping criteria. In our implementation we followed the fuzzification model described in (4) but with the simplification of assuming a single objective (Z), besides the aggregated violations of constraints (objective M in (4)). Further, when we use the SA algorithm for solving fuzzy optimization problems, the decision maker can select thresholds levels (α-cuts) as well as the tolerances/deviations (i.e.fuzzification) for each constraint parameter and/or for each objective function's coefficient and/or for each constraint's coefficient. In addition, we must point that this implementation was based on a first prototype explained in [2] but with modifications and additions, as for example the notion of *seed* [3].

As mentioned, we included in the algorithm implementation the notion of seed for the random numbers generation [22] to allow the generation of identical values (i.e. same random numbers), in different program executions but for the same example. With the seed we can repeat the a priori conditions for testing the same problem with different types of fuzzification (e.g. coefficients or resources or both). For our tests we used a seed value of 1.

For more details about the SA algorithm and its implementation, used to test the set of selected examples (after formulating them with the fuzzification model (4)), see [3].

3 Set of linear examples tested and discussion of the results

Considering that there are many methods proposed in the literature to solve fuzzy optimization problems we selected arbitrarily a set of linear and non-linear examples from the literature. We believe the set is a good sample to compare different results with the ones obtained with our approach. Instead of comparing the approach used in this work with other approaches our objective is to test a set of diverse linear and non-linear fuzzy optimization problems. We solve the set of examples with our approach and then we compare our results with both the crisp results and the fuzzy results of the authors (when we have them). This is a more understandable and simple way to discuss the benefits of our approach. We must point that in some of the examples we made some simplifications to enable comparisons between our results and the ones found in the literature. All modifications are explained with each example presentation.

In order to standardize the tests we also made some assumptions regarding the membership functions of the problems tested. In this work we always use linear triangular membership functions to solve the examples. According to the sign of the equation (e.g. bigger than, equal to, less than) we use the following notation: left-open-triangle=]value, tolerance]; right-open-triangle=[tolerance, value[; triangular= [leftTolerance, value, rightTolerance]. For the linear set of examples we followed the author's fuzzy parameters limits as close as possible. For the set of non-linear examples, instead of following the deviations (membership functions limits) considered by the author we simplified them to a 10% flexibilization either in the coefficients or in the resources limits or in the goals. This 10% tolerance is smaller than the one used by the authors but this strengths our claim of achieving better results.

The presentation and discussion of the examples (section 3 and 4) will follow the steps:

- First we depict the initial example (proposed by each author) and the fuzzified parameters that were considered in the respective method (when we have them).
- Second, we transform the example with our fuzzification model (4) but for reasons of space we only show this formulation for example 3.1. Besides the authors fuzzy parameters limits we also show our fuzzy parameters limits, if we performed some modification. We do not show our fuzzified limits for the non-linear examples because, as mentioned above, they were simplified to a 10% tolerance from the central value.
- Third, we solve the problem with the SA algorithm implementation, described in sub-section 2.3, and we discuss and compare the results we obtain with three thresholds (0.3, 0.6 and 0.9) with both the one provided by the author (denoted fuzzy-author) and the crisp result (when we have it). In some cases we added another threshold solution to enable us to compare our result with the similar one presented by an author.

It is important to remind that we will only show the mathematical transformation formulation that is performed in the examples using (4) for the first example (3.1.), because of two aspects: space considerations and because the transformation is quite straightforward. Hence, we will only show the original example formulation, the fuzzified parameters limits and the final results comparison.

Although our approach may consider all kinds of fuzzification, we solve each example with just the respective fuzzification proposed by the authors. This is the only way to make a meaningful comparison between results. As mentioned, in some cases we did simplifications but they are indicated and do not change the discussion.

In addition, we perform more than one simulation for each example, with different threshold values, to obtain three different scenarios: optimistic (high satisfaction level, 90%), average (average satisfaction level, 60%) and pessimist (low satisfaction level, 30%). These scenarios will provide more information for comparative purposes and allow us to neglect the small modifications done in the fuzzification of the examples. For each alternative test we also show the computational execution time that the SA implementation took to solve the problem. This will allow us to assess the computational effort of the fuzzification approach used.

3.1 Example by Tanaka, Ichihashi, Asai (in: [6])

In the case of Tanaka, Ichihashi and Asai method we used the example found in [6] instead of the original one [24] because it is a simpler version using the same method. The example is,

$$\max 25x_1 + 18x_2$$
$$s.t\ \tilde{a}_{11}x_1 + \tilde{a}_{12}x_2 \le 7\tilde{8}0$$
$$\tilde{a}_{21}x_1 + \tilde{a}_{22}x_2 \le 3\tilde{8}0$$
$$x_i \ge 0$$

where the fuzzy parameters were: $a_{11} = [12, 18]$; $a_{12} = [32, 36]$; $b_1 =]780, 850]$; $a_{21} = [19, 21]$; $a_{22} = [7, 13]$; $b_2 =]380, 480]$

Let us consider the following fuzzy parameters for the coefficients: $a_{11} = [12, 15, 18]$; $a_{12} = [31.96, 34, 36.04]$; $a_{21} = [19, 20, 21]$; $a_{22} = [7, 10, 13]$ and for the resources identical limits. Now transforming the example with our formulation (4) we obtain,

$$\max Z1 = 25x_1 + 18x_2$$
$$\begin{cases} \max Z2 = \min(\mu_{a_{11}}, \mu_{a_{12}}, \mu_{a_{21}}, \mu_{a_{22}}, \mu_{b_1}, \mu_{b_2}) \\ s.t.\ y_{11}x_1 + y_{12}x_2 \le 7\tilde{8}0 \\ \quad\quad y_{21}x_1 + y_{22}x_2 \le 3\tilde{8}0 \\ \quad\quad y_{11} = \tilde{1}5 \\ \quad\quad y_{12} = \tilde{3}4 \\ \quad\quad y_{21} = \tilde{2}0 \\ \quad\quad y_{22} = \tilde{1}0 \\ \quad\quad x_i, y_i \ge 0 \end{cases}$$

The solutions for our three thresholds, the crisp solution and the fuzzy solution of the author are:

solutions	#	Z	x_1	x_2	μ	Time
Crisp solution	1	577.736	9.660	18.672	—	—
Fuzzy Author	2	623.54	12.14	17.78	0.4	—
Fuzzy ours: α= 0.3	3	673.267	12.336	20.270	0.3151	15"
Fuzzy ours: α= 0.4	4	648.875	11.699	19.800	0.4144	16"
Fuzzy ours: α= 0.6	5	644.708	12.320	18.707	0.6337	17"
Fuzzy ours: α= 0.9	6	596.519	10.386	18.716	0.9024	17"

Observing the results above we see that with our method most of our solutions, #3, #4, #5 are better than solution #2 of Tanaka, Ichihashi and Asai. Only when we set a threshold of $\alpha = 0.9$ our solution is worse than the one by Tanaka, Ichihashi and Asai with $\mu = 0.4$ (596.519 vs. 623.54). Again the results clearly show the trade-off that happens with our approach: for better satisfaction of constraints and coefficients (less violation) we get worse values for our objective function.

The time to achieve a solution with the SA implementation is quite reasonable, for all our tested solutions (around 16 seconds).

3.2 Example by Carlsson and Korhonen [4]

Carlsson and Korhonen proposed an interesting method that considers a complete fuzzification of linear programming problems. These authors used the following example to illustrate their method,

$$\max [1, 1.5)x_1 + [1, 3)x_2 + [2, 2.2)x_3$$
$$\text{s.t.}$$
$$[3, 2)x_1 + [2, 0)x_2 + [3, 1.5)x_3 \leq [18, 22)$$
$$[1, 0.5)x_1 + [2, 1)x_2 + [1, 0)x_3 \leq [10, 40)$$
$$[9, 6)x_1 + [20, 18)x_2 + [7, 3)x_3 \leq [96, 110)$$
$$[7, 6.5)x_1 + [20, 15)x_2 + [9, 8)x_3 \leq [96, 110)$$
$$x_i \geq 0$$

where the intervals used for the coefficients and resources fuzzification are represented by exponential functions with the intervals shown in the example.

We formulated this example using the the open interval limit as the central point for the tolerances, in the same fashion as Lai & Hwang [6],

$$\max 1.5x_1 + 3x_2 + 2.2x_3$$
$$\text{s.t.}$$
$$2x_1 + 1.5x_3 \leq 22$$
$$0.5x_1 + 1x_2 \leq 40$$
$$6x_1 + 18x_2 + 3x_3 \leq 110$$
$$6.5x_1 + 15x_2 + 8x_3 \leq 110$$
$$x_i \geq 0$$

and our fuzzy parameters are: $C_1 = [1, 1.5, 2]$; $C_2 = [1, 3, 5]$; $C_3 = [2, 2.2, 2.4]$; $a_{11} = [1, 2, 3]$; $a_{13} = [0.51.5, 3]$; $b_1 =]18, 22]$; $a_{21} = [0, 0.5, 1]$; $a_{22} = [0, 1, 2]$; $b_2 =]10, 40]$; $a_{31} = [3, 6, 9]$; $a_{32} = [16, 18, 20]$; $a_{33} = [0, 3, 7]$; $b_3 =]96, 110]$; $a_{41} = [6, 6.5, 7]$; $a_{42} = [10, 15, 20]$; $a_{43} = [7, 8, 9]$; $b_4 =]96, 110]$.

As mentioned above, for comparative purposes we show the solutions obtained with our fuzzy approach (2 simulations for different thresholds), the crisp solution for the problem, and the respective two solutions from author method.

solutions	#	Z	x_1	x_2	x_3	μ	time
Crisp solution	1	12	0	0	6	—	—
Fuzzy Author 1	2	23.76	0	1.22	10.44	0.3	
Fuzzy Author 2	3	13.08	0	0	6.52	0.9	—
Fuzzy ours α= 0.3	4	34.110	4.220	3.030	7.220	0.33	1'24"
Fuzzy ours α= 0.6	5	29.938	8.733	2.666	2.693	0.68	8'5"
Fuzzy ours α= 0.9	6	17.663	2.723	2.970	2.119	0.95	23'36"

As can be observed in the results obtained, with our approach we obtain much better results for the similar cases, #2 and #3 of Carlsson and Korhonen versus ours #4 and #6. In particular, there is a big difference between our solution #4 and the equivalent Carlsson and Korhonen solution #2 (34.110-23.76= 10.35!), if we note that the allowed violations of the fuzzy parameters were similar $\mu = 0.33$ vs $\mu = 0.3$. The differences using both methods are quite significant and show that by simplifying the membership functions to triangular ones helped to obtain better results. The time to solve this problem is large, particularly for high thresholds (e.g. $\alpha = 0.9$), which mean having less violation of the fuzzy parameters. I.e. the S.A.algorithm implementation takes time to find a solution when we require better satisfaction levels (less flexibility) for completely fuzzified problems.

3.3 Example by Chanas [5]

Chanas used the following example (fuzzy resources) to test his method,

$$\max 3x_1 + 4x_2 + 4x_3$$
$$s.t. \ 6x_1 + 3x_2 + 4x_3 \leq 1\tilde{2}00$$
$$5x_1 + 4x_2 + 5x_3 \leq 1\tilde{5}50$$
$$x_i \geq 0$$

where the fuzzy parameters were $Z = [1600, 1750[$; $b_1 =]1200, 1300]$; $b_2 =]1550, 1750]$.

In our approach the formulation and fuzzy parameters used are identical and the solutions obtained were:

solutions	#	Z	x_1	x_2	x_3	μ	time
Crisp solution	1	1550	0	387.5	0	—	—
Fuzzy Author	2	1649.8	0	412.45	0	0.57	—
Fuzzy ours $\alpha= 0.3$	3	1691.732	0.792	416.918	5.421	0.3	16"
Fuzzy ours $\alpha= 0.4$	4	1656.010	2.929	408.313	3.492	0.451	13"
Fuzzy ours $\alpha= 0.5$	5	1646.578	3.090	405.108	4.219	0.5102	16"
Fuzzy ours $\alpha= 0.6$	6	1626.023	2.195	402.956	1.905	0.6080	14"

Observing the results we see that our solutions are better only for a $\mu < 0.5$. This implies that we need to violate the constraints a little more than with the Chanas parametric approach to obtain better results. However, Chanas method does not handle a complete flexibilization of the model and does not have the facility of setting a threshold satisfaction level. For example, with a smaller violation of the constraints ($\mu = 0.3$ versus $\mu = 0.57$) we obtain a better level for the objective function (1691.732 versus 1649.8). Of course, in our method, the higher the violation of the constraints the smaller the objective function is.

Chanas [5] also solved other examples using his parametric method such as Zimmermann multi-objective example (example 3.10). Since the results obtained were similar for both authors, we leave the comparison of those results for later.

3.4 Example by Lai & Hwang [6]

Lai and Hwang presented an example with fuzzy resources to discuss their parametric method,

$$\begin{aligned}
\max \quad & 4x_1 + 5x_2 + 9x_3 + 11x_4 \\
s.t. \quad & x_1 + x_2 + x_3 + x_4 \leq \tilde{15} \\
& 7x_1 + 5x_2 + 3x_3 + 2x_4 \leq \tilde{120} \\
& 3x_1 + 5x_2 + 10x_3 + 15x_4 \leq \tilde{100} \\
& x_i \geq 0
\end{aligned}$$

where the fuzzy parameters were $b_1 =]15, 18]$; $b_3 =]100, 120]$. Our formulation followed the same fuzzy parameters tolerances. The results obtained were:

solutions	#	Z	x_1	x_2	x_3	x_4	μ	time
Crisp solution	1	99.29	7.14	0	7.86	0	—	—
Fuzzy Authors $\tau=0.1$	2	101.28	7.283	0	8.017	0	0.1	—
Fuzzy Authors $\tau=0.3$	3	105.248	7.569	0	8.331	0	0.3	—
Fuzzy Authors $\tau=0.9$	4	117.19	8.427	0	9.273	0	0.9	—
Fuzzy ours: $\alpha= 0.1$	5	116.463	8.394	0.210	8.246	0.693	0.11	20"
Fuzzy ours: $\alpha= 0.3$	6	110.686	8.628	0.339	7.018	1.029	0.33	19"
Fuzzy ours: $\alpha= 0.9$	7	101.223	6.852	0.453	7.746	0.167	0.90	19"

Comparing results #2 and #3 from Lai and Hwang with our corresponding first two solutions (#5 and #6) we see that our method performs better than the authors method, i.e. we obtain higher objective function values. However we get a smaller value for solution #7 vs. #4 because our philosophy is that for higher satisfaction values of the fuzzy parameters (i.e. less violation of constraints) we should "pay" more. We do have a trade-off between better solutions and bigger violation of the fuzzy parameters.

Lai and Hwang also tested the same problem for bigger violations of the constraints, but since our method would always perform better (due to its generality) for lower thresholds (higher violation of constraints) we did not perform more comparisons.

3.5 Example by Sakawa [7]

Sakawa presented the following multiple objective example,

$$\min C_{11}x_1 - 4x_2 + x_3$$
$$\max -3x_1 + C_{22}x_2 + x_3$$
$$\text{equal } 5x_1 + x_2 + C_{33}x_3$$
$$\text{s.t. } a_{11}x_1 + a_{12}x_2 + 3x_3 \leq 12$$
$$x_1 + 2x_2 + a_{23}x_3 \leq b_2$$
$$x_i \geq 0$$

where the fuzzy parameters are $C_{11} = [0, 2, 2.5]$; $C_{22} = [-1.25, -0.75, -0.25]$; $C_{33} = [-0.25, 0, 1]$; $a_{11} = [0, 3, 4]$; $a_{12} = [0.5, 1, 1.5]$; $a_{23} = [0.5, 1, 1.5]$; $b_2 = [8, 12, 14]$.

In order to use our SA implementation we have to transform the problem into a single objective one. It should be noted that our fuzzification model (section 2.2.) allows solving multiple objective problems but the implemented algorithm solution is not yet prepared for this. Hence, we start by considering the following simplifications in the objective functions,

$$\max -C_{11}x_1 + 4x_2 - x_3$$
$$\max -3x_1 + C_{22}x_2 + x_3$$
$$\max 5x_1 + x_2 + C_{33}x_3$$

After we use a simple combination of the objectives to obtain a single objective and our formulation of the example becomes,

$$\max -\tilde{2}x_1 + 4x_2 - x_3 - 3x_1 + 0.\tilde{7}5x_2 + x_3 + 5x_1 + x_2 + \tilde{0}x_3$$
$$\text{s.t. } \tilde{3}x_1 + \tilde{1}x_2 + 3x_3 \leq 12$$
$$x_1 + 2x_2 + \tilde{1}x_3 \leq \tilde{8}$$
$$x_i \geq 0$$

where our fuzzy parameters are identical to the above except for $b_2 =]8, 14]$.
The solutions obtained were:

solutions	#	Z	x_1	x_2	x_3	μ	time
Crisp solution	1	17	0	4	0	—	—
Fuzzy Author	2	12.91(*)	2.3074	2.3443	1.921	0.65	—
Fuzzy ours: α= 0.3	3	34.729	0.196	5.756	0.000	0.3737	36"
Fuzzy ours α= 0.6	5	25.552	0.015	4.943	0.000	0.6773	1'34"
Fuzzy ours: α= 0.9	6	19.281	0.016	4.276	0.000	0.951	7'39"

(*) approximate solution calculated from the three O.F. in Sakawa example.

In this case it is more difficult to compare the results because we made some assumptions in the simplifications to obtain a single objective. However, if we consider that the decision maker wants to have the best possible value for all objectives and that the equality objective can be transformed into a maximizing objective, then our results are much better than the ones presented by Sakawa (for all solutions #3, #4, #5, #6). It should also be noted that there seems to be a correlation between the time to solve problems and higher threshold levels (less violation) when the problems include fuzzification of some coefficients (time=7'39" for #6). Our SA implementation algorithm behaviour decreases when there is an increase in both fuzzy parameters and higher satisfaction values for the violations (less flexibility).

3.6 Example by Delgado, Verdegay, Vila [9]

Delgado, Verdegay and Vila illustrated their method with the following example,

$$\begin{aligned}
\max\ & 5x_1 + 6x_2 \\
\text{s.t.}\ & \tilde{3}x_1 + \tilde{4}x_2 \leq \tilde{18} \\
& \tilde{2}x_1 + \tilde{1}x_2 \leq \tilde{7} \\
& x_i \geq 0
\end{aligned}$$

where the fuzzy parameters were: $a_{11} = [3, 2, 4]$; a_{12}=[4, 2.5, 5.5]; $b_1 = [18, 16, 19]$; $a_{21} = [2, 1, 3]$; $a_{22} = [1, 0.5, 2]$; $b_2 = [7, 6, 9]$. These authors then construct two alternative auxiliary problems to handle the fuzzy parameters with their parametric method. In this paper we only discuss the authors solutions for the first auxiliary problem.

Our fuzzy membership functions limits are: $a_{11} = [2, 3, 4]$; $a_{12} = [2.5, 4, 5.5]$; $b_1 =$]18, 19]; $a_{21} = [1, 2, 3]$; $a_{22} = [0.5, 1.25, 2]$; $b_2 =$]7, 9].

The solutions obtained were:

solutions	#	Z	x_1	x_2	μ	time
Crisp solution	1	28	2	3	—	—
Fuzzy Author $\tau= 0.3$	2	31.22	2.14	3.42	0.7(*)	—
Fuzzy Author $\tau= 0.6$	3	29.84	2.08	3.24	0.4(*)	—
Fuzzy Author $\tau= 0.9$	4	28.46	2.02	3.06	0.1(*)	—
Fuzzy ours $\alpha= 0.3$	5	34.36	2.581	3.576	0.3380	9"
Fuzzy ours $\alpha= 0.6$	6	31.77	0.656	4.749	0.6402	9"
Fuzzy ours $\alpha= 0.9$	7	29.63	1.337	3.658	0.9062	14"

(*) this corresponds to $(1 - \tau)$.

Comparing the results for the same τ and α we see that with our approach we always obtain better results than the parametric method proposed by Delgado, Verdegay and Vila. Our method is more flexible, hence it allows a wider search of space. In terms of time our algorithm performed quite well considering it has fuzzy coefficients and that solution #7 has high satisfaction value.

We also tested other examples by the same authors Delgado, Verdegay, Vila and Campos [25] [10] for their parametric method. Since their method did not change our results proved to be always better. For reasons of space we will not present these results.

3.7 Example by Zimmermann [11]

Zimmermann was the first author to propose a fuzzy method to deal with fuzzy resources. The example he presented to discuss his symmetrical method was,

$$\min 41400x_1 + 44300x_2 + 48100x_3 + 49100x_4$$
$$\text{s.t. } 0.84x_1 + 1.44x_2 + 2.16x_3 + 2.4x_4 \leq 1\tilde{7}0$$
$$16x_1 + 16x_2 + 16x_3 + 16x_4 \leq 1\tilde{3}00$$
$$x_i \geq 0$$

and the fuzzy parameters limits were: $b_1 = [160, 170[$, $b_2 = [1200, 1300[$ and $b_3 = [0, 6[$.

To solve this problem with our method we do not need to make any modification in the above formulation. Because we already tested problems with only fuzzy resources in this test we only used two thresholds, 0.3 and 0.6 because they are enough to draw conclusions.

The solutions obtained were:

solutions	#	Z	x_1	x_2	x_3	x_4	μ	time
Crisp solution	1	3,864,975	6	16.29	0	66.54	—	—
Fuzzy Author	2	3,988,250	17.411	0	0	66.54	—	—
Fuzzy ours $\alpha= 0.3$	3	3,701,349	4.058	11.743	2.777	58.647	0.3560	6"
Fuzzy ours $\alpha= 0.6$	4	3,759,867	5.540	14.401	0.200	58.716	0.6170	6"

As Zimmmermann comments in his book his solution is not very good for this example (the results are worse than the crisp solution). With our approach the results tested (#3 and #4) are both significantly better than the crisp and Zimmermann solutions (#1 and #2). This clearly shows the flexibility of our approach compared with the symmetric method of Zimmermann. In addition the computational time to solve this problem is quite small, which means that the SA algorithm quickly obtains a good solution.

3.8 Unfeasible crisp example (without crisp solution) [12]

This following example is an unfeasible crisp problem,

$$\begin{aligned}
\max\ & 40x_1 + 30x_2 \\
\text{s.t.}\ & 0.4x_1 + 0.5x_2 \leq 20 \\
& 0.2x_2 \leq 5 \\
& 0.6x_1 + 0.3x_2 \leq 21 \\
& x_1 \geq 30 \\
& x_2 \geq 15
\end{aligned}$$

However, when we fuzzify the resource limits or even the resource limits and the constraints coefficients we can find a feasible solution. We tested the two types of flexibilization (one test for resources and another for resources and coefficients) using the 10% tolerance from the preferred value (the given ones).

The solutions obtained for only fuzzified resources (B_i) are,

solutions	Z	x_1	x_2	B_1	B_2	B_3	B_4	B_5	μ	time
Crisp solution	No	No	No	—	—	—	—	—	—	—
Fuzzy Author	1570	28	15	—	—	—	—	—	—	—
Fuzzy ours α=0.1	1891.8	33.00	19.06	22.73	3.81	25.52	33.00	19.06	0.25	48"
Fuzzy ours α=0.3	1882.7	30.98	21.45	23.12	4.29	25.02	30.98	21.45	0.33	45"
Fuzzy ours α=0.6	1766.9	28.37	21.07	21.88	4.21	23.34	28.37	21.07	0.61	1'4"
Fuzzy ours α=0.9	1598.6	28.97	14.66	18.92	2.93	21.78	28.97	14.66	0.90	1'5"

The solutions for fuzzified resources and coefficients are,

solutions	Z	x_1	x_2	μ	time
Crisp solution	No	No	No	—	—
Fuzzy Author	1570	28	15	—	—
Fuzzy ours α=0.1	2180.05	32.61	24.26	0.14	18"
Fuzzy ours α=0.3	1999.69	33.14	19.06	0.40	18"
Fuzzy ours α=0.6	1841.28	30.22	19.79	0.62	19"
Fuzzy ours α=0.9	1621.35	29.12	14.62	0.90	20"

and the solutions for the fuzzy coefficients and fuzzy resources are now:

	A_{11}	A_{12}	C_{11}	C_{12}	C_{21}	C_{31}	C_{32}	B_1	B_2	B_3	B_4	B_5
α=0.1	43.59	32.22	0.37	0.46	0.20	0.61	0.30	22.80	6.14	24.19	24.58	30.03
α=0.3	42.13	31.65	0.39	047	0.21	0.57	0.31	21.77	3.95	24.98	33.14	19.06
α=0.6	40.84	30.70	0.39	0.51	0.21	0.58	0.29	21.74	4.06	22.30	30.22	19.79
α=0.9	40.47	30.29	0.40	0.50	0.20	0.59	0.30	19.13	2.91	21.68	29.12	14.62

The performance of the SA algorithm in this example is quite enlightening. The time to solve a more flexible problem is much smaller than when we are less flexible (just fuzzy resources). The reason is that with more flexibility we obtain many more possible solutions to choose from.

Another interesting aspect is that all solutions found in this work are better in terms of the objective function value, but in terms of violating constraint 4 (the one that is causing problems) for a small threshold we have more violation ($x_1 = 24.58$ versus $x_1 = 28$), i.e. we violate more for lower thresholds and this should not be the case. We believe the reason for this is due to the nature of the SA algorithm (sometimes it gets stuck in local optimum [2] [3]). For the other thresholds we obtain better results and less violations.

This example clearly shows the potential of using a fuzzy approach to obtain results for otherwise unfeasible problems.

4 Set of non-linear examples tested and discussion of the results

In this section we will present some non-linear examples from the literature. However, most of the examples selected are crisp and not from fuzzy authors because there are not many fuzzy methods that can handle non-linear problems. Some exceptions can be seen in the following books [14] [7, 15]. However, since most of the examples were too big, in this section the two fuzzy examples discussed are from Sakawas book [7].

In summary, the selected example set to be tested with our fuzzy approach is: (4.1) a non-linear peak-load pricing problem [13]; (4.2) a non-linear sales force allocation problem [13]; (4.3) and (4.4.) two non-linear fuzzy examples from Sakawa [7], one with only fuzzy goals and another with fuzzy goals and fuzzy coefficients.

For this set of non-linear problems we considered a simplification of using deviations of 10% for all resources coefficients or goals fuzzifications. I.e. the membership functions were constructed considering a fuzzification of 10% from the preferred value of the resource or coefficient or goal value to be fuzzified. In addition, for all our solutions we show the values obtained for the fuzzy parameters.

4.1 Crisp peak-load pricing example [13]

This is a crisp peak-load pricing problem that we fuzzified to assess what are the gains obtained by being flexible and how the fuzzy approach can handle non-linear problems. We tested the fuzzification of all the variables coefficients with the above mentioned 10% tolerances. The example formulation is,

$$\max 60P - 0.5P^2 + 0.2FP + 40F - F^2 - 10C$$
$$\text{s.t. } 60 - 0.5P + 0.1F \leq C$$
$$40 + 0.1P - F \leq C$$
$$F, P, C \geq 0$$

The solutions obtained for the three thresholds considered are:

solutions	Z	P	F	C	μ	time
Crisp solution	2202.3	70.31	26.53	27.5	—	—
Fuzzy ours α=0.1	2903.30	75.08	27.83	29.40	0.15	32"
Fuzzy ours α=0.3	2715.25	71.64	24.48	28.59	0.30	44"
Fuzzy ours α=0.6	2513.14	74.59	28.75	25.72	0.61	1'03"
Fuzzy ours α=0.9	2274.75	66.56	25.86	29.26	0.95	1'51"

And the fuzzy parameters solutions are:

	A_{11}	A_{12}	A_{13}	A_{14}	A_{15}	A_{16}	C_{11}	C_{12}	C_{13}	C_{21}	C_{22}	C_{23}
α=0.1	64.86	0.458	0.197	43.16	0.929	9.49	60.78	0.541	0.095	40.04	0.099	1.013
α=0.3	64.20	0.473	0.211	42.40	0.941	10.53	59.40	0.466	0.104	37.32	0.097	0.989
α=0.6	62.28	0.485	0.196	41.48	0.961	9.81	61.80	0.492	0.096	38.64	0.101	0.962
α=0.9	60.60	0.495	0.201	40.28	0.990	9.93	60.24	0.497	0.101	40.28	0.099	1.006

As can be observed, by being flexible the results improve considerably. Even considering a threshold of 90% (meaning that we only allow a violation of constraints of 10% or that we want a satisfaction for our constraints of 90%) we do obtain better results than the crisp solution, $Z = 2274.75$ versus crisp $Z = 2202.3$. All the other results for the objective function are much better than the crisp solution. It seems that the expense of considering small deviation on the coefficients pays off in term of obtaining higher profits.

If we consider a further fuzzification of the resources and/or goals we could have achieved even better results. However, we must point that these better results would be obtained at the expense of using more or less resources (depending on the constraint sign).

In terms of time to solve the problem this example is not very good for higher thresholds $\alpha = 0.6$ and 0.9 because the SA took more than one minute to solve a problem with 3 variables and 2 constraints. However, if we think it is a non-linear problem (hence more difficult to sole) the results are acceptable.

4.2 Crisp sales force allocation example [13]

This is a small sales force allocation crisp problem with only non-negativity constraints (it is a simplified version of a problem by Lodish et al in: Interfaces 18, 1 (1996): 5-20, presented in [13]). We should also point that the authors also show an integer version of the same problem, but here is not considered.

$$\max 200x_1^{0.5} + 150x_2^{0.75} + 180x_3^{0.6} + 300x_4^{0.3} - 50x_1 - 50x_2 - 50x_3 - 50x_4$$
$$\text{s.t. } x_i \geq 0$$

The solutions obtained for this problem with the different thresholds are:

solutions	Z	x_1	x_2	x_3	x_4	μ	time
Crisp solution	1125.876	4	25.6	6.85	2.3	—	—
Fuzzy ours α=0.1	1551.41	5.31	34.39	11.43	1.68	0.38	8"
Fuzzy ours α=0.3	1474.81	8.40	33.17	6.66	3.91	0.43	9"
Fuzzy ours α=0.6	1458.75	4.29	29.17	8.77	8.77	0.77	10"
Fuzzy ours α=0.9	1204.00	5.54	32.11	7.54	2.58	0.90	12"

And the results for the objective function fuzzified coefficients are:

	A_{11}	A_{12}	A_{13}	A_{14}	A_{15}	A_{16}	A_{17}	A_{18}
α=0.1	207.42	162.30	191.73	325.83	47.32	45.63	46.46	52.17
α=0.3	206.87	162.27	194.81	307.36	48.31	45.64	46.38	53.59
α=0.6	209.84	160.91	191.40	316.34	50.24	46.20	46.09	51.77
α=0.9	201.65	152.97	183.50	305.75	49.01	49.19	50.28	49.31

The results obtained for this example show a similar behaviour than in the previous example. All the solutions are better than the crisp solution; hence, we may say that flexibility is compensatory in terms of results. Obviously, as in the previous example, the higher the threshold (more satisfaction of constraints) the less flexible the problems are and lower results are obtained.

The curiosity of this example is that we are only fuzzifying the coefficients of the objective function and hence we can observe exactly what are the gains obtained for deviations on the coefficients (in each test with different thresholds).

Interesting enough because this example does not have constraints, besides the non-negativity ones, the time to solve the problem is quite small (maximum 12 seconds). It does show that the calculations of the constraints add a considerable weight to the SA algorithm.

4.3 Example 7.2. by Sakawa [7]

In this example of a non-linear problem the author only considered fuzzy goals,

$$\begin{aligned}
\max F_1 &= 2x_1^2 + 4(x_2 - 20)^2 + 3(x_3 - 15)^2 \\
\min F_2 &= (x_1 - 10)^2 + 2(x_2 - 25)^2 + 3(x_3 + 5)^2 \\
\text{equal } F_3 &= 3(x_1 + 15)^2 + 2(x_2 + 10)^2 + (x_3 + 20)^2 \\
\text{s.t. } & (x_1 + 5)^2 + (x_2 + 8)^2 + (x_3 - 10)^2 \leq 200 \\
& x_i \leq 10 \\
& x_i \geq 10
\end{aligned}$$

The fuzzy goals limits, proposed by the author, were: $F_1 =]950, 2200]$; $F_2 = [1900, 1750[$; $F_3 = [1300, 1900, 2500]$.

Our solutions and the fuzzy solution of the author are:

solutions	F_1	F_2	F_3	x_1	x_2	x_3	μ	time
Crisp solution	–	–	–	–	–	–	–	–
Fuzzy Author	2063.41	1646.96	1853.06	4.3348	0.0225	3.0358	–	–
Fuzzy ours α=0.1	2171.81	1418.50	1752.27	3.9191	0.0744	1.8058	0.50	42"
Fuzzy ours α=0.3	2145.54	1439.67	1809.61	4.2278	0.0478	2.3287	0.61	40"
Fuzzy ours α=0.6	2156.05	1439.69	1800.25	4.1909	0.0148	2.2438	0.68	43"
Fuzzy ours α=0.9	2094.10	1471.61	1853.98	4.3236	0.0567	3.0570	1	52"

All the results for objective functions F_1 and F_2 that we obtained are considerably better than the author ones. Of course the less flexibility we allow the worse results we obtain (F_1 with $\alpha = 0.9$ is 2094.10 is relatively better than the author $F_1 = 2063.41$). The more intriguing case is the equality objective, F_3, where only for the less flexible, F_1 with $\alpha = 0.9$, our results are better than the author one. We believe that the reason for this is that we only considered 10% tolerance on the memberships; hence, the values close to the preferred one have bigger membership values that the ones obtained by the author.

4.4 Example 7.6 by Sakawa [7]

In this example Sakawa considered the fuzzification of the parameters that are depicted in the example formulation, as well as the goals:

$$\begin{aligned}
\min F_1 &= (x_1 + 5)^2 + A_{11}x_2^2 + 2(x_3 - A_{12})^2 \\
\min F_2 &= A_{21}(x_1 - 45)^2 + (x_2 + 15)^2 + 3(x_3 + A_{22})^2 \\
\text{equal } F_3 &= A_{31}(x_1 + 20)^2 + A_{32}(x_2 - 45)^2 + (x_3 + 15)^2 \\
\text{s.t. } & B_1 x_1 + B_2 x_2 + B_3 x_3 \leq 100 \\
& x_i \leq 10 \\
& x_i \geq 10
\end{aligned}$$

The fuzzy parameters limits proposed by the author were: A_{11}=[3.8, 4, 4.3], $A_{12} = [48.5, 50, 52]$, $A_{21} = [1.85, 2, 2.2]$, $A_{22} = [18.2, 20, 22.5]$, $A_{31} =$

$[2.9, 3, 3.15]$, $A_{32} = [4.7, 5, 5.35]$, $B_{11} = [0.9, 1, 1.1]$, $B_{12} = [0.8, 1, 1.2]$, $B_{13} = [0.85, 1, 1.15]$. The author also fuzzified the goals and the limits proposed are: $f1 = [5400, 3300[$; $f2 = [6900, 3900[$, $f3 = [7800, 10000, 13300]$. We must remind the readers that we simplified these deviations to 10% for all fuzzy parameters.

The solutions given by the author, as well as our solutions were:

solutions	F_1	F_2	F_3	x_1	x_2	x_3	μ	time
Crisp solution	–	–	–	–	–	–	—	—
Fuzzy Author	4816.80	4526.11	10455.02	8.177	5.878	2.258	—	—
Fuzzy ours $\alpha=0.1$	3784.17	4407.86	11017.46	7.990	3.980	3.433	0.22	34"
Fuzzy ours $\alpha=0.3$	4242.07	4494.45	10780.07	7.841	5.064	2.390	0.66	45"
Fuzzy ours $\alpha=0.6$	4625.42	4512.46	10371.89	7.749	5.736	1.155	0.75	36"
Fuzzy ours $\alpha=0.9$	4787.88	4508.31	10350.00	7.868	5.526	1.785	0.90	53"

and our solutions, found for the fuzzy parameters are:

	A_{11}	A_{12}	A_{21}	A_{31}	A_{32}	B_{11}	B_{12}	B_{13}
$\alpha=0.1$	3.919	45.584	1.923	18.272	3.221	4.846	1.064	1.069	0.956
$\alpha=0.3$	4.072	46.959	1.934	19.377	3.175	5.027	1.058	1.021	1.077
$\alpha=0.6$	3.865	47.715	2.036	19.320	3.028	5.046	0.973	1.057	0.950
$\alpha=0.9$	4.004	49.219	1.967	19.622	2.974	4.979	1.001	1.014	1.016

As can be observed we obtained considerably better results for F_1 and F_2, for the three thresholds. Even for $\alpha = 0.9$ the results are better, particularly for $F_1 = 4787.88$ (our solution) versus $F_1 = 4816.8$ (author solution). In this example we also obtain better results for the equality objective, F_3, except for lower thresholds, $\alpha = 0.1$ and 0.3. In terms of the results obtained for the coefficients ours are quite close to the central value, given by the author, which represents the preferred value for that coefficient. This is due to a more restrict fuzzification of the coefficients (only 10% tolerance), than the author considered. However, we still obtain better results for the objective functions and close enough values for the fuzzy parameters. This was an interesting example to test because it included fuzzification of coefficients and goals.

In terms of time the SA took to solve the problem, even though it cannot be compared with other results (the solving time was not a consideration in the author proposal) it seems quite reasonable when compared with the previous example (4.3). For the more restrictive case, $\alpha = 0.9$, it took 53 seconds versus 52 seconds for the same threshold of the previous example. Further, comparing the times to solve the two examples (4.3 and 4.4) we can say that the SA solving time is more dependent on the dimension of the problem than in being more or less fuzzified.

5 Conclusions

This paper compared the results obtained by solving several examples with a fuzzy approach and with different methods proposed in the literature. Most of the results obtained with our approach are better than the ones shown by other methods. Even for some methods that showed better results, for similar satisfaction values, we could provide better results for lower satisfaction values since our method allows simulations with different thresholds.

The selected set of examples included seven linear examples, four non-linear and one unfeasible problem. We believe this set of examples provided a significant test-bed for discussion of the method used.

We also showed that the approach allows a way to study the trade-off between better objective function and worse satisfaction values for the fuzzy parameters (i.e. more violation is required) and vice versa. This characteristic, in combination with its generality, makes this method a very flexible method. We also pointed the main disadvantage of the fuzzification model used, because the flexibility and generality of the approach is gained by having a larger dimension formulation as well as a non-linear one.

Finally, we showed that using the simulated annealing algorithm for solving fuzzy optimization problems is a good solution technique for solving this type of problems. However, the implementation needs further improvements to allow using the full potential of multiple objective fuzzy optimization problems as well as other types of membership functions.

References

1. Pires, F.M., J. Pires, Moura, and R. Ribeiro (1996) Solving Fuzzy Optimisation problems: Flexible Approaches using Simulated Annealing. In Proceedings of the World Automation Congress (WAC96). Montpelier, France: TSI Press series.
2. Ribeiro, R.A. and F. Moura-Pires (1999), Fuzzy Linear Programming Via Simulated Annealing. Kybernetica, 35(1):57–67.
3. Varela, L.R. and R.A. Ribeiro (2001), Utilizao de Simulated Annealing em Optimizao Difusa (in Portuguese). Investigao Operacional, 21(2):205–231.
4. Carlsson, C. and P. Korhonen (1986) A Parametric Approach to Fuzzy Linear Programming. Fuzzy Sets and Systems, **20**:17–30.
5. Chanas, S. (1983) The Use of Parametric Programming in Fuzzy Linear Programming. Fuzzy Sets and Systems, **11**:243–251.
6. Lai, Y.-J. and C.-L. Hwang (1992) Fuzzy Mathematical Programming: Methods and Applications. Lecture Notes in Economics and Mathematical Systems. Vol. 39. Springer-Verlag.
7. Sakawa, M. (1993) Fuzzy Sets and Interactive Multiobjective Optimization. In: Applied Information Technology. M.G. Singh (Ed). Plenum Press.
8. Tanaka, H. and K. Asai (1984)Fuzzy Linear Programming Problems with Fuzzy Numbers. Fuzzy Sets and Systems, **13**:1–10.
9. Delgado, M., J.L. Verdegay, and M.A. Vila (1989) A General Model for Fuzzy Linear Programming. Fuzzy Sets and Systems, **29**:21–29.

10. Delgado, M., J.L. Verdegay, and M.A. Vila (1990) Relating Different Approaches to Solve Linear Programming Problems With Imprecise Costs. Fuzzy Sets and Systems, **37**:33–42.
11. Zimmermann, H.-J. (1996) Fuzzy Set Theory and its Applications. 3rd edition. Kluwer Academic publisher.
12. Pires, F.M. and R.A. Ribeiro(1998) A New Risk Function for Fuzzy Linear Programming. In: Proceedings of the World Automation Congress (WAC98),Hawaii. TSI Press.
13. Winston, W.L. and S.C. Albright (1997) Pratical Management Science: Spreadsheet Modeling Applications. Duxbury Press- International Tomhson Publishing Inc.
14. Lai, Y.-J. and C.-L. Hwang (1994) Fuzzy Multiple Objective Decision Making. Lectures Notes in Economics and Mathematical Systems. Springer-Verlag.
15. M. Delgado, J. Kacprzyk, J.L. Verdegay and M.A. Vila (Eds). (1994) Fuzzy Optimization. Studies in Fuzziness. Physica-Verlag.
16. Bellman, R.E. and L.A. Zadeh (1970) Decision-Making in a Fuzzy Environment. Management Science, Vol.17(4) :141–164.
17. Zimmermann, H.-J. (1978) Fuzzy Programming and Linear Programming with Several Objective Functions. Fuzzy Sets and Systems, **1**:45–55.
18. Kickert, W.J.M. (1978) Fuzzy Theories on Decision Making. Frontiers in Systems Research, Vol 3. Martinus Nijhoff Social Sciences Division.
19. Fedrizzi, M., J. Kacprzyk, and J.L. Verdegay, (1991). A Survey of Fuzzy Optimization and Mathematical Programming. In: Interactive Fuzzy Optimization, M. Fedrizzi, J. Kacprzyk, and M. Roubens (Eds). Springer-Verlag.
20. Delgado, M., J.L. Verdegay, and M.A. Vila (1994) Fuzzy Linear Programming: From Classical Methods to New Applications. In: Fuzzy Optimization, M. Delgado, J. Kacprzyk, J.L. Verdegay and M.A. Vila (Eds). Physica-Verlag. :111–134.
21. Ribeiro, R.A. and F.M. Pires (1999), Fuzzy Site Location Problems and Simulated Annealing. Series Studies in Locational Analysis, 13(June):61–76.
22. Reeves, C.R., Ed. (1995) Modern Heuristic Techniques for Combinatorial Problems. Advanced Topics in Computer Science. McGraw Hill.
23. Kirkpatrick, S., C.D. Gelatt, and M.P. Vecchi (1983) Optimization by Simulated Annealing. Science, Vol. 220(4598):671–680.
24. Tanaka, H., H. Ichihashi, and K. Asai (1985) Fuzzy Decision In Linear Programming Problems With Trapezoid Fuzzy Parameters. In: Management Decision Support Systems Using Fuzzy Sets and Possibility Theory, J. Kacprzyk and R. Yager (Eds), Verlag TUV Rheinland. :146–155.
25. Campos, L. and J.L. Verdegay (1989) Linear Programming Problems and Ranking of Fuzzy Numbers. Fuzzy Sets and Systems, **32**:1–11.

Multi-stage Supply Chain Network by Hybrid Genetic Algorithms

Mitsuo Gen[1] and Admi Syarif[1,2]

[1] Dept.of Industrial and Information System Engg.
Ashikaga Institute of Technology, Japan
[2] Dept. of Mathematics, Faculty of Mathematics and Sciences
Lampung University, Indonesia

Abstract. This research is concern with logistic system design considering production/distribution planning in the view of multi-stage structure. The design tasks of this problem involve the choice of the facilities (plants and distribution centers) to be opened or not and the distribution network design to satisfy the demand with minimum cost. This problem is known as one of the NP-hard problems. To solve this problem, a hybrid spanning tree-based genetic algorithm, hst-GA, is proposed. In order to improve the performance of the proposed method, we develop a local search technique called displacing Prüfer number and adopt the concept fuzzy logic controller (FLC) to dynamically control the GA parameters. The effectiveness of the proposed method is checked by comparing its computational experiment result with those of other traditional methods.

1 Introduction

To be competitive in the global manufacturing environment, the strategic logistic system design becomes one of the most important factors. The logistic system design problem is defined as [1]:

- the process of anticipating customer needs and wants; acquiring the capitals, material, people, technologies and information necessary to meet those need and want;
- optimizing the goods- or service-producing network to fulfill customer request;
- and utilizing the network customer request in a timely way .

Since one important concept of logistics is in the sense of management of goods-flows, it is also often defined as the art of bringing the right amount of the right product to the right place at the right time.

In the recent years, logistic network problems have been widely researched production/distribution models have been developed for design of supply chain and its logistic system. The common objective is to find strategic options for improving the efficiency in the production/distribution performance that meets the demand at minimum cost or fills demand for maximum profit. The simplest model related to multi-stage logistic system is known as multi-echelon location/allocation problems that have taken great interest of many researchers [2], [3]. An order plan model for multi-stage production system was introduced by Azevedo and Sousa [4]. Such a system processes products and materials through different production units that are part of logistics chain organized in several phases. In this case they try to determine, for each incoming order, an 'optimal' path (concerning cost) through the network. The discussion of the production/distribution planning in relation to the supply chain management concept was given by Sim *et. al* [5].

The Genetic Algorithm (GA) is one of a family of heuristic optimization techniques, which include simulated annealing, tabu search, and evolutionary strategies. It has been demonstrated to give the optimal solution for many diverse and difficult problems. For the specific engineering problems, however, GA cannot be guaranteed the optimality and sometimes can suffer from the premature convergence situation of its solution because of their fundamental requirements that are not using a priori knowledge and not exploiting local search information [6]. In order to overcome this problem, several hybridized techniques are introduced. The most popular technique is to combine the local search technique into GA process. Here, genetic search is used to perform global exploration among the population and local search is used to perform local exploitation around chromosomes.

Another important issue in GA is how to determine the best GA parameters that can bring us to the optimal solution. To this problem, fuzzy logic controller (FLC) is acceptable because it provides an algorithm that can convert linguistic control strategy based on expert knowledge into an automatic control strategy. The first authors introducing FLC to dinamically tune the GA parameters are Lee *et. al.*'s. They used the phenotypic diversity measure for the best, average and worst of fitness to automatically tune crossover probability (p_C) and mutation probability (p_M) [7]. Zeng *et. al.*'s FLC is designed to adjust p_C, p_M and the position for crossover operation by approximating the relationship between GA parameters and several measures in populations [8]. Wang *et. al.* suggested that the heuristic updating principle of the p_C and p_M based on the changes in the average fitness of the population [9]. It has been shown that adapting such GA parameters automatically can improve the quality of solution.

In this chapter, we consider a multi-stage logistic system problem modeled by 0-1 mixed linear programming model. In this model, we give the maximum number of facilities (plants and distribution center) to be opened as the constraints. These constraints increase the difficulty of the problem, yet the relevance for the real-world applicability also increases significantly. The design tasks of this problem involve the choice of facilities to be opened and the distribution network design to satisfy the customer demand with least cost. This kind of problem can be viewed as the combination of multiple-choice Knapsack problem with the transportation problem. So this problem is known to be NP-hard [10]. This fact describes that non-polynomial time must be spent to solve this problem with general algorithm. At this point, the evolutionary computation methods give a powerful support to find out the best heuristic solution. We assume that the logistic system process for this problem is organized according to a three sequence of stages as introduced by Yu [11]. The illustration of this problem is given in Figure 1.

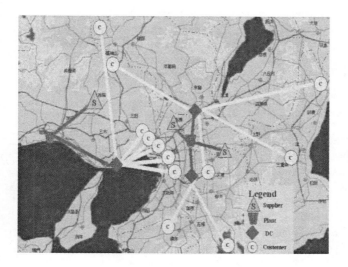

Fig. 1. The illustration of the problem

To solve the problem, here, we propose a novel technique called hybrid spanning tree-based genetic algorithm (hst-GA). We adopt the Prüfer number [12], [13] that is known to be an efficient way to represent various network problems [10], [14]. We briefly discuss the main characteristics of hst-GA. These characteristics include the representation of the solution, repairing procedure for infeasible chromosome, mechanism to create the initial population and genetic operations. Further, to improve the performance of the algorithm, we also develop a local search technique call displacing Prüfer number and adopt the automatic fine tuning for the crossover ratio and mutation ratio using FLC [9]. The effectiveness and efficiency of the proposed method are evaluated by comparing several numerical experiment results of the proposed method with those of traditional methods.

The rest of this paper is organized as follows: The mathematical formulation of this problem is given in Section 2. In Section 3, we describe the features of our method including the chromosome representation, the GA process, local search technique and FLC concept for auto-tuning the GA parameters. In Section 4, the overall procedure for the proposed method is given. Numerical experiments and comparison with the results of traditional algorithm to demonstrate the efficiency of the proposed method follow in Section 5. Finally, some concluding remarks are given in Section 6.

2 Mathematical Model

In this section, we shall present a comprehensive mathematical formulation that considers real-world factors and constraints. We assume that the number of customers and the number of source centers as well as their demand and capacity are known in advance. The number of potential plants and distribution centers as well as their maximum capacity is also known in advance. Our formulation uses the following notations:

Parameters
- a_i capacity of source center i ($i = 1, 2, \cdots, I$)
- b_j capacity of plant j ($j = 1, 2, \cdots, J$)
- c_k capacity of distribution center k ($k = 1, 2, \cdots, K$)
- d_l demand of customer l ($l = 1, 2, \cdots, L$)
- s_{ij} cost of shipping one unit of material product from source i to plant j.
- t_{jk} cost of shipping one unit product from plant j to distribution center k
- u_{kl} cost of shipping one unit product from distribution k to customer l
- f_j fixed cost for operating plant j
- g_k fixed cost for operating distribution center k
- P an upper limit on total DCs that can be opened
- W an upper limit on total plants that can be opened

Decision variables
- x_{ij} number of raw materials to be shipped from source center i to plant j
- y_{jk} number of products to be shipped from plant j to distribution center k
- z_{kl} number of products to be transported from DC k to customer l

$$w_j = \begin{cases} 1 & \text{if plant j is opened} \\ 0 & \text{otherwise} \end{cases} \qquad z_k = \begin{cases} 1 & \text{if DC k is opened} \\ 0 & \text{otherwise} \end{cases}$$

The problem is to choose the subset of plants and distribution centers to be opened and to design the distribution network strategy that can satisfy all capacity and demand requirement imposed by customers with minimum cost. We formulate the problem by using the following mixed integer linear programming model (MIP):

$$\min Z = \sum_{i=1}^{I}\sum_{j=1}^{J} s_{ij}x_{ij} + \sum_{j=1}^{J}\sum_{k=1}^{K} t_{jk}y_{jk} + \sum_{k=1}^{K}\sum_{l=1}^{L} u_{kl}z_{kl} \\ + \sum_{j=1}^{J} f_j w_j + \sum_{k=1}^{K} g_k z_k \quad (1)$$

s.t.
$$\sum_{j=1}^{n} x_{ij} \leq a_i, \; i = 1, 2, \cdots, I \tag{2}$$

$$\sum_{k=1}^{K} y_{jk} \leq b_j w_j, \; j = 1, 2, \cdots, J \tag{3}$$

$$\sum_{j=1}^{J} w_j \leq P, \tag{4}$$

$$\sum_{l=1}^{L} z_{kl} \leq c_k z_k, \; k = 1, 2, \cdots, K \tag{5}$$

$$\sum_{k=1}^{K} z_k \leq W, \tag{6}$$

$$\sum_{k=1}^{K} z_{kl} \geq d_l, \; l = 1, 2, \cdots, L \tag{7}$$

$$w_j, z_k = 0 \text{ or } 1, \forall \; j, \; k \tag{8}$$

$$x_{ij}, \; y_{jk} \text{ and } z_{kl} \geq 0, \; \forall \; i, \; j, \; k, \; l. \tag{9}$$

The constraint (2) ensures source capacity is enough. The constraints (3) and (5) are the capacity constraint for the plants and distribution centers, respectively. The constraints (4) and (6) ensure that the opened plants and opened distribution centers do not exceed the upper limit of opened plants and distribution centers, respectively. Constraint (7) ensures that all demands are met. In this model, we assume that the total capacity of source center and candidate facilities are sufficient to satisfy all customer demand. And we also can assume that the balance equation is satisfied, since we can convert an unbalance model into a balance one by introducing a dummy source center, facility or customer.

3 Design of the Algorithm

Nowadays, as the use of computer is rapidly increasing, many evolutionary computation methods for solving optimization problems have been introduced. Probably, among them, Genetic Algorithm (GA) is the most well known class of evolutionary algorithms. It has taken a lot of attention of researchers in the several years. Other variants of evolutionary algorithm such as *Genetic Programming, Evolutionary Strategies* or *Evolutionary Programming* are less popular, though very powerful too. The first author used GA for solving linear and nonlinear transportation/ distribution problems was Michalewicz [15] . In this method, he represent each chromosome of the problem by using two dimensional matrix.

3.1 Spanning Tree-based Genetic Algorithm

The use of Prüfer number to represent the transportation graph was first introduced by Gen and Cheng [10]. The Prüfer number here is randomly generated $m + n - 2$ digit number in the range $[1, m + n]$ to uniquely represent a distribution network with m sources and n destinations. In our previous work [16], however, we found that his technique only works when the number of source center and the demand center are almost the same. For relatively large size problem, where the difference between the number of source center and the number of demand center is large, their method cannot work. The reason is because it is very difficult to generate the chromosome that satisfy their feasibility criteria. To handle such kind of problems, it is necessary to develop a repairing procedure for the infeasible chromosome.

3.1.1 Representation of the chromosome

In this paper, each chromosome is represented by using five parts of numbers. The first part is a J digits 0-1 variable to determine the opened/closed plants. The second part is a K digits 0-1 variable to shows whether the distribution center is opened or not. The last three parts are three Prüfer numbers representing the sub-tree $I - J$, sub-tree $J - K$ and sub-tree $K - L$. As the first step in generating the chromosome, we should generate the two 0-1 variables and check whether or not the number of opened plants or distribution centers exceeds the given upper limit. If the number of opened plants or distribution centers exceed the maximum, then closed plants or distribution centers with minimum capacities. Also, we should also check the total capacity of the opened plants and distribution centers to satisfy the customer demands. If the total capacity is less then the total demand then open one of closed facilities with maximum capacities to satisfy the demand. The three Prüfer numbers are generated after this step. For the sub-tree $I - J$, denote the source center $1, 2, \cdots, I$ as the component of set $S = \{1, 2, \cdots, I\}$ and define plant $1, 2, \cdots, J$ as the component of the set $D = \{I+1, I+2, \cdots, I+J\}$. Obviously this distribution graph has $I + J$ nodes which means that we need $I+J-2$ digits Prüfer number in the range $[1, I+J]$ to uniquely represent this sub-tree $I - J$. When generating the Prüfer number, it is also feasible for us to generate an infeasible Prüfer number that cannot be adapted to generate the transportation sub-tree. To this reason, we need to check its feasibility by using the following feasibility condition.

Feasibility criteria for Prüfer number

Let R_i denotes the number of appearances of node i in the Prüfer number $P(T)$ and L_i be the number of connection of the node i for all $i \in S \cup D$. Here, $L_i = R_i + 1$. The Prüfer number P(T) is said to be feasible if

$$\sum_{i=1}^{I} L_i = \sum_{i=I+1}^{I+J} L_i. \tag{10}$$

The Prüfer number that does not satisfy the above criteria is called infeasible Prüfer number. There are two ways to handle with this infeasible Prüfer number: First is to reject that Prüfer number and generate a new one. However, this technique will take a large computational time especially when the difference of the source number and destination number is very large. Another way is to develop the repairing procedure for the infeasible Prüfer number. Here, we design the feasibility check and repairing procedure for the Prüfer number to be decoded into spanning tree as follows:

step 1: Generate the Prüfer number randomly.
step 2: Determine R_i for $i = S \cup D$ from $P(T)$
step 3: $L_i = R_i + 1$
step 4: If $\sum_{i \in S} L_i > \sum_{i \in D} L_i$, then select one digit in $P()$ which contains node i $(i \in S)$ and replace it with the number j $(j \in D)$. Otherwise select one digit in P(T) which contains node i $(i \in D)$ and replace it with the number j $(j \in S)$ then go to Step 1.

Next, we can also generate two other Prüfer numbers representing $J - K$ and $K - L$ sub-trees by using the similar way. These two Prüfer numbers consist of $J + K - 2$ and $K + L - 2$ digits, respectively. After checking for their feasibility, the Prüfer numbers can be decoded into spanning trees in order to determine the distribution pattern in each stage. The procedure of encoding the network problem into Prüfer number and also decoding Prüfer number into the network graph is given in [10]. The process of generating the feasible chromosome is done *pop_size* times in order to generate the initial population. As an example, let us the problem that has 4 suppliers, 3 feasible plants, 3 feasible distribution centers and 4 customers. We decide 2 plants and 2 distribution centers as the upper limit of opened plants and opened distribution centers for this problem. Figure 2 gives an illustration of a feasible chromosome representation:

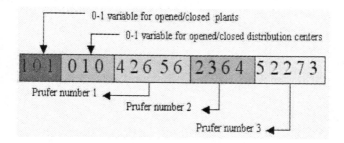

Fig. 2. The illustration of chromosome representation

The decoding procedure for the Prüfer numbers is used to find the distribution pattern of this chromosome and compute its fitness value. We illustrate the decoding of the chromosome using the first Prüfer number in the above representation in the following Figure 3. The decoding procedures of the second and third Prüfer numbers are similar.

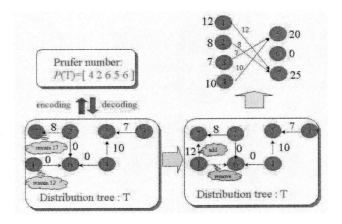

Fig. 3. The illustration of the decoding Prufer number

3.1.2 Genetic Operations
● **Crossover**

In the genetic algorithms, the aim of crossover operation is to provide a powerful exploration in the neighborhood of the solution. It has been shown in many literatures that the crossover operation technique plays very important role at improving the performance of the genetic algorithm in both of speed and accuracy. It is generally done by exchanging the information of two parents to reproduce offspring. In this paper, we employ a one-cut-point crossover, which randomly selects a one-cut-point and exchanges the right parts of two parents to generate offspring as shown in Figure 4.

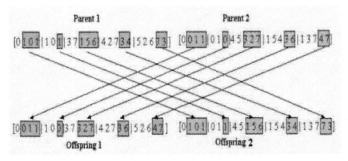

Fig. 4. The illustration of crossover process

Mutation

Modifying one or more of the gene values of an existing individual, mutation creates new individual to increase the variability of the population. Here, we use inversion that can guarantee to generate feasible chromosome when the parents are feasible.

Evaluation and Selection

As in nature, it is necessary to provide driving mechanism for better individual to survive. Evaluation is to associate each chromosome with a fitness value that shows how good it based on its achievement of the objective function. The higher fitness value of an individual, the higher its chances for survival for the next generation. Selection is to select individuals to be parents in the next generation according their fitness value. So the evaluation and selection play a very important role in the evolutionary process. For this problem, the evaluation procedure can be operated as follows:

procedure: Evaluation
step 1: Convert Prüfer numbers into spanning trees
step 2: Calculate the total cost and the total fixed cost of trees according to the objective function
step 3: Repeat the procedure on all individuals.

As to selection, we adopt the elitist selection strategy [11] in which duplicate chromosomes are prohibited.

3.2 Local Search Technique

Though it has been shown that GA is an effective way to find the solution for many difficult optimization problems, when searching the space of possible solutions, we often reach a local optimum and find it difficult to make further progress. To help move out of a local optimum and carefully search the near optimal region, many researchers reported that combining a GA and a local search technique into a hybrid approach often produces certain benefits [6]. In contrast to the local search techniques that are simple and computationally efficient, global search techniques explore the global search space without using local information about promising search direction. Consequently, they are less likely to be trapped in local optimal solution, but their computational cost is higher. The distinction between local search techniques and global search techniques is referred to as the exploitation - exploration trade-off [6].

It has been shown by [18] that changing one element of Prufer number can change its corresponding tree topology dramatically. To remedy this problem, we develop a simple local search operation called displacing Prufer number that can keep the similarity of the chromosome. The concept of displacing Prüfer number is stated as follows: select one digit in the Prüfer number randomly; And, exchange it with the first Prüfer number. The Prüfer number resulted by this operation is guaranteed to be feasible since the appearance number for each node remains the same.

In hybridizing GA using local search techniques, most common form is to incorporate a local search technique in GA loop [6]. It is also noted by Lo [18] that, according to the result obtained by the well-known Add and Drop searching heuristics, changing only one element in every iteration of the search process always leads to a globally optimal solution. Thus our technique can significantly improve the quality of newly found chromosome.

3.3 Fuzzy Logic Controller

Generally the behavior of GAs depends on many uncertain factors. One of the important factors is a balance between exploitation and exploration in the search space [6]. The balance is strongly affected by design strategy for GA parameters such as population size, maximum generation, crossover probability and mutation probability. Fuzzy Logic Controller (FLC) provides an algorithm than can convert linguistic control strategy based on expert knowledge into an automatic control strategy. The main idea of FLC is to dynamically change the GA parameters based on the information in the previous generations such as the average fitness of the population.

Here, we apply the Wang *et. al* [9] technique to automatically adjust the crossover ratio and mutation ratio. Let $\triangle \overline{f}(V; t-1)$ be the difference in the average fitness value at continuous two generation $(t-1,$ and $t)$, the input variables of the FLC are $\triangle \overline{f}(V; t-1), \triangle \overline{f}(V; t)$. Based on a number of experimental data and domain expert opinion, we use the fuzzy decision table as given in Table 1. For simplicity, the input values are respectively normalized into integer values in the range [-4.0,4.0]. After scaling $\triangle \overline{f}(V; t-1)$ and $\triangle \overline{f}(V; t)$, assign these values to the corresponding indexes i and j in the fuzzy decision table as given in Table 2. The membership functions of fuzzy all input and output linguistic variables are illustrated in Figure 5.

Table 1. Fuzzy decision table

		$\Delta f(t-1)$							
		NR	NL	NM	NS	ZE	PS	PM	PL	PR
$\Delta f(t)$	NR	NR	NL	NL	NM	NM	NS	NS	ZE	ZE
	NL	NL	NL	NM	NM	NS	NS	ZE	ZE	PS
	NM	NL	NM	NM	NS	NS	ZE	ZE	PS	PS
	NS	NM	NM	NS	NS	ZE	ZE	PS	PS	PM
	ZE	NM	NS	NS	ZE	PM	PS	PS	PM	PM
	PS	NS	NS	ZE	ZE	PS	PS	PM	PM	PL
	PM	NS	ZE	ZE	PS	PS	PM	PM	PL	PL
	PL	ZE	ZE	PS	PS	PM	PM	PL	PL	PR
	PR	ZE	PS	PS	PM	PM	PL	PL	PR	PR

where

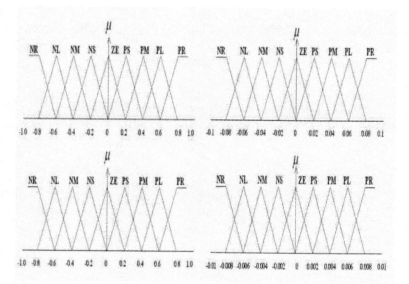

Fig. 5. The Membership function for input-output FLC variables

Table 2. Control action value

$z(i,j)$		-4	-3	-2	-1	0	1	2	3	4
	-4	-4	-3	-3	-2	-2	-1	-1	0	0
	-3	-3	-3	-2	-2	-1	-1	0	0	1
	-2	-3	-2	-2	-1	-1	0	0	1	1
	-1	-2	-2	-1	-1	0	0	1	1	2
j	0	-2	-1	-1	0	2	1	1	2	2
	1	-1	-1	0	0	1	1	2	2	3
	2	-1	0	0	1	1	2	2	3	3
	3	0	0	1	1	2	2	3	3	4
	4	0	1	1	2	2	3	3	4	4

NR-Negative larger,   NL -Negative large
NM - Negative medium, NS - Negative Small
ZE - Zero,   PS - Positive small
PM - Positive medium, PL - Positive large
PR - Positive larger.

Using the result of the above process, the changes of crossover probability and mutation probability are computed as follows:

$$\triangle c(t) = \alpha \cdot z(i,j), \qquad \triangle m(t) = \beta \cdot z(i,j)$$

where α and β is a given value to regulate an increasing and decreasing range for crossover probability and mutation probability (e.g. α=0.02, β=0.002).

The crossover probability and mutation probability are computed by the following equation:

$$p_C(t) = \triangle c(t) + p_C(t-1)$$
$$p_M(t) = \triangle m(t) + p_M(t-1)$$

where $p_C(t)$ and $p_M(t)$ are crossover probability and mutation probability at generation t, respectively.

4 Overall Procedure

In this section, we propose the overall procedures of the proposed algorithm, hst-GA. At first, we describe the procedure of displacing Prufer number method and FLC, as follows:

4.1 Displacing Prufer number

The local search using displacing Prufer number is implemented in each generation of GA. The detailed procedure for this local search technique is as follows:

procedure: Local search
   begin
        $local \longleftarrow$ FALSE;
        select a optimum choromosome (v_c) in current GA loop;
        repeat
           generate new chromosome (v_n) by exchanging the content of first digit
               with the other digit of each Prufer number;
           evaluate the new chromosome (v_n)
           if $f(v_c) > f(v_n)$ **then**
               $\underline{v}_c \longleftarrow v_n$
           else $local \longleftarrow$ TRUE
        end
        until $local$
   end

4.2 Fuzzy Logic Controller

The procedures of the proposed FLC are used to dynamically regulate the crossover probability and mutation probability. That is, these procedures are to consider the average fitness considered by continuous generation of GA. And the values obtained regulate the crossover probability and mutation probability for the next generation. The detailed procedures are as follows:

procedure: Auto-tuning to strategy parameters
step 1: The input variables of the FLC for GA parameters are the changes in average fitness at continuous three generation $(t-2, t-1, \text{ and } t)$ as follows:

$$\triangle \overline{f}(V; t-1), \triangle \overline{f}(V; t)$$

step 2: After scaling $\triangle \overline{f}(V; t-1)$ and $\triangle \overline{f}(V; t)$, assign these values to the indexes i and j corresponding to the control actions in the fuzzy decision table [9]
step 3: Calculate the changes of crossover probability and mutation probability as follows:

$$\triangle c(t) = 0.02 \cdot z(i, j), \quad \triangle m(t) = 0.002 \cdot z(i, j)$$

step 4: Update the crossover probability and mutation probability by the following equation:

$$p_C(t) = \triangle c(t) + p_C(t-1)$$
$$p_M(t) = \triangle m(t) + p_M(t-1)$$

step 5: Return to optimal search procedure.

Overall procedure of the proposed algorithm, hst-GA, is as follows:

procedure: Overall procedure of hst-GA
```
begin
    t ← 0;
    use the populations obtained by random search for
        initializing P(t);
    evaluate P(t);
    while (t ≤ max_gen) do
        combine P(t) using genetic operators to yield C(t);
        evaluate C(t);
        perform local search
        dynamically regulate p_C(t) and p_M(t) using FLC
        select P(t + 1) from P(t) and C(t) using elitist strategy;
        t ← t + 1;
    end
end
```

5 Numerical Examples

We implement our proposed method using Visual C language. To see the efficiency and the effectiveness of the method, we compare the computational results of the proposed method with those of matrix-based GA, and two other traditional spanning tree-based GA. All algorithms were run on the same Pentium 500 PC. As the initial value for the GA parameters, we set $p_C = 0.4$ and $p_M = 0.2$. Three different size of test problems are used in the computational experiment and each of them were run 10 times.

In Figure 6, we show the comparative convergence behaviour of the methods in the generations for test problem 1. The average results of the numerical experiments and the computational time of the proposed algorithm in comparison to the other traditional methods are summarized in Table 3 and Table 4 respectively.

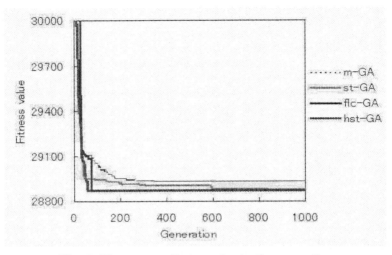

Fig. 6. The average fitness value in the generation

Table 3. The average fitness value in the numerical experiments

Problem	pop_size	max_gen	m-GA	st-GA[a]	flc-GA[b]	hst-GA
1	25	750	28942	28920	28895	28870
	30	1000	28924	28875	28870	28870
2	250	2500	35542572	35379480	34216845	33423673
	300	3000	35511341	34626132	33973273	33284735
3	250	4000	601962	601237	600678	596021
	300	6000	601094	600336	598693	595273

[a] Proposed method without FLC and local search
[b] Proposed method without local search.

6 Conclusion

In this paper we proposed a hybrid spanning tree-based genetic algorithm (hst-GA) approach to find the best production/distribution design in multi stages logistics system. We utilize the Prüfer number that is known to be an efficient way to represent various network problems. Even though the

Table 4. The average computational time

Problem	pop_size	max_gen	m-GA	st-GA[a]	flc-GA[b]	hst-GA
1	25	750	1.6	1.4	1.5	1.8
	30	1000	2.6	2.7	2.4	2.9
2	250	2500	103.6	93.3	97.6	106.2
	300	3000	216.4	167.2	186.7	198.1
3	250	4000	579.6	322.4	351.3	376.4
	300	6000	947.5	546.3	592.6	617.3

[a] Proposed method without FLC and local search
[b] Proposed method without local search

structure of the proposed methods is very simple, experimental results show that our algorithm not only can give better heuristic solution in almost all of the time but also has better performance in the sense of computational time and required memory for computation than those of matrix-based genetic algorithm (m-GA). So we believe this method will be an efficient and robust method to solve this kind of multi-stage logistics chain design problems.

Acknowledgments

This research was supported by the International Scientific Research Program, the Grant-in-Aid for Scientific Research (No.10044173) by the Ministry of Education, Science and Culture, the Japanese Government. The work of the second author was also supported by Development Undergraduate Education Project, The University of Lampung, Indonesia (Indonesian Government and World Bank Project).

References

1. Tilanus, B. (1997) Introduction to Information System in Logistics and Transportation, in "*Information System in Logistics and Transportation*", ed. Tilanus, Pergamon, Elsevier Science Ltd.
2. Ro, H. and Tcha, D. (1984). A branch and bound algorithm for two level uncapacitated facility location problem with some side constraint, *European Journal of Operational Research*, Vol. 18, pp. 349-358.
3. Pirkul, H., and Jayaraman, V., (1998) A Multy Commodity, Multy Plant, Capacitated Location Alocation Problem: Formulation and Efficient Heuristic Solution, *Computers and Operations Research*, Vol. 25, No. 10 pp. 869-878.
4. Azevedo, AL. and Sousa, J. P., (2000) Order Planning for Networked Make-to-order Enterprises - A Case Study, *Journal of Operational Research Society*, Vol. 51, pp. 1116-1127
5. Sim, E., Jang, Y., and Park, J., (2000) A Study on the Supply Chain Network Design Considering Multy-level, Multy-product, Capacitated Facility, *Proc. of Korean Supply chain Management Society*
6. Lee, C. Y., Yun, Y. S. and Gen, M. (2002) Reliability Optimization Design for Complex System by Hybrid GA with Fuzzy Logic Controller and Local Search, *IEICE Trans. Fundamental*, to appear.

7. Lee, M. and Takagi, H. (1993) Dynamic control of genetic algorithm using fuzzy logic techniques, *Proc. of the 5th Inter. Conf. on Genetic Algorithm*, pp. 76-83.
8. Zeng, X. and Rabenasolo, B. (1997) A Fuzzy Logic Based Design for Adaptive Genetic Algorithms, *Proc. of the 5th European Congress on intelligent Techniques and Soft Computing*, pp. 660-664.
9. Wang, P. T., Wang, G. S. and Hu, Z. G. (1997) Speeding up the Search Process of Genetic algorithm by Fuzzy Logic, *Proc. of the 5th European Congress on intelligent Techniques and Soft Computing*, pp. 665-671.
10. Gen, M. and Cheng, R. (1997) *Genetic Algorithms and Engineering Design*, John Wiley & Sons, New York.
11. Yu, H. (1997). ILOG in the supply chain, *ILOG Technical Report*.
12. Prufer, H. (1918) Neuer bewis eines saizes uber permutationen, *Arch. Math. Phys*, Vol, 27, pp. 742-744.
13. Dossey, J., Otto, A., Spence, L. and Eynden, C. (1993) *Discrete Mathematics*, Harper Collins.
14. Gen, M. and R. Cheng (2000) *Genetic Algorithms and Engineering Optimization*, John Wiley & Sons, New York.
15. Michalewicz, Z. (1994) *Genetic Algorithms + Data Structures = Evolution Programs*, 2nd ed., Springer-Verlag, New York.
16. Syarif, A. and Gen, M. (2002) Solving Exclusionary Side Constrained Transportation Problem by Using A Hybrid Spanning Tree-based Genetic Algorithm, *Journal of Intelligent Manufacturing*, to appear.
17. Palmer, C. and Kershenoaum, A. (1995) An approach to a problem in network design using genetic algorithms, *Networks*, Vol. 26, pp. 151-163.
18. Lo, C. C. and Chang. W. H. (2000), A multiobjective hybrid genetic algorithm for the capacitated multipoint network design problem, *IEEE Transaction on System, Man and Cybernetic*, Vol. 30, No. 3, pp.461-470.

Fuzzy Evolutionary Approach for Multiobjective Combinatorial Optimization: Application to Scheduling Problems

Imed Kacem, Slim Hammadi, and Pierre Borne

Laboratoire d'Automatique et Informatique de Lille, Ecole Centrale de Lille, Cité Scientifique, Villeneuve d'Ascq, 59651, France.

1 Introduction

Most combinatorial problems are complex and very hard to solve. That is why, lots of methods focus on the optimization according to a single criterion. The combining of several criteria presents additional complexity and new problems.

In this paper we propose a fuzzy evolutionary methodology to solve multiobjective combinatorial optimization problems and we apply it to Flexible Job-shop Scheduling Problem (FJSP). We show how Fuzzy Logic (FL) could be useful to make heuristics, such as Evolutionary Algorithms (EAs), more robust and more efficient to solve combinatorial problems such as FJSP.

This chapter is organized as follows: in the second section, we present the state of the art concerning the general techniques used for solving the Multiobjective Optimization Problems (MOPs). Then, the fuzzy evolutionary approach will be described in the third section. The fourth section focuses on the illustration of the suggested approach by applying it to solve FJSP and highlights some practical aspects of the application of such an approach for solving hard combinatorial problems. Finally, the last section deals with concluding remarks and shows the efficiency of the hybridization of FL and EAs through different numerical experiments.

2 Multiobjective Optimization: The State Of The Art

In most cases, a MOP can be described, without loss of generality, using the following formulation (Eq. 1):

$$Minimize_{x \in \Omega} \left(f_1(x), f_2(x), ..., f_L(x) \right) \qquad (1)$$

where x is a possible solution for the considered problem, Ω is the feasible solution space and $f_q(.)$ is the qth objective function (for $1 \leq q \leq L$). In the majority of combinatorial problems, we have to simultaneously optimize a set of conflicting objective functions. The literature presents many possible

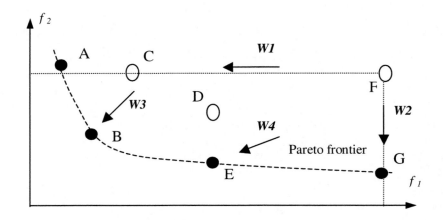

Fig. 1. Pareto-optimality approach

considerations and techniques that can be useful to evaluate solutions [1]. Mainly, we can distinguish two classes: the Pareto-optimality approaches and the non Pareto-optimality approaches.

2.1 Pareto-optimality Approach

It is one of the important approaches used in the MOP [2]. Pareto-optimality is expected in MOPs to provide flexibility and a large set of choices for the decision-maker. The optimality notion in the Pareto approaches can be formulated as follows:

- the Pareto-optimal set is constituted of nondominated solutions.
- solution is nondominated, if it is not dominated by any other one.
- x dominates y ($x \in \Omega$ and $y \in \Omega$) if: $\forall\ 1 \leq q \leq L, f_q(x) \leq f_q(y)$ and at least one index r exists such that $f_r(x) < f_r(y)$.

As an example, in Fig. 1, we consider a 2 objective function case. Solutions C, D, and F are dominated and {A, B, E, G} is the Pareto-optimal set of solutions. We can notice that Pareto-optimal set (Pareto frontier) is constituted of several nondominated solutions. The main objective of such an approach is to find all the elements of this set in order to give more choices to the decision-maker [3].

2.2 Non Pareto-optimality Approaches

The literature presents a large set of different approaches.

The first variant is applied in a lot of works and it is based on the aggregation of the different criteria in a single objective. As an example, a well-known fitness function has been proposed to measure the global quality of solutions according to predefined preferences (Eq. 2):

$$F_{fitness}(x) = \sum_{q=1}^{q=L} w_q.f_q(x) \qquad (2)$$

$F_{fitness}(.)$ is the weighted sum of objective functions, where w_q is the weight of the qth objective function such that $w_q \in [0,1] \, \forall 1 \leq q \leq L$, and $\sum_{q=1}^{q=L} w_q = 1$. We note $W = (w_1, w_2, ..., w_L)^T$ the weight vector of the different criteria. Depending on the studied problem, the literature presents many other expressions and formulations for aggregating the different objectives. But, in most cases, it is difficult to design such a formulation especially when relations between objectives are conflicting and non linear [2]. Thus, a careful study of such relations is necessary before drawing any formulation.

In 1993, Fonseca and Fleming proposed a slightly different variant. It was based on the ranking of a solution according to the number of solutions dominating it [4] [5].

In [6], we can find an application of another method for solving MOPs. It is the optimization by phases. It is based on alternating the objectives iteratively. In other words, we construct a set of solutions in optimizing it according to one objective function in the current iteration. Then, the obtained set will be optimized according to the next criterion in the next iteration, thus, the current objective will represent a constraint to respect in the optimization of the next one.

Recently, Evolutionary Algorithms (EAs) have been proposed to solve MOPs [2] [7] [8] [9] [10]. These algorithms are applied by many users in different areas of engineering framework, computer science and operation research. The main advantage of such an optimization technique consists in its special modular aspect: it operates on data without using preliminary knowledge on the processed problem and it evolves from an initial population of possible solutions iteratively bringing a global improvement according to one predefined criterion [11] [12]. Therefore, the EAs can represent a suitable technique for implementation of the multiobjective optimization based on the aggregative approaches. In addition, the literature shows that EAs are successfully used to construct efficient Pareto-optimality approaches:

1) Schaffer proposed to consider the L objective functions $f_q(.)$ and L subpopulations P_q ($\forall \, 1 \leq q \leq L$). In each subpopulation P_q, we use the function $f_q(.)$ to optimize solutions [13]. As an example, $W1$ and $W2$, represent the search directions used in 2-objectives case by such a method. Thus, the considered approach can easily find solutions A and G but it represents the inconvenience of being unable to find other Pareto solutions such as E or B (Fig. 1).

2) Ishibuchi and Murata proposed to aggregate different objective functions as shown in Eq. 2 and generate the weights randomly when evaluation is needed in the selection step [3]. In other words, they proposed to vary the search directions randomly. This approach is interesting to obtain various Pareto solutions because we can generate various sets of weight vectors such as $W3$ and $W4$ as shown in Fig. 1.

3) Recently, Leung and Wang have proposed a very interesting approach based on the application of Uniform Design to overcome some disadvantages of Ishibuchi and Murata's method such as irregularity of the random approach [14]. In addition, they showed how Uniform Design can be useful for finding Pareto solutions scattered uniformly.

In this chapter, we present a new methodology based on the hybridization of EAs and FL. Such a methodology can be used to find a single preferred solution according to some predefined performances. In addition, it can also be applicable to find a set of Pareto solutions when the user cannot express his/her preferences.

3 Fuzzy Evolutionary Approach

In this section, we propose a new approach for MOPs. Two methodological levels will be considered.

The first one concerns the multiobjective evaluation problem. It defines a set of practical considerations that could be useful to formulate an adapted fitness function based on a fuzzy aggregation of different objective functions. Such a formulation can also integrate some predefined preferences to find a particular solution according to these preferences and it remains applicable for finding a set of Pareto-optimal solutions.

The second level concerns the resolution method. It will be started by reducing the search space (Localization stage) in order to accelerate the convergence. In addition, this Localization stage will enable us to generate a control model that can be useful to build an evolutionary approach with high performance.

3.1 Multiobjective Evaluation

In this part, we are interested in evaluating and comparing the solutions according to several criteria. Generally, these criteria present non linear and complex relations between them and do not obviously have the same importance for the decision-makers. Thus, many considerations must be done to take all these difficulties into account. In such a way, the proposed multiobjective evaluation will be based on the three following steps:

1) First step: determination of lower-bounds

For each objective function $f_q(.)$, we can (in a large class of combinatorial problems) find or estimate a possible lower-bound f_q^* (Eq. 3), such that:

$$\forall\, x \in \Omega, \quad f_q(x) \geq f_q^* \tag{3}$$

The lower-bounds can be very useful to evaluate and judge the obtained solutions precisely.

2) Second step: resolution of the scale problem (hmogenization of objective functions)

In most cases, the values of objective functions can belong to different intervals with variable lengths. Thus, to be efficient, the fitness function (described in Eq. 2) must be homogenized in order to avoid that some objectives be dominated by the others. For example, if $f_1(x) \in [100, 1000]$ and $f_q(x) \in [0,1]$ ($\forall\, 2 \leq q \leq L$), then the first objective can dominate the others.

To avoid such an inconvenient that can disregard some objective functions and amplify the others, we can use a simple application of fuzzy logic based on the following steps:

- to each feasible solution x, we associate a vector $f(x) \in [f_1^*, +\infty[\, \times\, [f_2^*, +\infty[\, \times\, ...\, \times\, [f_L^*, +\infty[$. This vector will characterize by its components the L objectives to be optimized: $f(x) = (f_1(x), f_2(x), ..., f_L(x))^T$,
- in particular, let H a chosen heuristic and f_q^H the best value of the qth objective function given by the considered heuristic,
- for each vector $f(x)$, we propose a fuzzification of its components $f_q(x)$ according to their positions in the intervals $[f_q^*, f_q^H + \varepsilon_q]$ where ε_q is a little positive value designed to avoid the problem of dividing by zero (when $f_q^* = f_q^H$) and formulated as shown in Eq. 4.

$$\varepsilon_q = 0.01.f_q^* \quad \text{if} \quad f_q^* = f_q^H; \quad \text{else} \quad \varepsilon_q = 0 \tag{4}$$

The two considered fuzzy subsets are the following ones:

- G_q: the subset of the good solutions according to the qth objective.
- B_q: the subset of the bad solutions according to the qth objective.

The fuzzification is applied using the membership function as shown in Fig. 2

Thus, to each vector $f(x)$, we associate two vectors $\widetilde{f}_G(x)$ and $\widetilde{f}_B(x)$ described in Eq. 5 and Eq. 6 such that:

$$\widetilde{f}_G(x) = (\mu_1^G(f_1(x)), \mu_2^G(f_2(x)), ..., \mu_L^G(f_L(x)))^T \tag{5}$$

$$\text{and} \quad \widetilde{f}_B(x) = (\mu_1^B(f_1(x)), \mu_2^B(f_2(x)), ..., \mu_L^B(f_L(x)))^T \tag{6}$$

where the membership functions are described in Eq. 7:

$$\mu_q^G(f_q(x)) = \frac{f_q^H - f_q(x) + \varepsilon_q}{f_q^H - f_q^* + \varepsilon_q} \quad \text{if} \quad f_q(x) \in [f_q^*, f_q^H + \varepsilon_q];$$

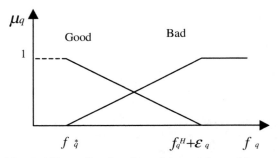

Fig. 2. FL application for solving scale problem

$$\mu_q^G(f_q(x)) = 0 \text{ if } f_q(x) \geq f_q^H + \varepsilon_q \text{ and } \mu_q^B(f_q(x)) = 1 - \mu_q^G(f_q(x)) \quad (7)$$

Thereafter, the quality of each solution x is characterized by the vector $\widetilde{f}_G(x)$ of which all the components are homogeneous since they belong to the same interval $[0,1]$.

3) Third step: formulation of the fuzzy fitness function

By combining Eq. 2 and Eq. 7, we can now reduce the multiobjective optimization to the minimization of the following global criterion (Eq. 8):

$$f_g(x) = \sum_{q=1}^{q=L} w_q . \mu_q^G(f_q(x)) \quad (8)$$

Depending on the way of the determination of the different weight values and using the proposed fuzzy fitness function, we can now construct two possible approaches. The first one is an aggregative approach using Eq. 8 to find a single solution according to the decision-maker's preferences. The second one is a Pareto approach in which we apply FL to compute dynamically the weights of the different objective functions when the decision-maker is not able to express his/her preferences.

Non Pareto-optimal Approach: In this paragraph, we deal with the case where the decision-maker is able to express his/her preferences. Such preferences are generally of a subjective nature and could be well modelled using FL. In this way, we propose a simple application of FL for computing the different weights of objective functions according to the choices evoked by the decision-maker. Two cases are considered:

- First case: the user gives his preferences by the attribution of a note N_q to each objective function $f_q(.)$. We calculate the corresponding weight w_q according to the following formula (Eq. 9):

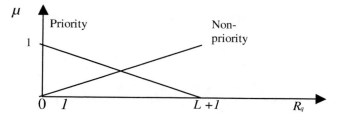

Fig. 3. Membership function for optimized objectives

$$w_q = \frac{N_q}{\sum_{q=1}^{q=L} N_q} \quad \forall 1 \leq q \leq L \qquad (9)$$

- Second case: the user evokes his preferences by a relative ranking of criteria. In this case, to each criterion, we associate $R_q \in \{1, 2, ..., L\}$ that represents the rank of $f_q(.)$ according to the user. We propose a simple application of FL to calculate the different weights of objective functions according to theirs ranks. Let us consider the primary proposal shown in Fig. 3. According to its rank, a criterion can be in one of the fuzzy subsets: P (the subset of the priority criteria) and $\tilde{P}$ (the subset of the non-priority criteria) according to the following membership function formulated in Eq. 10.

$$\mu^P(R_q) = \left(1 - \frac{R_q}{L+1}\right) \quad \forall\, 1 \leq q \leq L \qquad (10)$$

However, we can easily notice that this priority is not *uniform* in all cases. In fact, we can be confronted to situations where an urgency intensifies the priority of certain criteria. To integrate this concept of intensification, we propose to introduce a small modification on the preceding function $\mu^P(.)$. Thus, it will be replaced by the function $\tilde{\mu}^P(.)$ described in Eq. 11.

$$\tilde{\mu}^P(R_q) = \left(1 - \frac{R_q}{L+1}\right)^\rho \quad \forall\, 1 \leq q \leq L \qquad (11)$$

with ρ is the degree of the priority intensity.
If ρ increase, the values of the function $\tilde{\mu}^P(.)$ drop for all $R_q (1 \leq q \leq L)$ and especially for badly-ranked criteria. We conclude that the rising of ρ involves a significant reduction in the priority of the badly-ranked criteria, that is why we use this coefficient ρ.

Thus, we can allot to each criterion a note which takes account these priorities and their intensities (Eq. 12):

$$N_q = \tilde{\mu}^P(R_q) = \left(1 - \frac{R_q}{L+1}\right)^\rho \quad \forall\ 1 \leq q \leq L \qquad (12)$$

Therefore, the calculation of the weights w_q becomes possible by using the same formula suggested in Eq. 9.

In [15], others considerations based on fuzzy concepts could be used to differently compute these weights.

Pareto-optimality Approach: In this paragraph, we try to help the decision maker when he/she cannot clearly give a particular preference of some objective functions. In such a dubious state, we propose to find a set of Pareto-optimal solutions without according any privilege to some particular search directions. This approach will be based on an EA in which we use the fitness function $f_g(.)$ (formulated by Eq. 8) for selection process. The weights w_q ($1 \leq q \leq L$) will dynamically be computed using a specific fuzzy rule. The idea is to measure the mean quality of solutions according to each criterion at each iteration and to compute the different weights according to this quality.

Let $\overline{f}_q^k$ the mean of the qth objective function values of solutions at the kth iteration of the EA (Eq. 13):

$$\overline{f}_q^k = \frac{\sum_{x \in P_k} f_q^k(x)}{cardinal(P_k)} \qquad (13)$$

where P_k is the population of the solutions at the kth iteration of the EA.

At the kth iteration of the EA, we evaluate the quality of solutions according to the qth objective function. The evaluation is done using the membership functions shown in Fig. 4. So, two fuzzy subsets could be distinguished:

- $Near_q$: the subset of solutions near the lower-bound value f_q^* according to the qth objective.
- Far_q: the subset of solutions far from the lower-bound f_q^* value according to the qth objective.

As shown in Fig. 4, the membership functions could be formulated as follows (Eq. 14):

$$\mu_{q,k}^{Far}\left(\overline{f}_q^k\right) = \frac{\overline{f}_q^k - f_q^*}{\overline{f}_q^0 - f_q^* + \varepsilon_q'} \text{ if } \overline{f}_q^k \in \left[f_q^*, \overline{f}_q^0 + \varepsilon_q'\right] \text{ ; else } \mu_{q,k}^{Far}\left(\overline{f}_q^k\right) = 1 \quad (14)$$

where ε_q' is a little positive value designed to avoid the problem of dividing by zero (when $\overline{f}_q^0 = f_q^*$) and formulated as shown in Eq. 15.

$$\varepsilon_q' = 0.01.f_q^* \quad \text{if} \quad \overline{f}_q^0 = f_q^*; \quad \text{else} \quad \varepsilon_q' = 0 \qquad (15)$$

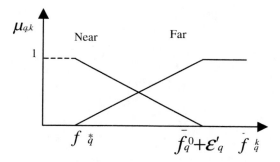

Fig. 4. Membership functions for criteria values

The computation of the different weights w_q^{k+1} will be achieved using the following fuzzy rule:

IF ($\overline{f}_q^k$ near f_q^*) THEN (w_q^{k+1} ↘),

IF ($\overline{f}_q^k$ far f_q^*) THEN (w_q^{k+1} ↗).

The application of the preceding rule could be formulated as shown in Eq. 16.

$$w_q^1 = \frac{1}{L} \; ; \; w_q^{k+1} = \frac{\mu_{q,k}^{Far}\left(\overline{f}_q^k\right)}{\sum_{q=1}^{q=L} \mu_{q,k}^{Far}\left(\overline{f}_q^k\right)} \quad \forall \; 1 \leq q \leq L, \forall \; 1 \leq k \leq Q-1 \quad (16)$$

where Q is the total number of iterations.

The different weight vectors ($W^1, W^2, .., W^k, .., W^Q$) will dynamically be computed (by evolving from the kth generation P_k to the next P_{k+1}) according to the distance between the lower-bounds and the mean of the individuals of the generation P_k (depicted by a black closed circles in Fig. 5). The aim is to investigate the possible gains and improvements of the solutions by giving the priority to the optimization of the objective functions which the mean of values is far from the optimal value (or the lower-bound value). Thereafter, using such a fuzzy technique, it would be possible to control the direction search in order to construct a final set of solutions near the optimal one (see in Fig. 5). This final set will be used for selecting Pareto-optimal solutions and the others will be rejected [16].

3.2 Evolutionary Optimization

In this part, we are interested in the resolution of the problem itself. To obtain a high solution quality, many considerations must be done. For this purpose, we propose a method based on the two following stages:

- Localization stage;
- Controlled evolutionary optimization stage.

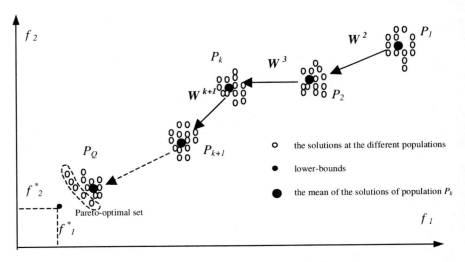

Fig. 5. Fuzzy dynamic control of search directions

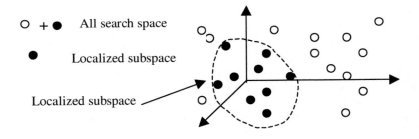

Fig. 6. Search space and Localization approach

Localization Stage: In spite of the spectacular evolution of the computer technology, the combinatorial problems remain very hard to solve. This difficulty is due to their high complexity. In order to make the resolution more efficient, we propose to reduce the search space in a subspace where the chance to obtain good solutions is increased (see in Fig. 6). This Localization technique will obviously depend on the problem that we process and may be difficult to be carried out but it can certainly make the optimization more efficient and can ensure a good level of satisfaction. The set of solutions obtained in the Localization stage will after be used as an initial population by the Controlled Evolutionary Approach. This can be useful to reduce the computing time spent in the evolutionary optimization stage starting from a set of good solutions [6].

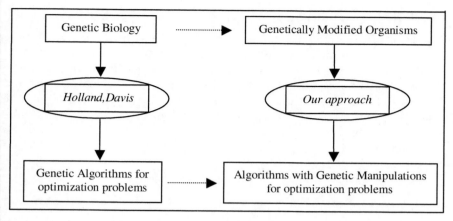

Fig. 7. GMO concept in controlled evolutionary optimization

Controlled Evolutionary Optimization Stage: In such an evolutionary optimization, we use the set of solutions obtained in the Localization stage as an initial population. The generation of new individuals will be controlled using the control model given in the preceding stage. In addition, we make genetic operators able to contribute in this optimization. These operators of mutation present a new way of application of evolutionary algorithms: it is the way of "genetic manipulations". In genetic biology, these manipulations allow us to generate Genetically Modified Organisms (GMO). The proposed method is inspired of such a principle and intervenes in the construction phase of the new chromosomes by applying the "artificial mutations" in order to accelerate the convergence and insure a high quality of final solutions [6](see in Fig. 7).

4 Application: Case Of Flexible Job-shop Scheduling Problem (FJSP)

A job shop scheduling is a process-organized manufacturing facility; its main characteristic is a great diversity of jobs to be performed [17] [18]. A job shop produces goods (parts); these parts have one or more alternative process plans. Each process plan consists of a sequence of operations; these operations require resources and have certain (predefined) durations on machines. The task of planning, scheduling, and controlling the work is very complex, and a perfect knowledge of the problem is necessary to assist in these tasks [18] [19].

The flexible job-shop scheduling problem is to select a sequence of operations together with assignment of start/end times and resources for each operation. The main considerations to be taken into account are the cost of having idle machine and labor capacity, the cost of carrying in-process inventory, and the need to meet certain completion due dates.

4.1 FJSP Formulation

The problem is to organize the execution of N jobs on M machines. The set of machines is noted U. Each job J_j represents a number of n_j non preemptable ordered operations (precedence constraint). The execution of the ith operation of job J_j (noted $O_{i,j}$) requires one resource or machine selected from a set of available machines. The assignment of the operation $O_{i,j}$ to the machine M_k entails the occupation of this machine during a processing time called $d_{i,j,k}$. Thus, to each FJSP, we can associate a table D of processing times such that: $D = \{d_{i,j,k} \in IN^* \mid 1 \leq j \leq N; 1 \leq i \leq n_j; 1 \leq k \leq M\}$ (see the example in Fig. 8(a)).

In this problem, we make the following hypotheses:

- all machines are available at $t = 0$ and each job J_j can be started at $t = r_j$,
- at a given time, a machine can only execute one operation: it becomes available to other operations once the operation which is currently assigned to is completed (resource constraints),
- to each operation $O_{i,j}$, we associate an earliest starting time $r_{i,j}$ calculated by the following formula (Eq. 17):

$$r_{1,j} = r_j \text{ and } r_{i+1,j} = r_{i,j} + \gamma_{i,j} \; \forall \; 1 \leq i \leq n_j - 1, \; \forall \; 1 \leq j \leq N \quad (17)$$

where $\gamma_{i,j} = \min_k (d_{i,j,k}) \quad \forall \; 1 \leq i \leq n_j, \forall \; 1 \leq j \leq N$.

The FJSPs present two difficulties. The first one is to assign each operation $O_{i,j}$ to a machine M_k (selected from the set U). The second one is the computation of the starting time $t_{i,j}$ and the completion time $tf_{i,j}$ of each operation $O_{i,j}$.

The considered objective is to minimize the following criteria described in Eq. 18, Eq. 19 and Eq. 20:

- the makespan:

$$Cr_1 = \max_j \{tf_{n_j,j}\} \quad (18)$$

- the workload of the most loaded machine:

$$Cr_2 = \max_k \{W_k\} \quad (19)$$

where W_k is the workload of M_k,

- the total workload of machines:

$$Cr_3 = \sum_k W_k \quad (20)$$

4.2 Resolution

As proposed in the third section, the fuzzy evaluation is started by the determination of a lower-bounds set for the considered criteria. For FJSP, such a set is formulated by Kacem et al in the following theorem.

Theorem 1. Cr_1^*, Cr_2^* and Cr_3^* are respectively lower-bounds for the considered criteria Cr_1, Cr_2 and Cr_3 as shown in Eq. 21:

$$Cr_1^* = \max\left(\max_j\left(r_j + \sum_i \gamma_{i,j}\right), \widetilde{E}\left(\frac{R_M + \sum_j \sum_i \gamma_{i,j}}{M}\right), \theta_{k_0,\widetilde{N}}\right),$$

$$Cr_2^* = \max\left(\widetilde{E}\left(\frac{\sum_j \sum_i \gamma_{i,j}}{M}\right), \delta_{k_0,\widetilde{N}}\right) \quad \text{and} \quad Cr_3^* = \sum_j \sum_i \gamma_{i,j} \qquad (21)$$

where $\widetilde{E}(x)$ is the first integer superior or equal to x; $\widetilde{N} = \widetilde{E}(N_t/M)$ with $N_t = \sum_{j=1}^{j=N} n_j$; $\delta_{k_0,\widetilde{N}} = \min_k\left(D_{\widetilde{N}}^k\right)$ with $D_{\widetilde{N}}^k$ is the sum of the $\widetilde{N}$ shortest processing times of the operations that we can execute on the machine M_k ; R_M is the sum of the M little values of the starting times $(r_{i,j})$; $\theta_{k_0,\widetilde{N}} = \min_k\left(\min_{1\leq z\leq N_t-\widetilde{N}+1}\left(r_{i_z,j_z} + d_{i_z,j_z,k} + \min_{C'_{z,\widetilde{N}} \in E'_{z,\widetilde{N}}}\left(\Delta^k\left(C'_{z,\widetilde{N}}\right)\right)\right)\right)$; $\Delta^k\left(C'_{z,\widetilde{N}}\right)$ is the sum of the processing times of the operations of $C'_{z,\widetilde{N}}$ on M_k ; $C'_{z,\widetilde{N}}$ is an element of $E'_{z,\widetilde{N}}$ and $E'_{z,\widetilde{N}}$ is the set of the combinations of $\left(\widetilde{N}-1\right)$ operations chosen among the operations of V_z; V_z is a part of the operations set defined as follows: $V_z = \{O_{i_{z+1},j_{z+1}}, O_{i_{z+2},j_{z+2}}, ..., O_{i_{N_t},j_{N_t}}\}$ for $z \in \{1,2,...,N_t - \widetilde{N}+1\}$ where $\{r_{i_1,j_1}, r_{i_2,j_2}, ..., r_{i_{N_t},j_{N_t}}\}$ are ranged in the ascending order.

Proof. For proof, see [20].

Thus, using Eq. 8 (by respectively exchanging f_q, f_q^*, and f_q^H by Cr_q, Cr_q^*, and Cr_q^H), we can reduce the multiobjective optimization to the minimization of the following global criterion (Eq. 22):

$$Cr_g = \sum_{q=1}^{q=3} w_q \cdot \mu_q^G(Cr_q) \quad \text{such that} \quad \sum_{q=1}^{q=3} w_q = 1 \qquad (22)$$

where Cr_q^H is the best value of the qth criterion given by the Approach by Localization (AL), the chosen heuristic H that we present in the next paragraph.

After the formulation of the fitness function, the optimization stage is started by applying the AL. It will represent the heuristic H to use. This approach enables us to solve the problem of resource allocation taking into

(a). Processing times table (D)

		M1	M2	M3	M4
J_1	O 1,1	1	3	4	1
	O 2,1	3	8	2	1
	O 3,1	3	5	4	7
J_2	O 1,2	4	1	1	4
	O 2,2	2	3	9	3
	O 3,2	9	1	2	2
J_3	O 1,3	8	6	3	5
	O 2,3	4	5	8	1

(b). A

$O_{i,j}$	$A_{i,j}$
O 1,1	1
O 2,1	1
O 3,1	2
O 1,2	2
O 2,2	4
O 3,2	3
O 1,3	4
O 2,3	4

(c). A^{ch}

$O_{i,j}$	$U_{i,j}$
O 1,1	{1,4}
O 2,1	{1,3,4}
O 3,1	{1, 2, 3}
O 1,2	{2,3}
O 2,2	{1, 2, 4}
O 3,2	{2, 3, 4}
O 1,3	{3, 4}
O 2,3	{4}

Fig. 8. AL and assignment solutions

account the processing times and minimize the total workload of the machines (localization of the most interesting zones of the search space before scheduling). In fact, the AL makes it possible to build a set E of assignments ($E = \{A^z \mid 1 \leq z \leq cardinal(E)\}$) and balance the workloads of the machines.

Each assignment A^z is represented in a table (the same length that D), $A^z = \{A^z_{i,j} \mid 1 \leq j \leq N; 1 \leq i \leq n_j\}$. For each i and j, the value of $A^z_{i,j}$ is belonging to $\{1, 2, .., M\}$. The value "$A^z_{i,j} = k$" means that $O_{i,j}$ is assigned to M_k (see in Fig. 8(b)).

Starting from the assignment set E, the AL enables us to conceive a model of chromosomes that suits the problem [6]. This model is going to serve us in the construction of new individuals in order to integrate the good properties contained in the schemata. The objective is to make genetic algorithms more efficient and more rapid in constructing the solution by giving the priority to the reproduction of individuals respecting model generated by the schemata and not from the whole set of chromosomes.

The idea is to determine (for each operation) the subset $U_{i,j} \subseteq U$ of possible machines where the probabilities of having a good schedule are raised [6]. As an example, for the FJSP shown in Fig. 8(a), we obtain the schemata A^{ch} as follows in Fig. 8(c). Such an approach also makes it possible to compute the starting time of each operation thanks to a modular algorithm (called "Scheduling Algorithm") and to construct a set of feasible solutions taking into account precedence and resource constraints [21]. This set of solutions will be used in the next stage (when we apply genetic algorithms) as an initial population.

Now, starting from the solution set given by the AL, we apply an advanced genetic algorithm in order to enhance solution quality. The conceptual details of this algorithm such that coding, crossover operator, and mutation operators are presented in the following paragraphs:

1) Coding: Operations List Coding (OLC)

It consists in representing the schedule in the same assignment table A, we only add a third column for the starting and completion times $(t_{i,j}, tf_{i,j})$.

> - Select randomly 2 parents x^1 and x^2 ;
> - Select randomly 2 integers j and j' such that $j \leq j' \leq N$;
> - Select randomly 2 integers i and i' such that $i \leq n_j$ and $i' \leq n_{j'}$ (in the case where $j=j'$, $i \leq i'$) ;
> - The individual e^1 receives the same assignments from parent x^1 for all operations between row (i,j) and row (i',j');
> - The remainder of assignments for e^1 is obtained from x^2;
> - The individual e^2 receives the same assignments from parent x^2 for all operations between row (i,j) and row (i', j');
> - The remainder of assignments for e^2 is obtained from x^1;
> - Calculate the starting and completion times according to the algorithm "Scheduling Algorithm".

Fig. 9. Crossover algorithm

Cases $A_{i,j}$ are unchanged (see the example in Fig. 10(a)). Other interesting coding possibilities are proposed for solving the considered problem [21] [22] [23].

2) Crossover operator:

This operator is designed in order to explore the search space and to offer more diversity by exchanging information between two individuals. One algorithm illustration is presented in Fig. 9. As an illustration, we treat the job-shop problem D already presented in Fig. 8(a) (where $r_1 = r_2 = r_3 = 0$) and we use the same instance to explain all the proposed genetic operators.

Example 1. we suppose that x^1 and x^2 are randomly chosen and $j = 1, j\prime = 2, i = 2, i\prime = 2$; (as shown in Fig. 10(a) and (b)). Thus, the chromosomes obtained are presented in Fig. 10(c) and (d).

3) Mutation operators:

The objective of our research is to minimize the makespan and workloads of machines. It would therefore be interesting to make genetic operators able to contribute in this optimization. As suggested in the third section, these operators of mutation are inspired by the novel biologic concept of GMO and intervene in the construction phase of the new chromosomes by applying the "artificial manipulations" in order to accelerate the convergence and to insure a high quality of final solutions. As an example, we propose two operators of artificial mutation:

- Operator of mutation reducing the $\zeta_j = \sum_{i=1}^{i=n_j} \sum_k \{d_{i,j,k} \mid A_{i,j} = k\}$ (See in Fig. 11 for algorithm.):

Example 2. for the schedule x shown in Fig. 12(a), the job J_1 has the most raised value of the ζ_j ($\zeta_1 = 6$ units of time). We therefore have to cover the list of its operations to reduce this duration. For operation $O_{1,1}$, the

(a). x^1 :

$O_{i,j}$	$A_{i,j}$	$t_{i,j}, tf_{i,j}$
O 1,1	1	0, 1
O 2,1	4	5, 6
O 3,1	3	6, 10
O 1,2	3	0, 1
O 2,2	1	1, 3
O 3,2	2	3, 4
O 1,3	4	0, 5
O 2,3	4	6, 7

(b). x^2 :

$O_{i,j}$	$A_{i,j}$	$t_{i,j}, tf_{i,j}$
O 1,1	4	0, 1
O 2,1	3	3, 5
O 3,1	1	5, 8
O 1,2	2	0, 1
O 2,2	1	1, 3
O 3,2	4	3, 5
O 1,3	3	0, 3
O 2,3	2	3, 8

(c). e^1 :

$O_{i,j}$	$A_{i,j}$	$t_{i,j}, tf_{i,j}$
O 1,1	1	0, 1
O 2,1	3	1, 3
O 3,1	1	3, 6
O 1,2	2	0, 1
O 2,2	1	1, 3
O 3,2	2	3, 4
O 1,3	4	0, 5
O 2,3	4	5, 6

(d). e^2 :

$O_{i,j}$	$A_{i,j}$	$t_{i,j}, tf_{i,j}$
O 1,1	4	0, 1
O 2,1	4	1, 2
O 3,1	3	4, 8
O 1,2	3	0, 1
O 2,2	1	1, 3
O 3,2	4	3, 5
O 1,3	3	1, 4
O 2,3	2	4, 9

Fig. 10. Crossover operator illustration

- Select randomly an individual x ;
- Choose the job J_j whose ζ_j is the longest ;
- $i=1; r = 0$;
- **WHILE** ($i \leq n_j$ and $r = 0$)
  - Find K_0 such that $A_{i,j} = K_0$;
  - **FOR** ($k=1, k \leq M$)
    IF ($d_{i,j,k} < d_{i,j,k0}$) **Then** $\{A_{i,j}=k;\ r=1\ ;\}$
    End IF
    End FOR
  - $i=i+1$;

End WHILE
- Calculate starting and completion times according to the algorithm "Scheduling Algorithm".

Fig. 11. First mutation algorithm

(a). x:

$O_{i,j}$	$A_{i,j}$	$t_{i,j}, tf_{i,j}$
O 1,1	4	0, 1
O 2,1	3	3, 5
O 3,1	1	5, 8
O 1,2	2	0, 1
O 2,2	2	1, 4
O 3,2	4	4, 6
O 1,3	3	0, 3
O 2,3	4	3, 4

(b). x':

$O_{i,j}$	$A_{i,j}$	$t_{i,j}, tf_{i,j}$
O 1,1	4	0, 1
O 2,1	4	1, 2
O 3,1	1	2, 5
O 1,2	2	0, 1
O 2,2	2	1, 4
O 3,2	4	4, 6
O 1,3	3	0, 3
O 2,3	4	3, 4

Fig. 12. First mutation algorithm illustration

- Select randomly an individual x ;
- Find the most loaded machine M_{k1};
- Find the less loaded machine M_{k2};
- Choose randomly an operation $O_{i,j}$ such that $A_{i,j}=k_1$;
- Assign this operation to the less loaded machine: $A_{i,j}= k_2$;
- Calculate the starting and completion times according to the algorithm "Scheduling Algorithm".

Fig. 13. Second mutation algorithm

processing time $d_{1,1,4}$ corresponds to the minimum of processing times. On the other hand, operation $O_{2,1}$, can be assigned to the machine M_4 instead of the machine M_3 (because $d_{2,1,4} < d_{2,1,3}$) and thereafter we reduce the ζ_1 to 5 units of time and the makespan to 6 units instead of 8. The obtained chromosome $x\prime$ after mutation is presented in Fig. 12(b).

- Mutation balancing workloads of machines:

Such an operator is based on assigning a random chosen operation (already assigned to the most loaded machine) to the less loaded machine [23] [24] as shown in Fig. 13.

Remark 1. The mutation and crossover probabilities are fixed in a classic way (crossover probability: $P_c = 0.90$; mutation probability: $P_m = 0.10$).

Remark 2. The used selection mechanism gives the priority to the reproduction of the best individuals according to their values of global criterion Cr_g.

Remark 3. The Controlled Genetic Algorithm will be stopped when the maximum number of the iterations Q is reached.

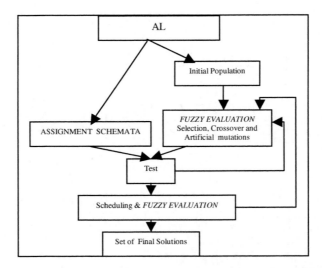

Fig. 14. Fuzzy Controlled Evolutionary Approach for FJSP

In conclusion, the suggested application can be summarized by the flowchart described in Fig. 14.

The evolution of generations will be controlled on two levels. On the first one, we apply artificial mutations (genetic manipulations) in order to reduce the blind aspect of genetic operators and we test if the new individual respects the model imposed by the assignment schemata. On the second level, we use the fuzzy multi-criteria evaluation to decide if the new individual represents an interesting solution to be selected in the next step or not (see the flowchart shown in Fig. 14).

4.3 Numerical Experiments

Computational experiments are carried out to evaluate the efficiency of our approach with a large set of representative problem instances based on practical data.

As proposed in the third section and depending on the use of the global criterion, we considered two approaches. The first one is an aggregative approach in which we compute the different weights of objectives according to the decision-maker's preferences. The second is a Pareto-optimal approach. In such an approach, we apply a fuzzy dynamic control for computing the different weights.

Based on the preceding consideration, our numerical experiments could be classified in two classes: some simulations of the first approach and others of the second one.

Table 1. Some simulation results of the aggregative approach

Instances I_h $_{h\leq 10}$	Weights w_1 w_2 w_3	Lower-bounds Cr_1^* Cr_2^* Cr_3^*	Obtained results Cr_1 Cr_2 Cr_3	Distance $\Delta(\%)$
I_1	0.1 0.1 0.8 0.1 0.5 0.4 0.8 0.2 0 0.6 0.1 0.3 0.79 0.01 0.2	16 7 32	18 8 32 18 7 33 16 9 40 16 10 36 16 11 36	2.7 2.5 5.7 8 3
I_2	.511 .322 .167	15 9 60	15 11 61	7.4
I_3	.511 .322 .167	23 10 91	24 11 94	5.9
I_4	0.58 0.31 0.11 0.22 0.03 0.75	7 5 41	7 5 45 8 7 42	1 6
I_5	0.5 0.5 0	17 13 63	17 13 63	0
I_6	0.5 0.5 0	26 26 126	26 26 126	0
I_7	0.5 0.5 0	51 51 252	51 51 252	0
I_8	0.5 0.5 0	16 13 63	16 13 63	0
I_9	0.5 0.5 0	38 38 189	38 38 189	0
I_{10}	0.5 0.5 0	63 63 315	63 63 315	0

Simulation Of The Aggregative Approach: In order to evaluate the suggested approach, many series of instances are tested. As an example, the lecturer could consult the simulation results of some instances at the web address: http://www.ec-lille.fr/~kacem/tests.pdf. These results are summarized in Table 1. For each instance and each weighting, we present the lower-bound values and the objective function values of the best obtained solution. In addition, we calculate the global criterion Cr_g and we define $\Delta(\%)$ to measure the distance between the obtained solution and the lower-bounds as shown in Eq. 23.

$$\Delta(\%) = (1 - Cr_g).100 \tag{23}$$

As can be seen in Table 1, the results show that the solutions obtained are generally near the optimal values (according to the chosen weights). In the case of "Parallel Machines Problem" (a particular case of the FJSP described by the instances I_5, I_6, I_7, I_8, I_9, I_{10}), we obtain no distance between final solutions and lower-bounds. In the general case, the small distance that could be obtained is due to the non homogeneous nature of the criteria and the difficulty of such a conflicting multiobjective optimization.

Simulation Of The Fuzzy Evolutionary Pareto Approach: In this case, the different weights of objectives are computed using the fuzzy rules already presented in the third section. The aim is to find a set of Pareto-

Table 2. Some simulation results of the Pareto approach

Instances $I_{h\ h\leq 10}$	Lower-bounds Cr_1^* Cr_2^* Cr_3^*	Obtained results Cr_1 Cr_2 Cr_3
I_1	16 7 32	18 8 32 18 7 33 16 9 35 16 10 34
I_2	15 9 60	15 11 61 16 10 66 16 12 60 17 10 64 18 10 63
I_3	23 10 91	24 11 91 23 11 95
I_4	7 5 41	7 5 45 8 7 41 8 5 42
I_5	17 13 63	17 13 63
I_6	26 26 126	26 26 126
I_7	51 51 252	51 51 252
I_8	16 13 63	16 13 63
I_9	38 38 189	38 38 189
I_{10}	63 63 315	63 63 315

optimal solutions of high quality (in other words, very near the lower-bound values).

The application of the suggested approach on the instance set $\{I_1, I_2, I_3, I_4, I_5, I_6, I_7, I_8, I_9, I_{10}\}$ (already presented in the preceding paragraph) confirmed its efficiency. So, for instances I_5, I_6, I_7, I_8, I_9, and I_{10}, we obtain the optimal solution. Then, the Pareto frontier is reduced to a single point representing the optimality. For the other instances, the values of the different criteria of the obtained Pareto-optimal solutions are presented in Table 2.

Remark 4. the different Pareto-optimal solutions obtained for the studied instances $\{I_1, I_2, I_3, I_4, I_5, I_6, I_7, I_8, I_9, I_{10}\}$ could be consulted at the following web address: http://www.ec-lille.fr/~kacem/testsPareto.pdf.

5 Discussions And Conclusions

In this paper, we propose a new fuzzy evolutionary methodology for solving MOPs and we showed the efficiency of making cooperate FL and EAs. Such a methodology presents the advantage of regrouping several conflicting criteria to be optimized simultaneously. Moreover, the modular aspect of the above

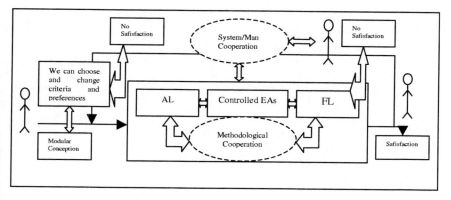

Fig. 15. Cooperative decision system

approach could simplify the resolution of many MOPs such as FJSP. In fact, the suggested method can integrate the decision-maker's preferences in order to find a single solution according to those preferences. In addition, when preferences cannot be obtained, the decision system can help the user by finding a set of Pareto-optimal solutions. In a more general way, we can build efficient decision system using these forms of cooperation and interaction. In fact, the last figure (Fig. 15) presents the two cooperative aspects that can be introduced to solve complex problems (like scheduling and planning):

1) first level: this level is only related to the conceptual and methodological aspects of resolution system which must make cooperate many optimization techniques (Genetic Algorithms, Heuristics, Fuzzy Logic) to ensure the solution efficiency [7],

2) second level: it concerns the interaction "Decision System/Man". In fact, we need to implement modular and flexible conception to contribute in the correct functioning of the production system and to make easier this interaction.

References

1. Zitzler, E., Deb, K., Thiele, L., Coello, C., Corne, D., 2001. Evolutionary Multi-Criterion Optimization. Lecture Notes in Computer Science, Vol. 1993, Springer Verlag.
2. Sarker, R., Abbas, H.A., Newton, C., 2001. Solving multiobjective optimization problems using evolutionary algorithm. Proceedings of International CIMCA Conference, July 9-11, 2001, Las Vegas, Nevada, USA.
3. Ishibuchi, H., Murata, T., 1998. A Multiobjective Genetic Local Search Algorithm and Its Application to Flowshop Scheduling. IEEE/SMC Transactions, Part C, Vol 28, August 1998, pp 392-403.
4. Fonseca, C.M., Fleming, P.J., 1998, (a). Multiobjective Optimization and Multiple Constraint Handling with Evolutionary Algorithms: Part I: A Unified

Formulation. IEEE/SMC Transactions, Part A, Vol 28, January 1998, pp 26-37.
5. Fonseca, C.M., Fleming, P.J., 1998, (b). Multiobjective Optimization and Multiple Constraint Handling with Evolutionary Algorithms: Part II: Application Example. IEEE/SMC Transactions, Part A, Vol 28, January 1998, pp 38-47.
6. Kacem, I., Hammadi, S., Borne, P., 2002, (a). Approach by Localization and Multiobjective Evolutionary Optimization for Flexible Job-Shop Scheduling Problems. IEEE Transactions on Systems, Man, and Cybernetics, Part C: Reviews and Applications, Vol 32, N°1, 2002.
7. Bachelet, V., Hafidi, Z., Preux, P., Talbi, E-G., 1998. Vers la coopération des métaheuristiques. Calculateurs parallèles, Vol.9, N.2.
8. Davis, L., 1990. Handbook of Genetic Algorithms. Van Nostrand Reinhold, New-York, USA.
9. Portmann, M-C., 2000. Study on Crossover Operators Keeping good schemata for some scheduling problems. Genetic and Evolutionary Computation Conference, July 8-12, 2000, Las Vegas, USA.
10. Ghedjati, F., 1994. Résolution par des heuristiques dynamiques et des algorithmes génétiques du problème d'ordonnancement de type job-shop généralisé. Ph.D Thesis, University of Paris VI, December 15, 1994, FRANCE.
11. Michalewicz, Z., 1992. Genetic Algorithms + Data Structures = Evolution Programs. Springer Verlag.
12. Fogel, D.B., 1994. An Introduction to Simulated Evolutionary Optimization. IEEE Transactions on Neural Networks, Vol. 5, N°1, Jan 1994, pp 3-14.
13. Schaffer, J.D., 1985. Multiobjective Optimization with Vector Evaluated Genetic Algorithms. in Proc. 1st Int. Conf. Genetic Algorithms, 1985, pp 93-100.
14. Leung, Y. W., Wang, Y., 2000. Multiobjective Programming Using Uniform Design and Genetic Algorithm. IEEE/SMC Transactions, Part C, Vol 30, August, 2000, pp 293-303.
15. Gzara, M., 2001. Méthode coopérative d'aide multicritère à l'ordonnancement flou. Ph.D. Thesis, Ecole Centrale de Lille et Université de Lille 1. FRANCE.
16. Kacem, I., Hammadi, S., Borne, P., 2002 (b). Pareto-optimality Approach for Flexible Job-shop Scheduling Problems: Hybridization of Evolutionary Algorithms and Fuzzy Logic. Journal of Mathematics and Computers in Simulation, Elsevier.
17. Carlier, J., 1989. An algorithm for solving the job shop problem. Management science, Vol. 35, pp164-176.
18. Hillier, F.S., Lieberman, G.J., 1967. Introduction to Operations Research. Holden-Day, San Fransisco, CA.
19. Dauzère-Pérès,S., Paulli,J. 1997. An integrated approach for modelling and solving the general multiprocessor job-shop scheduling problem using tabu search. Annals of Oper. Res., 70, pp. 281-306.
20. Kacem, I., Hammadi, S., Borne, P., 2002, (c). Bornes Inférieures pour les Problèmes d'Ordonnancement des Job-shop Flexibles. Proceedings of CIFA'02, July 2002, Nantes, FRANCE.
21. Kacem, I., Hammadi, S., Borne, P., 2001, (a). Approach by Localization and Genetic Manipulations Algorithm for Flexible Job-shop Problems. Proceedings of International IEEE Conference on Systems, Man, and Cybernetics, October 7-10, 2001, pp 2599-2604, Tucson, Arizona, USA.

22. Kacem, I., Hammadi, S., Borne, P., 2001, (b). Direct Chromosome Representation and Advanced Genetic Operators for Flexible Job-shop Problems. Proceedings of International CIMCA Conference, pp 123-131, July 9-11, Las Vegas, Nevada, USA.
23. Mesghouni, K., 1999. Application des algorithmes évolutionnistes dans les problèmes d'optimisation en ordonnancement de production. Ph.D Thesis, University of Lille1, January 5, 1999, FRANCE.
24. Kacem, I., Hammadi, S., Borne, P., 2001, (c). Multiobjective Optimization for Flexible Job-Shop Scheduling Problem: Hybridization of Genetic Algorithms with Fuzzy Logic. Proceedings of IFDICON'2001, European Workshop on Intelligent Forecasting, Diagnosis and Control, 24-28 June, 2001, Santorini, GREECE.

Fuzzy Sets based Heuristics for Optimization: Multi-objective Evolutionary Fuzzy Modeling

Fernando Jiménez, Antonio F. Gómez Skarmeta, Gracia Sánchez, and José M. Cadenas

Dpto. Ingeniería de la Información y las Comunicaciones, Universidad de Murcia, 30071 Espinardo, Murcia, Spain

Abstract. Combining fuzzy systems with evolutionary computation has provided very fruitful research results in the last decades. Both technologies are suitable in optimization environment where the decision maker may prefer to obtain satisfactory solutions according to his preferences rather than optimal ones, or when uncertainty appears in the optimization problem or there are not appropriate techniques known to solve it. We deal a wide classical field: Fuzzy Modeling. Interpretability aspects of fuzzy models have received quite some attention in recent years and may be obtained by using transparent rule-structures and well characterized fuzzy membership functions. Moreover, model compactness is important for the interpretability and is related to the number of rules and fuzzy sets. Besides these two criteria, the model accuracy should always be taken into account. In this way, several criteria appear in fuzzy modeling and then multi-objective evolutionary algorithms are suitable to capture several non-dominated solutions in a single run of the algorithm. We describe a multi-objective neuro-evolutionary algorithm that considers all three objectives. The algorithm applies an accuracy criterium and a transparency criterium, based on fuzzy set similarity, while compactness is achieved by a specific technique, incorporated ad hoc within the evolutionary alrorithm. Results are shown for an approximation problem studied before by other authors.

1 Introducction

In recent years, fuzzy modeling, as a complement to conventional modeling techniques, has become an active research topic and has found successful applications in many areas. Evolutionary Algorithms (EAs) [4] have been applied to learn both the antecedent and consequent part of fuzzy rules, and models with both fixed and varying number of rules have been considered [25,9]. Also, EAs have been combined with other techniques like fuzzy clustering [7,5] and neural networks [10,17]. This has resulted in many complex algorithms and, as recognized in [23] and [18], often the transparency and compactness of the resulting rule base are not considered to be of importance; accuracy aspect prevails and interpretability aspects are partly ignored. In such cases, the fuzzy model becomes a black-box, and one can question the rationale for applying fuzzy modeling instead of other techniques.

Transparency and model interpretability for data-based fuzzy models received quite some interest in recent literature [15,13,2,14]. Based on this literature, we introduce three important criteria to be optimized in fuzzy model identification: compactness, transparency and accuracy. Compactness is related to the size of the model, i.e. the number of rules, the number of fuzzy sets and the number of inputs for each rule, while transparency is related to linguistic interpretability of the fuzzy sets and locality of the rules [24,18]. Often one is interested in the local behavior of the global nonlinear model. Such information may be obtained by constraining the model structure during identification or by using these criteria by multi-objective optimization techniques, like multi-objective EAs [1,11]. Most evolutionary approaches to multi-objective fuzzy modeling consist of multiple EAs, usually designed to achieve a single task each, which are then applied sequentially. In these cases, each EA optimizes the model attending to a single criterion separately, which is an impediment for global search. Simultaneous optimization of all criteria is more appropriate. Therefore, others used approaches based on classical multi-objective techniques in which multiple objectives are aggregated into a single function to be optimized [5,16]. In this way, the EA obtains a compromise solution that consists of the weighted criteria.

Another promising method to handle multi-criteria optimization problems is the multi-objective EA based on the Pareto optimality notion, in which all objectives are optimized simultaneously to find multiple non-dominated solutions in a single run of the EA [3,6,20]. The solution for a such a multi-objective optimization problem is a set of so-called Pareto optimal solutions. A Pareto-based multi-objective EA incorporates the Pareto concept to identify multiple solutions through a single run of the algorithm. This practically leaves obsolete the classical tendency to aggregate the different objectives using a weight vector or similar approach to obtain a single function which is then optimized. In that case, the method often requires several executions of the EA with different weights, in order to identify one Pareto solution in each run. In this aspect, multi-objective optimization, based on the Pareto-optimality concept, distinguishes itself from related optimization methods, like gradient techniques, simulated annealing or neural networks.

Pareto-based multi-objective evolutionary approaches can also be considered from the fuzzy modeling perspective [8,12]. The advantage of the the classical optimization approach with aggregated objectives, a single solution is obtained without further interaction with the decision maker. However, it may often be difficult to define a good aggregation function. Then, if the solution cannot be accepted, new runs of the EA are required until a satisfying solution is found. The advantages of the Pareto approach are that no aggregation function need to be defined, and that the decision maker can choose the most appropriate solution according to the current decision environment at the end of the EA run. Moreover, if the decision environment changes, it is not always necessary to run the EA again. Another solution may be chosen out of the family of non-dominated solutions that has already been obtained.

In this paper, we propose a multi-objective neuro-EA to find multiple non-dominated solutions for fuzzy modeling problems. In section 2, the fuzzy model is defined and in section 3 and 4, techniques to improve transparency and compactness of rule set and training are approached respectively. The criteria taken into account (accuracy, tranparency and compactness) are discussed in section 5. The main components of the multi-objective neuro-EA are described in section 6. Section 7 shows the results obtained for a test problem. Section 8 concludes the paper and indicates lines for future research.

2 Fuzzy model identification

We consider rule-based models of the Takagi-Sugeno (TS) type [22]. The rule consequents are taken to be linear functions of the inputs:

$$R_i : \textbf{If } x_1 \textbf{ is } A_{i1} \textbf{ and} \ldots x_n \textbf{ is } A_{in} \textbf{ then} \\ \hat{y}_i = \zeta_{i1}x_1 + \ldots + \zeta_{in}x_n + \zeta_{i(n+1)} \quad (1)$$

where $i = 1, \ldots, M$, and $\mathbf{x} = (x_1, \ldots, x_n)$ is the input vector, $\hat{y}_i$ is the output of the ith rule, A_{ij} are fuzzy sets defined in the antecedent space by membership functions $\mu_i : \Re \to [0,1]$, $\zeta_{ij} \in \Re$ are the consequent parameters, and M is the number of rules. The total output of the model is computed by aggregating the individual contributions of each rule:

$$\hat{y} = \sum_{i=1}^{M} p_i(\mathbf{x})\hat{y}_i \quad (2)$$

where $p_i(\mathbf{x})$ $i = 1, \ldots, M$ is the normalized firing strength of the ith rule:

$$p_i(\mathbf{x}) = \frac{\prod_{j=1}^{n} \mu_{A_{ij}}(x_j)}{\sum_{k=1}^{M} \prod_{j=1}^{n} \mu_{A_{kj}}(x_j)} \quad (3)$$

where $i = 1, \ldots, M$.

Each fuzzy set A_{ij} is described by an asymmetric gaussian membership function.

$$\mu_{A_{ij}}(x_j) = \begin{cases} \exp\left(-\frac{(c_{ij}-x_j)^2}{2\sigma_{lij}^2}\right) & if \ x_j < c_{ij} \\ \exp\left(-\frac{(x_j-c_{ij})^2}{2\sigma_{rij}^2}\right) & if \ x_j \geq c_{ij} \end{cases} \quad (4)$$

where $i = 1, \ldots, M$ and $j = 1, \ldots, n$.

This fuzzy model is defined by a radial basis function neural network. The number of neurons in the hidden layer of an RBF neural network is equal to the number of rules in the fuzzy model. The firing strength of the ith neuron in the hidden layer matches the firing strength of the ith rule in the fuzzy model. We apply an asymmetric gaussian membership function defined by three parameters, the center c, left variance σ_l and rigth variance σ_r. Therefore, each neuron in the hidden layer has these three parameters that define its firing strength value. The neurons in the output layer perform the computations for the first order linear function described in the consequents of the fuzzy model, therefore, the ith neuron of the output layer has the parameters ζ_i that correspond to the linear function defined in the ith rule of the fuzzy model.

3 A technique to improve transparency and compactness of the fuzzy rule sets

Automated approached to fuzzy modeling often introduce redundancy in terms of several similar fuzzy sets that describe almost the same region in the domain of some variable. According to some similarity measure, two or more similar fuzzy sets can be merged to create a new fuzzy set representative for the merged sets. This new fuzzy set substitutes the ones merged in the rule base. On the other hand, if there are two fuzzy sets which are similar, but not very similar, the best approach is to split the fuzzy sets, so that their similarity improves. The merging-splitting process is repeated until fuzzy sets for each model variable cannot be merged, i.e., they are not similar. This process may results in several identical rules, which are removed from the rule set.

We consider the following similarity measure between two fuzzy sets A and B:

$$S(A,B) = \max\left\{\frac{A \cap B}{A}, \frac{A \cap B}{B}\right\} \quad (5)$$

If $S(A,B) > \theta_1$ (we use $\theta_1 = 0.9$), fuzzy sets A and B are merged in a new fuzzy set C as follows:

$$c_C = \alpha c_A + (1-\alpha)c_B \quad (6)$$

$$\sigma_{lC} = \frac{c_C - \min\{c_A - n_\sigma \sigma_{lA}, c_B - n_\sigma \sigma_{lB}\}}{n_\sigma} \quad (7)$$

$$\sigma_{rC} = \frac{\max\{c_A + n_\sigma \sigma_{rA}, c_B + n_\sigma \sigma_{rB}\} - c_C}{n_\sigma} \quad (8)$$

where $n_\sigma = 3$ and $\alpha \in [0,1]$ determines the influence of A and B on the new fuzzy set C:

$$\alpha = \frac{\sigma_{rA} + \sigma_{lA}}{\sigma_{rA} + \sigma_{lA} + \sigma_{rB} + \sigma_{lB}} \quad (9)$$

If $\theta_2 < S(A,B) < \theta_1$ (we use $\theta_2 = 0.6$), fuzzy sets A and B are split as follows:

$$\sigma_C^l = \sigma_C^l(1-\beta) \quad (10)$$

$$\sigma_C^r = \sigma_C^r(1-\beta) \tag{11}$$

where $\beta \in [0,1]$ denotes the amount of splitting between A and B; in our algorithm, we use $\beta = 0.1$.

4 Training of the RBF neural networks

The RBF neural networks associated with the fuzzy models can be trained with a gradient method to obtain more accuracy. However, in order to maintain the transparency and compactness of the fuzzy sets, only the consequent parameters are trained. The training algorithm incrementally updates the parameters based on the currently presented training pattern. The network parameters are updated by applying the gradient descent method to an error function. The error function for the th training pattern is given by the MSE function defined in (15). The update rule is:

$$\zeta_{ij}^{new} = \zeta_{ij}^{old} - \eta \frac{\partial MSE}{\partial \zeta_{ij}} \tag{12}$$

where $i = 1, \ldots, M$, $j = 1, \ldots, n+1$ and η is the learning rate. This rule is applied a number of epochs. Our algorithm use a value of $\eta = 0.01$ and a number of 10 epochs. The detailed derivation of $\Delta \zeta_{ij} = \frac{\partial MSE}{\partial \zeta_{ij}}$ is the following:

$$\Delta \zeta_{ij} = -(\hat{y}_k - y_k)p_i(\mathbf{x})x_j \tag{13}$$

$$\Delta \zeta_{i(n+1)} = -(\hat{y}_k - y_k)p_i(\mathbf{x}) \tag{14}$$

where $i = 1, \ldots, M$, $j = 1, \ldots, n$ and $p_i(\mathbf{x})$ is the firing value for the ith rule defined as in Eq. 3.

5 Criteria for fuzzy modeling

Identification of fuzzy models from data requires de presence of multiple criteria in the search process. In multi-objective optimization, the set of solutions is composed of all those elements of the search space for which the corresponding objective vector cannot be improved in any dimension without degradation in another dimension. These solutions are called *non-dominated* or *Pareto-optimal*. Given two decision vectors **a** and **b** in the universe U, **a** is said to *dominate* **b** if $f_i(\mathbf{a}) \leq f_i(\mathbf{b})$, for all objective functions f_i and $f_j(\mathbf{a}) < f_j(\mathbf{b})$, for at least one objective function f_j, for minimization. A decision vector $\mathbf{a} \in U$ is said to be *Pareto-optimal* if no other decision vector dominates **a**.

In the search for an acceptable fuzzy model, we consider three main criteria: (i) accuracy, (ii) transparency, and (iii) compactness. It is necesary to define quantitative measures for these criteria by means of appropiate objective functions which define the complete fuzzy model identificacion.

The accuracy of a model can be measured with the *mean squared error*:

$$MSE = \frac{1}{K}\sum_{k=1}^{K}(y_k - \hat{y}_k)^2 \qquad (15)$$

where y_k is the actual output and $\hat{y}_k$ is the desired output for the kth input vector, respectively, and K is the number of data samples.

Many measures are possible for the second criterion, transparency. Nevertheless, in this paper we only consider one of the most significant, *similarity*, as a first starting point. The similarity S among distinct fuzzy sets in each variable of the fuzzy model can be expressed as follows:

$$S = \max_{\substack{i,\,j,\,k \\ A_{ij} \neq B_{kj}}} S(A_{ij}, B_{kj}) \qquad (16)$$

whiere $i = 1,\ldots,M$, $j = 1,\ldots,n$ and $k = 1,\ldots,M$.

This is an aggregated similarity measure for the fuzzy rule-based model with the objective to minimize the maximum similarity between the fuzzy sets in each input domain.

Finally, measures for the third criterion, the compactness, are the number of rules, M and the number of different fuzzy sets L of the fuzzy model. We assume that models with a small number of rules and fuzzy sets are compact.

6 Multi-objective neuro-evolutionary algorithm

The main characteristics of the Multi-Objective Neuro-Evolutionary Algorithm are the following:

1. The proposed algorithm is a Pareto-based multi-objective EA for fuzzy modeling, i.e., it has been designed to find, in a single run, multiple non-dominated solutions according to the Pareto decision strategy. There is no dependence between the objective functions and the design of the EAs, thus, any objective function can easily be incorporated. Without loss of generality, the EA minimizes all objective functions.
2. The EA has a variable-length, real-coded representation. Each individual of a population contains a variable number of rules between 1 and max, where max is defined by a decision maker.
3. The initial population is generated randomly with a uniform distribution within the boundaries of the search space, defined by the learning data and model constraints.
4. Constraints with respect to the fuzzy model structure are satisfied by incorporating specific knowledge about the problem. The initialization procedure and variation operators always generate individuals that satisfy these constraints.

5. The EA searchs for among rule sets treated with the technique described in section 3 and trained as defined in section 4, i.e, all individuals in the population have been treated with technique 3 to mantain transparency and trained (after initialization and variation), which is an added ad hoc technique for transparency, compactness and accuracy.
6. Chromosome selection and replacement are achieved by means of a variant of the preselection scheme. This technique is, implicitly, a niche formation technique and an elitist strategy. Moreover, an explicit niche formation technique has been added to maintain diversity respect to the number of rules of the individuals. Survival of individuals is always based on the Pareto concept.
7. The EAs variation operators affect at the individuals at different levels: (i) the rule set level, (ii) the rule level, and (iii) the parameter level.

Representation of solutions An individual for this problem is a rule set of M rules defined by the weights of the RBF neural network. With n input variables, we have for each individual the following parameters:

- centers c_{ij}, left variances σ_{lij} and right variances σ_{rij}, $i = 1, \ldots, M$, $j = 1, \ldots, n$
- coefficients for the linear function of the consequent ζ_{ij}, $i = 1, \ldots, M$, $j = 1, \ldots, n+1$

Initial population The initial population is generated randomly. The number of individuals with M rules, for all $M \in [1, max]$, must be between $minNS$ and $maxNS$ to ensure diversity respect to the number of rules, where $minNS$ and $maxNS$, with $0 \leq minNS \leq PS/max \leq maxNS \leq PS$ (PS is the population size), are the minimum and maximum niche size respectively (see next section).

To generate an individual with M rules, the procedure is as follows: for each fuzzy number A_{ij} ($i = 1, \ldots, M, j = 1, \ldots, n$), three random real values from $[l_j, u_j]$ are generated and sorted to satisfy the constraints $c_{lij} \leq c_{ij} \leq c_{rij}$ ($\sigma_{lij} = (c_{ij} - c_{lij})/n_\sigma$ and $\sigma_{rij} = (c_{rij} - c_{ij})/n_\sigma$). Parameters ζ_{ij} ($i = 1, \ldots, M, j = 1, \ldots, n+1$) are real values generated at random from $[l, u]$. After, the individual is simplified according to the procedure described in a previous section.

Variation operators As already said, an individual is a set of M rules. A rule is a collection of n fuzzy numbers (antecedent) plus $n+1$ real parameters (consequent), and a fuzzy number is composed of three real numbers. In order to achieve an appropiate exploitation and exploration of the potential solutions in the search space, variation operators working in the different levels of the individuals are necessary. In this way, we consider three levels of variation operators: rule set level, rule level and parameter level.

Rule set level variation operators

- **Rule Set Crossover**: This operator interchanges rules. Given two parents $I_1 = (R_1^1 \ldots R_{M_1}^1)$ and $I_2 = (R_1^2 \ldots R_{M_2}^2)$, two children are produced: $I_3 = (R_1^1 \ldots R_a^1 R_1^2 \ldots R_b^2)$ and $I_4 = (R_{a+1}^1 \ldots R_{M_1}^1 R_{b+1}^2 \ldots R_{M_2}^2)$, where $a = round(\alpha \cdot M_1)$ and $b = round((1-\alpha) \cdot M_2)$. The number of rules of the children is between M_1 and M_2.
- **Rule Set Increase Crossover**: This operator increases the number of rules of the two children as follows: the first child contains all M_1 rules of the first parent and $\min\{\max - M_1, M_2\}$ rules of the second parent; the second child contains all M_2 rules of the second parent and $\min\{\max - M_2, M_1\}$ rules of the first parent.
- **Rule Set Mutation**: This operator deletes or adds, both with equal probability, one rule in the rule set. For deletion, one rule is randomly deleted from the rule set. For rule-addition, one rule is randomly generated, according to the initialization procedure described, and added to the rule set.

Rule Level Variaton Operators

- **Rule Arithmetic Croossover**: Performs arithmetic crossover of two random rules. Given two parents $I_1 = (R_1^1 \ldots R_i^1 \ldots R_{M_1}^1)$ and $I_2 = (R_1^2 \ldots R_j^2 \ldots R_{M_2}^2)$, this operator produces two children $I_3 = (R_1^1 \ldots R_i^3 \ldots R_{M_1}^1)$ and $I_4 = (R_1^2 \ldots R_j^4 \ldots R_{M_2}^2)$, with $R_i^3 = \alpha R_i^1 + (1-\alpha) R_j^2$ and $R_j^4 = \alpha R_j^2 + (1-\alpha) R_i^1$, where i,j are random indexes from $[1, M_1]$ and $[1, M_2]$ respectively.
- **Rule Uniform Crossover**: Performs uniform crossover of two random rules. Given two parents $I_1 = (R_1^1 \ldots R_i^1 \ldots R_{M_1}^1)$ and $I_2 = (R_1^2 \ldots R_j^2 \ldots R_{M_2}^2)$, this operator produce two children $I_3 = (R_1^1 \ldots R_i^3 \ldots R_{M_1}^1)$ and $I_4 = (R_1^2 \ldots R_j^4 \ldots R_{M_2}^2)$, where R_i^3 and R_j^4 are obtained with the *uniform* crossover.

Parameter level variation operators

- **Arithmetic Crossover**: Given two parents, and one rule of each parent randomly chosen, this operator performs an arithmetic cross of the fuzzy numbers corresponding to a random input variable or the consequent parameters.
- **Non-Uniform Mutation**: This operator changes the value of one of the antecedent fuzzy sets of a random fuzzy number, or a parameter of the consequent ζ, of a randomly chosen rule. The new value of the parameter is generated at random within the constraints given by a non-uniform mutation.
- **Uniform Mutation**: Similar to former, but within the constraints given by an uniform mutation.

- **Small Mutation**: Similar to former, but within the constraints given by an small mutation. The small mutation produced an small change in the individual and it is suitable for fine tuning of the real parameters.

Selection and generational replacement In each iteration, the neuro-EA executes the following steps:

1. Two individuals are picked at random from the population.
2. These individuals are crossed and mutated to produce two offspring.
3. Performs technique 3 in the offspring.
4. Performs the training in the offspring.
5. The first offspring replaces the first parent and the second offspring replaces the second parent only if:
   - the offspring is betther than the parent and
   - the number of rules of the offspring is equal to the number of rules of the parent, or the niche count of the parent is greater than $minNS$ and the niche count of the offspring is smaller than $maxNS$.

An individual I is better than another individual J if I dominates J. The niche count of an individual I is the number of individuals in the population with the same number of rules as I. The preselection scheme is an implicit niche formation technique to maintain diversity in the population because an offspring replaces an individual similar to itself (one of their parents). Implicit niche formation techniques are more appropiate for fuzzy modeling than explicit techniques, such as sharing function, which can provoke an excessive computational time. However, we need and additional mechanism for diversity with respect to the number of rules of the individuals in the population. The added explicit niche formation technique ensures that the number of individuals with M rules, for all $M \in [1, max]$, is greater or equal to $minNS$ and smaller or equal to $maxNS$. Moreover, the preselection scheme is also an elitist strategy because the best individual in the population is replaced only by a better one.

Optimization model and decision process After preliminary experiments in which we have checked different optimization models, the following remarks can be maded:

1. Instead of minimizing of the number of rules M we have decided to search for rules sets with a number of rules within an interval $[1, max]$ where a decision maker can feel comfortable. The explicit niche formation technique ensures the EA always contains a minimum of representative rule sets for each number of rules in the populations. Then, we do not minimize the number of rules during the optimization, but we will take it into account at the end of the run, in a posteriori decision process applied to the last population.

2. It is very important to note that a very transparent model will be not accepted by a decision maker if the model is not accurate. In most fuzzy modeling problems, excessively low values for similarity hamper accuracy, for which these models are normally rejected. Alternative decision strategies, as *goal programming*, enable us to reduce the domain of the objective functions according to the preferences of a decision maker. Then, we can impose a goal g_S for similarity, which stop minimization of the similarity in solutions for which goal g_S has been reached.
3. The measures M (number of rules) and L (number of different fuzzy sets) are reduced by the technique of section 3. So, we do not define an explicit objective function to minimize M and L.

According to the previous remarks, we finally consider the following optimization model:

$$Minimize\ f_1 = MSE \\ Minimize\ f_2 = \max(g_s, S) \qquad (17)$$

At the end of the run, we consider the following a posteriori decision process applied to the last population to obtain the final compromise solution:

1. Identify the set $X^* = \{x_1^*, \ldots, x_p^*\}$ of non-dominated solutions according to:

$$Minimize\ f_1 = MSE \\ Minimize\ f_2 = S \qquad (18) \\ Minimize\ f_3 = M$$

2. Choose from X^* the most accurate solution x_i^*; remove x_i^* from X^*;
3. If solution x_i^* is not accurate enough or there is no solution in the set X^* then STOP (no solution satisfies);
4. If solution x_i^* is not transparent or compact enough then go to step 2;
5. Show the solution x_i^* as output.

Computer aided inspection shown in Figure 3 can help in decisions for steps 3 and 4.

7 Experiments and results

We consider the modelling of the rule base given in [21] and shown in Fig. 1. The corresponding surface is shown in Fig. 2. In [19] a model with four rules was identified from sampled data ($N = 546$) by the supervised clustering algorithm, which was initialized with 12 clusters. This model was optimized using a Genetic Algorithm to result in a MSE of 1.6.

Table 1 shows results obtained with the Pareto based multi-objective neuro-EAs. The following values for the parameters were used in the simulations: $PS = 100$, $minNS = 5$, $maxNS = 30$, cross probability 0.9, mutation probability 0.9, $g_S = 0.25$, and $max = 5$.

We finally choose a compromise solution (4-rules fuzzy model). Figure 3 shows the local model, the surface generated by the model, fuzzy sets for each variable and the prediction error.

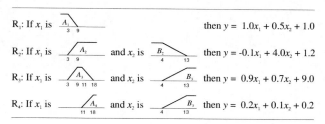

R_1: If x_1 is A_1 [3 9] then $y = 1.0x_1 + 0.5x_2 + 1.0$

R_2: If x_1 is A_2 [3 9] and x_2 is B_2 [4 13] then $y = -0.1x_1 + 4.0x_2 + 1.2$

R_3: If x_1 is A_3 [3 9 11 18] and x_2 is B_3 [4 13] then $y = 0.9x_1 + 0.7x_2 + 9.0$

R_4: If x_1 is A_4 [11 18] and x_2 is B_3 [4 13] then $y = 0.2x_1 + 0.1x_2 + 0.2$

Fig. 1. Modeling of the rule base given in [21]

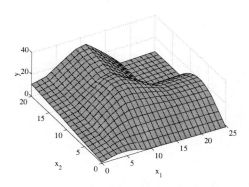

Fig. 2. Real surface for the example in [21].

Table 1. Non-dominated solutions according to (18) obtained with the multi-objective neuro-EA.

M	L	MSE	S
1	2	125.391	0.0
2	4	25.606	0.348
3	5	2.187	0.349
4	5	1.017	0.349
5	5	0.910	0.350

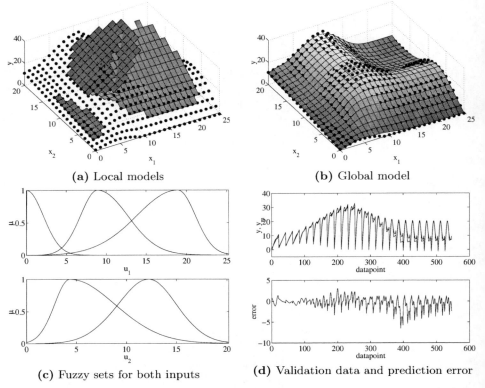

Fig. 3. Accurate, transparent and compact fuzzy model for the example in [21].

8 Conclusions and future research

In this paper we present a Pareto-based multi-objective neuro-evolutionary algorithm to obtain interpretable fuzzy models. Criteria such as accuracy, transparency and compactness have been propose and are taken into account in the optimization process. Some of these criteria have been partially incorporated into the EA by means of ad hoc techniques. Advantages of gaussian fuzzy sets arise with the possibility of training the RBF neural networks associated with the fuzzy models in order to obtain more accuracy. In addition, several new ideas to reduce computatinal load and improve the global search capabilities, have been incorporated in the evolutionary algorithm. An implicit niche formation technique (preselection) in combination with other explicit techniques with low computational costs have been used to maintain diversity. These niche formation techniques are appropriate in fuzzy modeling if excessive amount of data are required. Elitism is also implemented by means of the preselection technique. A goal based approach has been proposed to help to obtain more accurate fuzzy models. The main difference between the proposed EAs and other approaches for fuzzy modeling is the reduced complexity because we use a single EA for generating, tuning and simplification of the fuzzy model. Moreover, human intervention is only required at the end of the run in choosing one of the multiple non-dominated solutions.

Results obtained are good in comparison with other iterative techniques reported in literature, with the advantage that the proposed techniques identifies a set of alternative solutions.

In future works, the proposed algorithm will be used in real world applications by means of research projects, particularly under PANDORA Project. This proposal designs, implements and evaluates a platform for the experimentation with mobile robots based on the use of behaviour component learning and it fusion by means of intelligent techniques.

Acknowledgement

The authors thank the Ministerio de Ciencia y Tecnología (MCyT) and the Fondo Europeo de Desarrollo Regional (FEDER) for the support given to this work under the project TIC2001-0245-C02-01.

References

1. Coello, C.A. (2002). *List of references on multi-objective evolutionary optimization.* http://www.lania.mx/~ccoello/EMOO/EMOObib.html.
2. Cordón, O., Herrera, F. (2000). *A proposal for improving the accuracy of linguistic modeling.* IEEE Transactions on Fuzzy Systems, 8(3):335-344.
3. Fonseca, C.M., Fleming, P.J. (1995). *An overview of evolutionary algorithms in multiobjective optimization*, Evolutionary Computation, 3(1):1-16.
4. Goldberg, D.E. (1989). *Genetic Algorithms in Search, Optimization, and Machine Learning.* Addison-Wesley, 1989.
5. Gómez-Skarmeta, A.F., Jiménez, F. (1999). *Fuzzy modeling with hybrid systems.* Fuzzy Sets and Systems, 104:199-208.
6. Horn, J., Nafpliotis, N (1993). *Multiobjective optimization using the niched pareto genetic algorithm.* IlliEAL Report No. 93005.
7. Hwang, H.S. (1998). *Control strategy for optimal compromise between trip time and energy comsumption in a high-speed railway.* IEEE Transactions on Systems, Man and Cybernetics, Part A: Systems and Humans, 28(6):791-802.
8. Ishibuchi, H., Murata, T., Trksen, I. (1997). *Single-objective and two-objective genetic algorithms for selecting linguistic rules for pattern classification problems.* Fuzzy Sets and Systems, 89:135-150.
9. Ishibuchi, H., Nakashima, T., Murata, T. (1999). *Performance evaluation of fuzzy classifier systems for multidimensional pattern classification problems.* IEEE Transactions on Systems, Man, and Cubernetics - Part B: Cybernetics, 29/5):601-618.

10. Jagielska, I., Matthews, C., Whitfort, T. (1999). *An investigation into the application of neural networks, fuzzy logic, genetic algorithms, and rough sets to automated knowledge acquisition for classification problems.* Neurocomputing, 24:37-54.
11. Jiménez, F., Verdegay, J.L., Gómez-Skarmeta, A.F. (1999). *Evolutionary techniques for constrained multi-objective optimization problems.* Proceeding of the Genetic and Evolutionary Computation Conference (GECCO-99), Workshop on Multi-Criterion Optimization Using Evolutionary Methods, Orlando, Florida, USA, pp. 115-116.
12. Jiménez, F., Gómez-Skarmeta, A.F., Roubos, H., Babuska, R. (2001). *Accurate, Transparent, and Compact Fuzzy Models for Function Approximation and Dynamic Modeling Through Multi-objective Evolutionary Optimization.* Lectures Notes in Computer Science, Evolutionary Multi-Criterion Optimization, 1993:653-667.
13. Jin, Y. (2000). *Fuzzy Modeling of High-Dimensional Systems.* Transactions on Fuzzy Systems: Complexity Reduction and Interpretability Improvement. 8:212-221.
14. Johansen, T.A., Shorten, R., Murray-Smith, R. (2000). *On the interpretation and identification of dynamic Takagi-Sugeno fuzzy models.* IEEE Transactions on Fuzzy Systems, 8(3):297-313.
15. H. Pomares, H., Rojas, I., Ortega, J., Gonzalez, J., Prieto, A. (2000). *A systematic approach to a self-generating fuzzy rule-table for function approximation.* IEEE Transactions on Systems, Man, and Cybernetics – Part B: Cybernetics, 30(3):431-447.
16. Roubos, J.A., Setnes, M. (2001). *Compact and transparent fuzzy models and classifiers through iterative complexity reduction.* IEEE Transactions on Fuzzy Systems, 9(4):516-524.
17. Russo, M. (1998). *FuGeNeSys - a fuzzy genetic neural system for fuzzy modeling.* IEEE Transactions on Fuzzy Systems, 6(3):373-388.
18. Setnes, M., Babuška, R., Verbruggen, H.B. (1998). *Rule-Based Modeling: Precision and Transparency.* IEEE Transactions on Systems, Man and Cybernetics, Part C: Applications & Reviews, 28:165-169.
19. Setnes, M., Roubos, J.A. (1999). *Transparent Fuzzy Modeling using Fuzzy Clustering and GA's.* 18th International Conference of the North American Fuzzy Information Processing Society, 198-202.
20. Srinivas, N., Deb, K. (1995). *Multiobjective optimization using nondominated sorting in genetic algorithms.* Evolutionary Computation, 2(3),221-248.
21. Sugeno, M., Kang, G.T. (1998). *Structure identification of fuzzy model.* Fuzzy Sets and Systems, 28:15-23.
22. Takagi, T., Sugeno, M. (1985). *Fuzzy identification of systems and its application to modeling and control.* IEEE Transactions on Systems, Man and Cybernetics, 15:116-132.
23. Valente de Oliveira, J. (1999). *Semantic constraints for membership function optimization.* IEEE Transactions on Fuzzy Systems, 19(1):128-138.
24. Valente de Oliveira, J. (1999). *Semantic constraints for membership function optimization.* IEEE Transactions on Fuzzy Systems, 19(1):128-138.
25. Wang, C.H., Hong, T.P., Tseng, S.S. (1998). *Integrating fuzzy knowledge by genetic algorithms.* Fuzzy Sets and Systems, 2(4):138-149.

An Interactive Fuzzy Satisficing Method for Multiobjective Operation Planning in District Heating and Cooling Plants through Genetic Algorithms for Nonlinear 0-1 Programming

Masatoshi Sakawa[1], Kosuke Kato[1], Satoshi Ushiro[1,2], and Mare Inaoka[1]

[1] Hiroshima University, Higashi-Hiroshima, 739-8527, Japan
[2] Shinryo Corporation, Shinjyuku, Tokyo, 160-0004, Japan

Abstract. In recent years, operation planning techniques of district heating and cooling (DHC) plants have drawn considerable attention as a result of development of cooling load prediction methods for DHC systems. In the present paper, we formulate an operation planning problem of a DHC plant as a nonlinear 0-1 programming problem. Furthermore, in order to reflect actual decision making situations more appropriately, we reformulate it as a multiobjective nonlinear 0-1 programming problem. For the formulated multiobjective nonlinear 0-1 programming problem, we attempt to derive a satisficing solution for the decision maker through an interactive fuzzy satisficing method, where single-objective nonlinear 0-1 programming problems are repeatedly solved. Realizing that operation planning problems in actual plants may involve hundreds of variables and there does not exist an efficient exact solution method for nonlinear 0-1 programming problems, we propose an approximate solution method using genetic algorithms for the nonlinear 0-1 programming problem.

1 Introduction

District heating and cooling (DHC) systems have been actively introduced as an energy supply system in urban areas for the purpose of saving energy, saving space, inhibiting air-pollution or preventing city disaster. In a DHC system, cold water and steam used for air-conditioning at all facilities in a certain district are made and supplied by a DHC plant.

Since there exist a number of large-size freezers and boilers in a DHC plant, the control under an operation plan for these instruments on the basis of the amount of cold water and steam, called cooling load, is important for stable and economical management of a DHC system.

In recent years, with the improvement of cooling load prediction methods for DHC systems [1,2], the needs of the formulation of an operation planning problem of a DHC plant as a mathematical programming one and the development of solution methods to the problem has been increasing [3,4]. The authors also published literatures about the formulation of the operation planning of DHC plants and the proposition of practical solution methods based on genetic algorithms to the formulated problems. In these papers, while only the minimization of the running cost is treated as an objective, no other objective such as the minimization of the pollution is not considered.

Under these circumstances, in this paper, we formulate an operation planning problem of a DHC plant as a multiobjective nonlinear 0-1 programming problem to reflect the requirement that not only the running cost but the surplus of cold water and steam should be minimized. For the formulated problem, after incorporating fuzzy goals of the decision maker for the objective functions, we propose an interactive fuzzy satisficing method through genetic algorithms to derive a satisficing solution for the decision maker from the extended Pareto optimal solution set. An illustrative numerical example based on an actual plant data shows the feasibility of the proposed method.

2 Operational Planning of a DHC Plant

2.1 Structure of a DHC plant

In a DHC plant, cold water, steam and electricity are generated by running many instruments using gas and electricity.

Relations among instruments in a DHC plant are depicted in Fig. 1. From Fig. 1, it can be seen that steam required for heating and cold water required for cooling are generated by running four absorbing freezers ($DAR_1, \ldots, DAR_4$ where $DAR_1 \sim DAR_2$ are the same type), six turbo freezers ($ER_1, \ldots, ER_6$ where $ER_1 \sim ER_4$ are the same type) and three boilers ($BW_1, \ldots, BW_3$ where $BW_1 \sim BW_2$ are the same type) using gas and electricity in this DHC plant, where pumps and cooling towers are connected with the corresponding freezers.

For such a DHC plant, we consider the operational plan to minimize the cost of gas and electricity under the condition that the demand for cold water and steam must be supplied by running boilers and freezers.

2.2 Formulation

Given the (predicted) amount of the demand for cold water C_{load}^t and that for steam S_{dist}^t at time t, the operation planning problem of the DHC plant can be summarized as follows.

(I) The problem contains 14 decision variables. They are all 0-1 variables as $x_1^t, \ldots, x_{10}^t$ which indicate whether each of four absorbing freezers and six turbo freezers is running or not, $y_1^t, \ldots, y_3^t$ which indicate whether each of three boilers is running or not and z^t which indicates whether some condition holds or not.

(II) The freezer output load rate $P = C_{\text{load}}^t / C^t$, the ratio of the (predicted) amount of the demand for cold water $C_{\text{load}}(t)$ to the total output of running freezers $C^t = \sum_{i=1}^{10} a_i x_i^t$, must be less than or equal to 1.0, i.e.,

$$C^t \geq C_{\text{load}}^t \quad (1)$$

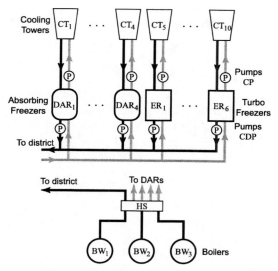

Fig. 1. Structure of a DHC plant.

where a_i denotes the rating output of the ith freezer. This constraint means that the sum of output of running freezers must exceed the (predicted) amount of the demand for cold water.

(III) The freezer output load rate $P = C_{\text{load}}^t / C^t$ must be greater than or equal to 0.2, i.e.,

$$0.2 \cdot C^t \leq C_{\text{load}}^t, \tag{2}$$

which means that the sum of output of running freezers must not exceed five times the (predicted) amount of the demand for cold water.

(IV) The boiler output load rate $Q = (S_{\text{DAR}}^t + S_{\text{dist}}^t)/S^t$, the ratio of the (predicted) amount of the demand for steam to the total output of running boilers $S^t = \sum_{j=1}^{3} f_j y_j^t$, must be less than or equal to 1.0, i.e.,

$$-S_{\text{DAR}}^t + S^t \geq S_{\text{dist}}^t \tag{3}$$

where f_j is the rating output of the jth boiler and S_{DAR}^t is the total amount of steam used by absorbing freezers at time t, defined as

$$S_{\text{DAR}}^t = \sum_{i=1}^{4} \theta(P) \cdot S_i^{\max} \cdot x_i \tag{4}$$

where $S_i^{\max}$ is the maximal steam amount used by the ith absorbing freezers. Furthermore, $\theta(P)$ denotes the rate of use of steam in an absorbing freezer, which is a nonlinear function of the freezer output load rate P. For the sake of simplicity, in this paper, we use the following piecewise linear approximation.

$$\theta(P) = \begin{cases} 0.8775 \cdot P + 0.0285 \,, & P \leq 0.6 \\ 1.1125 \cdot P - 0.1125 \,, & P > 0.6 \end{cases} \quad (5)$$

This constraint means that the sum of output of running boilers must exceed the (predicted) amount of the demand for steam.

(V) The boiler output load rate $Q = (S_{\text{DAR}}^t + S_{\text{dist}}^t)/S^t$ must be greater than or equal to 0.2, i.e.,

$$-S_{\text{DAR}}^t + 0.2 \cdot S^t \leq S_{\text{dist}}^t. \quad (6)$$

This constraint means that the sum of output of running boilers must not exceed five times the (predicted) amount of the demand for steam.

(VI) The minimizing objective function $J(t)$ is the energy cost which is the sum of the gas bill G^t and the electricity bill E^t.

$$J(t) = \text{G}_{\text{cost}} \cdot G^t + \text{E}_{\text{cost}} \cdot E^t \quad (7)$$

where G_{cost} and E_{cost} denote the unit cost of gas and that of electricity. The gas bill G^t is defined by the gas amount consumed in the rating running of a boiler g_j, $j = 1, 2, 3$ and the boiler output load rate Q.

$$G^t = \left(\sum_{j=1}^{3} g_j y_j\right) \cdot Q \quad (8)$$

On the other hand, E^t is defined as the sum of electricity amount consumed by turbo freezers, accompanying cooling towers and pumps.

$$E^t = E_{\text{ER}}^t + E_{\text{CT}}^t + E_{\text{CP}}^t + E_{\text{CDP}}^t$$
$$= \sum_{i=5}^{10} \xi(P) \cdot E_i^{\max} \cdot x_i^t + \sum_{i=1}^{10} c_i x_i^t + \sum_{i=1}^{10} d_i x_i^t + \sum_{i=1}^{10} e_i x_i^t \quad (9)$$

where $E_i^{\max}$ is the maximal electricity amount used by ER_i, c_i, d_i and e_i are the electricity amount of cooling tower and two kinds of pumps.

In the above equation, $\xi(P)$ denotes the rate of use of electricity in a turbo freezer, which is a nonlinear function of the freezer output load rate P. For the sake of simplicity, in this paper, we use the following piecewise linear approximation.

$$\xi(P) = \begin{cases} 0.6 \cdot P + 0.2 \,, & P \leq 0.6 \\ 1.1 \cdot P - 0.1 \,, & P > 0.6 \end{cases} \quad (10)$$

Accordingly, the operation planning problem is formulated as the following nonlinear 0-1 programming problem.

Problem $P(t)$
minimize

$$J(\boldsymbol{x}^t, \boldsymbol{y}^t, z^t) = \text{G}_{\text{cost}} \cdot \left(\sum_{j=1}^{3} g_j y_j^t \right) \cdot Q$$

$$+ \text{E}_{\text{cost}} \cdot \left\{ z^t \cdot \Xi_1(P) + (1 - z^t) \cdot \Xi_2(P) + \sum_{i=1}^{10} c_i x_i^t + \sum_{i=1}^{10} d_i x_i^t + \sum_{i=1}^{10} e_i x_i^t \right\}$$

subject to

$$(1 - z^t) \cdot (-C^t + C_{\text{load}}^t) \leq 0 \quad (11)$$
$$z^t \cdot (0.2 \cdot C^t) + (1 - z^t) \cdot (0.6 \cdot C^t) \leq C_{\text{load}}^t \quad (12)$$
$$z^t \cdot (-0.6 \cdot C^t + C_{\text{load}}^t) \leq 0 \quad (13)$$
$$z^t \cdot \Theta_1(P) + (1 - z^t) \cdot \Theta_2(P) - S^t \leq -S_{\text{dist}}^t \quad (14)$$
$$-z^t \cdot \Theta_1(P) - (1 - z^t) \cdot \Theta_2(P) + 0.2 \cdot S^t \leq S_{\text{dist}}^t \quad (15)$$
$$x_i^t \in \{0,1\}, \ i = 1, \ldots, 10, \ y_j^t \in \{0,1\}, \ j = 1, \ldots, 3, \ z^t \in \{0,1\} \quad (16)$$

where

$$C^t = \sum_{i=1}^{10} a_i x_i^t$$

$$S^t = \sum_{j=1}^{3} f_j y_j^t$$

$$P = \frac{C_{\text{load}}^t}{C^t}$$

$$\Theta_1(P) = \sum_{i=1}^{4} (0.8775 \cdot P + 0.0285) \cdot S_i^{\max} \cdot x_i^t$$

$$\Theta_2(P) = \sum_{i=1}^{4} (1.1125 \cdot P - 0.1125) \cdot S_i^{\max} \cdot x_i^t$$

$$\Xi_1(P) = \sum_{i=5}^{10} (0.6 \cdot P + 0.2) \cdot E_i^{\max} \cdot x_i^t$$

$$\Xi_2(P) = \sum_{i=5}^{10} (1.1 \cdot P - 0.1) \cdot E_i^{\max} \cdot x_i^t$$

$$Q = \{ z^t \cdot \Theta_1(P) + (1 - z^t) \cdot \Theta_2(P) + S_{\text{dist}}^t \}$$

and $z^t = 1$, $z^t = 0$ mean $P \leq 0.6$, $P > 0.6$, respectively. In the following, let $\boldsymbol{\lambda}^t = \left((\boldsymbol{x}^t)^T, (\boldsymbol{y}^t)^T, z^t \right)^T$ and let Λ^t, $\ell_i^t(\boldsymbol{\lambda}^t)$ and r_i^t, $i = 1, \ldots, 5$ represent the feasible region of $P(t)$, left sides and right ones of constraints (11) $\sim$ (15), respectively.

The problem $P(t)$ can be exactly solved by complete enumeration since it includes only fourteen 0-1 variables. However, a multi-period operation planning made by pasting K solutions to $P(t), P(t+1), \ldots, P(t+K-1)$ solved independently at each time t together often becomes such an unnatural one that the switching of instruments occurs frequently. Since the starting and stopping of instruments need more electricity and manpower than the continuous running does, the additional cost for these operations should be took into account in optimizing a multi-period operation planning problem.

Thus, in this paper, we formulate an extended operation planning problem in consideration of the starting and stopping of instruments. To be more specific, we deal with the following problem $P'(t, K)$ gathering K periods.

<u>Extended problem $P'(t, K)$</u>

$$\underset{\boldsymbol{\lambda}(t,K) \in \Lambda(t,K)}{\text{minimize}} \quad J'(\boldsymbol{\lambda}(t, K)) = \sum_{\tau=t}^{t+K-1} \left[J(\boldsymbol{\lambda}^\tau) + \sum_{j=1}^{13} \phi_j \left| \lambda_j^\tau - \lambda_j^{(\tau-1)} \right| \right] \quad (17)$$

where let $\boldsymbol{\lambda}(t, K) = \left((\boldsymbol{\lambda}^t)^T, \ldots, (\boldsymbol{\lambda}^{t+K-1})^T \right)^T$, $\Lambda(t, K) = \Lambda^t \times \cdots \times \Lambda^{t+K-1}$ and ϕ_j is the cost of switching of the jth instrument.

Furthermore, in order to reflect the requirement in a practical operation planning, we reformulate the extended problem as the following multiobjective nonlinear 0-1 programming problem aiming to simultaneously minimize the running cost which is the sum of the energy cost and the switching cost, the surplus of cold water and the surplus of steam.

<u>Multiobjective extended problem $\boldsymbol{P'}(t, K)$</u>

minimize

$$J_1(\boldsymbol{\lambda}(t, K)) = \sum_{\tau=t}^{t+K-1} \left[J(\boldsymbol{\lambda}^\tau) + \sum_{j=1}^{13} \phi_j \left| \lambda_j^\tau - \lambda_j^{(\tau-1)} \right| \right]$$

minimize

$$J_2(\boldsymbol{\lambda}(t, K)) = \sum_{\tau=t}^{t+K-1} (C^\tau - C_{\text{load}}^\tau)$$

minimize

$$J_3(\boldsymbol{\lambda}(t, K)) = \sum_{\tau=t}^{t+K-1} (z^\tau \cdot \Theta_1(P) + (1 - z^\tau) \cdot \Theta_2(P) - S^\tau - S_{\text{dist}}^\tau)$$

subject to $\quad \boldsymbol{\lambda}(t, K) \in \Lambda(t, K)$ \hfill (18)

where the objective functions $J_1(\boldsymbol{\lambda}(t, K))$, $J_2(\boldsymbol{\lambda}(t, K))$ and $J_3(\boldsymbol{\lambda}(t, K))$ mean the running cost, the difference between the production amount of cold water and its predicted demand and the difference between the production amount of steam and its predicted demand, respectively.

In order to consider the imprecise nature of the decision maker's judgments for each objective function $J_i(\cdot)$, if we introduce the fuzzy goals such as "$J_i(\cdot)$ should be substantially less than or equal to p_i", the multiobjective extended problem can be rewritten as:

$$\underset{\boldsymbol{\lambda}(t,K)\in \Lambda(t,K)}{\text{minimize}} \quad (\mu_1(J_1(\boldsymbol{\lambda}(t,K))), \mu_2(J_2(\boldsymbol{\lambda}(t,K))), \mu_3(J_3(\boldsymbol{\lambda}(t,K)))) \quad (19)$$

where $\mu_i(J_i(\cdot))$ are membership functions to quantify the fuzzy goals.

Since the problem (19) is regarded as a fuzzy multiobjective decision making problem, there rarely exist a complete optimal solution that simultaneously optimizes all objective functions. As a reasonable solution concept for the fuzzy multiobjective decision making problem, M. Sakawa et al. [5,6] defined M-Pareto optimality on the basis of membership function values by directly extending the Pareto optimality in the ordinary multiobjective programming problem.

Definition 1 (M-Pareto optimal solution). $\boldsymbol{\lambda}^*(t,K) \in \Lambda(t,K)$ is said to be an M-Pareto optimal solution if and only if there does not exist another $\boldsymbol{\lambda}(t,K) \in \Lambda(t,K)$ such that $\mu_i(J_i(\boldsymbol{\lambda}(t,K))) \geq \mu_i(J_i(\boldsymbol{\lambda}^*(t,K)))$ for all i and $\mu_j(J_j(\boldsymbol{\lambda}(t,K))) > \mu_j(J_j(\boldsymbol{\lambda}^*(t,K)))$ for at least one j.

In general, there exist so many M-Pareto optimal solutions to a fuzzy multiobjective decision making problem, it is not easy for the decision maker to obtain a satisficing solution for him. Thereby, we introduce an interactive fuzzy satisficing method, developed by M. Sakawa et al. [5,6] as a tool to generate a candidate for the satisficing solution which is M-Pareto optimal.

3 An Interactive Fuzzy Satisficing Method

An interactive fuzzy satisficing method proposed by M. Sakawa et al. can derive a satisficing solution for the decision maker by eliciting the local preference information from it through the interaction with it. Here, we summarize the algorithm as follows.

Step 1 Calculate the (approximate) minimum $\underline{J}_i$ and the (approximate) maximum $\overline{J}_i$ of each of objective functions $J_i(\boldsymbol{\lambda}(t,K))$, $i = 1, 2, 3$ from the multi-period operation planning made by pasting together K solutions to $P(t), P(t+1), \ldots, P(t+K-1)$ solved by the complete enumeration independently at each time t.

Step 2 Ask the decision maker to specify membership functions $\mu_i(J_i(\boldsymbol{\lambda}(t,K)))$, $i = 1, 2, 3$ based on individual approximate minima and maxima in (1), and set the initial reference membership levels $\bar{\mu}_i$, $i = 1, 2, 3$.

Step 3 For the reference membership levels $(\bar{\mu}_1, \bar{\mu}_2, \bar{\mu}_3)$, solve the corresponding augmented minimax problem as:

$$\operatorname*{minimize}_{\boldsymbol{\lambda}(t,K)\in\Lambda(t,K)} \max_{i=1,2,3}\left\{\Big(\bar{\mu}_i - \mu_i(J_i(\boldsymbol{\lambda}(t,K)))\Big) + \rho \sum_{l=1}^{3}\Big(\bar{\mu}_l - \mu_l(J_l(\boldsymbol{\lambda}(t,K)))\Big)\right\} \quad (20)$$

where, ρ is a sufficiently small positive number.

Step 4 If the decision maker is satisfied with the solution in step 3, which is guaranteed to be M-Pareto optimal, stop. Otherwise, ask the decision maker to update the reference membership levels $\bar{\mu}_i$, $i = 1, 2, 3$ in consideration of the current membership function values and objective function values, and return to step 3.

Since the problem (20) in step 3 is a large-scale nonlinear 0-1 programming problem which involves K times as many variables as $P(t)$ does, we propose a practical approximate solution method through genetic algorithms.

4 Genetic Algorithms for Nonlinear 0-1 Programming

Genetic algorithms (GAs) were initially proposed by Holland, his colleagues and his students at the University of Michigan in the early 1970's, as search algorithms based on the mechanism of natural selection and natural genetics, and have recently attracted considerable attention as a methodology for search, optimization, and learning [7–10].

For multidimensional 0-1 knapsack problems, Sakawa et al. [11] proposed genetic algorithms using double string representation and a decoding algorithm to decode an individual to a feasible solution and showed its effectiveness [12,13]. Furthermore, they [14] proposed genetic algorithms with double strings based on reference solution updating for general 0-1 programming problems involving positive and negative coefficients.

In this paper, we propose a new genetic algorithm using 0-1 string representation corresponding to $\boldsymbol{\lambda}(t, K)$, based on the decoding algorithm using a reference solution and the reference solution updating in [14].

4.1 Coding and decoding

In genetic algorithms for optimization problems, a solution $\boldsymbol{\lambda}(t, K)$ and an individual **S** are generally called a phenotype and a genotype, respectively. A mapping from a phenotype space to a genotype space is called coding, while a mapping from a genotype space to a phenotype space is called decoding. In this paper, we adopt an individual using only '0' and '1' corresponding to a solution $\boldsymbol{\lambda}(t, K)$ to $P'(t, K)$, illustrated in Fig. 2. In an individual **S**, $\mathbf{s}^\tau$ is a subindividual corresponding to a solution $\boldsymbol{\lambda}^\tau$ at time τ, and the basic decoding is represented by $\boldsymbol{\lambda}^\tau = \mathbf{s}^\tau$, $\tau = t, \ldots, t + K - 1$, i.e., $\boldsymbol{\lambda}(t, K) = \mathbf{S}$.

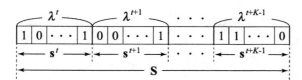

Fig. 2. Individual.

It should be noted that all phenotypes are not always feasible when the problem is constrained. From the viewpoint of search efficiency and feasibility, decoding such that all phenotypes are feasible seems desirable.

Sakawa et al. [11,14] proposed genetic algorithms using double string representation and a decoding algorithm to decode an individual to a feasible solution for multidimensional 0-1 knapsack problems and more general 0-1 programming problems involving positive and negative coefficients. Here, we propose a new decoding algorithm to decode an individual to a feasible solution for nonlinear 0-1 programming problems $P'(t, K)$, based on the decoding algorithm using a reference solution in [14].

In [14], a feasible solution used as a template in decoding, called a reference solution, must be obtained in advance, In order to find a feasible solution to the extended problem $P'(t, K)$, we solve a maximizing problem with the following objective function

$$G(\boldsymbol{\lambda}(t, K)) = \exp\left[-\theta \sum_{\tau=t}^{t+K-1} \sum_{i=1}^{5} R\left(\frac{\ell_i^\tau(\boldsymbol{\lambda}^\tau) - r_i^\tau}{|r_i^\tau|}\right)\right]$$

where

$$R(\xi) = \begin{cases} \xi, & \xi \geq 0 \\ 0, & \xi < 0 \end{cases}$$

and θ is a positive constant. To be more explicit, we find a feasible solution $\boldsymbol{\lambda}^0(t, K)$ to an optimization problem to maximize the following objective function $G(\boldsymbol{\lambda}(t, K))$

$$\text{maximize } G(\boldsymbol{\lambda}(t, K)), \quad \boldsymbol{\lambda}(t, K) \in \Lambda(t, K), \tag{21}$$

by a genetic algorithm which uses a fitness function $f(\mathbf{S}) = G(\boldsymbol{\lambda}(t, K))$ and the basic decoding. Then, let the obtained feasible solution be a reference function in decoding, i.e., $\bar{\boldsymbol{\lambda}} = \boldsymbol{\lambda}^0(t, K)$. If no feasible solution can be found in a prescribed number of generations, we judge the problem is inconsistent.

Using the feasible solution $\boldsymbol{\lambda}^0(t, K)$, the decoding algorithm to decode each of individuals to the corresponding feasible solution to $P'(t, K)$ is summarized as follows. In the algorithm, g_j^τ, $j = 1, \ldots, 14$, $\tau = t, \ldots, t+K-1$ denotes the jth gene of a subindividual $\mathbf{s}^\tau$ of an individual $\mathbf{S}$, and $\bar{\boldsymbol{\lambda}} = \left((\bar{\boldsymbol{\lambda}}^t)^T, \ldots, (\bar{\boldsymbol{\lambda}}^{t+K-1})^T\right)^T$ is a reference solution.

Decoding algorithm using a reference solution

Step 1 Let $\tau := t$.
Step 2 Let $j := 1$, $l := 0$ and $\boldsymbol{\lambda}^\tau := \left(\mathbf{0}^T, \mathbf{0}^T, g_{14}^\tau\right)^T$.
Step 3 Let $\lambda_j^\tau := g_j^\tau$.
Step 4 If $\ell^\tau(\boldsymbol{\lambda}^\tau) \leq \boldsymbol{r}^\tau$, let $l := j$, $j := j+1$ and go to step 5. Otherwise, let $j := j + 1$ and go to step 5.
Step 5 If $j > 14$, go to step 6. Otherwise, return to step 3.
Step 6 If $l > 0$, let $\tau := \tau + 1$ and go to step 11. Otherwise, go to step 7.
Step 7 Let $j := 1$ and $\boldsymbol{\lambda}^\tau := \bar{\boldsymbol{\lambda}}^\tau$.
Step 8 Let $\lambda_j^\tau := g_j^\tau$.
Step 9 If $\ell^\tau(\boldsymbol{\lambda}^\tau) \leq \boldsymbol{r}^\tau$, let $j := j + 1$ and go to step 10. Otherwise, let $\lambda_j^\tau := \bar{\lambda}_j^\tau$, $j := j + 1$ and go to step 10.
Step 10 If $j > 14$, let $\tau := \tau + 1$ and go to step 11. Otherwise, go to step 8.
Step 11 For instruments of the same type, revise $\boldsymbol{\lambda}^\tau$ so that they will be used in numerical order. Concretely, revise $\boldsymbol{\lambda}^\tau$ as $\lambda_j^\tau := 1$ $(1 \leq j \leq n_1)$, $\lambda_j^\tau := 0$ $(n_1 + 1 \leq j \leq 2)$, $\lambda_j^\tau := 1$ $(5 \leq j \leq n_2 + 4)$, $\lambda_j^\tau := 0$ $(n_2 + 5 \leq j \leq 8)$, $\lambda_j^\tau := 1$ $(11 \leq j \leq n_3 + 10)$, $\lambda_j^\tau := 0$ $(n_3 + 11 \leq j \leq 12)$, where $n_1 = \sum_{j=1}^{2} \lambda_j^\tau$, $n_2 = \sum_{j=5}^{8} \lambda_j^\tau$, $n_3 = \sum_{j=11}^{12} \lambda_j^\tau$. Then, go to step 12.
Step 12 If $\tau > t + K - 1$, let $\boldsymbol{\lambda} := \left((\boldsymbol{\lambda}^t)^T, \ldots, (\boldsymbol{\lambda}^{t+K-1})^T\right)^T$ and stop. Otherwise, return to step 2.

If phenotypes $\boldsymbol{\lambda}$ obtained by the decoding strongly depend on the reference solution, the global search may be impossible without updating the reference solution. Consequently, we propose the following algorithm for updating the reference solution, where the phenotype which is farthest from the reference solution is substituted for the current reference solution when the average distance between the reference solution and a phenotype in the current population is less than a sufficiently small positive constant η. In the algorithm, N is the population size, $\boldsymbol{\lambda}(r)$ is the phenotype of the rth individual, $\bar{\boldsymbol{\lambda}}$ is the reference solution.

Algorithm for reference solution updating

Step 1 Let $r := 1$, $r_{\max} := 1$, $d_{\max} := 0$ and $d_{\text{sum}} := 0$.
Step 2 After calculating $d_r := \sum_{\tau=t}^{t+K-1} \sum_{j=1}^{14} \left|\lambda_j^\tau(r) - \bar{\lambda}_j^\tau\right|$, let $d_{\text{sum}} := d_{\text{sum}} + d_r$.
If $d_r > d_{\max}$ and $J'(\boldsymbol{\lambda}(r)) < J'(\bar{\boldsymbol{\lambda}})$, let $d_{\max} := d_r$, $r_{\max} := r$, $r := r + 1$ and go to step 3. Otherwise, let $r := r + 1$ and go to step 3.
Step 3 If $r > N$, go to step 4. Otherwise, return to step 2.
Step 4 If $d_{\text{sum}}/(N \cdot n) < \eta$, replace the reference solution $\bar{\boldsymbol{\lambda}}$ with $\boldsymbol{\lambda}(r_{\max})$ and stop. Otherwise, stop without updating the reference solution.

It should be noted that most of individuals may be decoded in the neighborhood of the reference solution when constraints are tight. To avoid such a situation, every P generations, we replace the reference solution with a feasible solution obtained by solving the maximizing problem (21) through a genetic algorithm which uses a fitness function $f(\mathbf{S}) = G(\boldsymbol{\lambda}(t,K))$ and the basic decoding.

4.2 Evaluation

In this paper, the fitness $f(\mathbf{S})$ of an individual $\mathbf{S}$ is defined as

$$\hat{f} = 1.0 + k\rho - \max_{i=1,2,3}\left\{\left(\bar{\mu}_i - \mu_i(J_i(\boldsymbol{\lambda}(t,K)))\right) + \rho\sum_{l=1}^{k}\left(\bar{\mu}_l - \mu_l(J_l(\boldsymbol{\lambda}(t,K)))\right)\right\} \tag{22}$$

where $\underline{J}$ and $\overline{J}$ are the minimal objective function value and the maximal one of the operation plan obtained by connecting K solutions to $P(\tau)$, $\tau = t,\ldots,t+K-1$ solved by complete enumeration, $J'(\boldsymbol{\lambda})$ denotes the cost of the operation plan for the phenotype $\boldsymbol{\lambda}$ decoded from $\mathbf{S}$, dis is the average Hamming distance between the individual $\mathbf{S}$ and its phenotype $\boldsymbol{\lambda}$:

$$\text{dis} = \frac{1}{K \cdot 14}\sum_{\tau=t}^{t+K-1}\sum_{j=1}^{14}\left|\lambda_j^\tau - \mathbf{S}_j^\tau\right|.$$

Namely, $\hat{f}$ indicates the cost performance of the individual normalized by the difference between $\overline{J}$ and $\underline{J}$, while $(1 - \text{dis}) \cdot \hat{f}_i$ is the evaluation value in consideration of $(1 - \text{dis})$, which is the degree of similarity between the individual $\mathbf{S}$ and its phenotype $\boldsymbol{\lambda}$. Thereby, the dependence of the fitness $f(\mathbf{S})$ on the degree of the similarity increases with decreasing β. In this paper, β is determined as

$$\beta = \begin{cases} 0.2 & , I < I_{\max}/5, \\ I/I_{\max} & , I \geq I_{\max}/5 \end{cases} \tag{23}$$

where I and $I_{\max}$ denote the generation counter and the maximal search generation.

4.3 Scaling

When the variance of fitness in a population is small, it is often observed that the ordinary roulette wheel selection does not work well because there is little difference between the probability of a good individual surviving and that of a bad one surviving. In order to overcome this problem, the linear scaling technique is adopted in this paper.

4.4 Reproduction

In [11], Sakawa et al. suggested that elitist expected value selection is more effective than the other five reproduction operators (ranking selection, elitist ranking selection, expected value selection, roulette wheel selection, elitist roulette wheel selection). Accordingly, in the present paper, we adopt elitist expected value selection, which is the combination of expected value selection and elitist preserving selection.

4.5 Crossover

In order to preserve the feasibility of offspring generated by crossover, we use the one-point crossover where the crossover point is chosen from among $K-1$ end points of K subindividual $\mathbf{s}^\tau$, $\tau = t, \ldots, t+K-1$, as shown in Fig. 3.

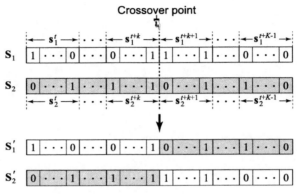

Fig. 3. An illustration of crossover.

4.6 Mutation

As the mutation operator, we use the mutation of bit-reverse type, illustrated in Fig. 4.

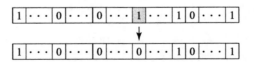

Fig. 4. An example of mutation.

4.7 Proposed genetic algorithm

Following the preceding discussions, the proposed genetic algorithm for operation planning of a DHC plant can be summarized as follows.

Proposed genetic algorithm

Step 0 Determine values of the parameters used in the genetic algorithm: the population size N, the generation gap G, the probability of crossover p_c, the probability of mutation p_m, the probability of inversion p_i, the minimal search generation $I_{\min}$, the maximal search generation $I_{\max}(>I_{\min})$, the scaling constant c_{mult}, the convergence criterion ε, the parameter for reference solution updating η, and set the generation counter t at 0.

Step 1 Generate the initial population consisting of N individuals.

Step 2 Decode each individual (genotype) in the current population and calculate its fitness based on the corresponding solution (phenotype).

Step 3 If the termination condition is fulfilled, stop. Otherwise, let $t := t+1$ and go to step 4.

Step 4 Carry out linear scaling.

Step 5 Apply reproduction operator using elitist expected value selection.

Step 6 Apply one-point crossover.

Step 7 Apply mutation of bit-reverse type. Return to step 2.

5 Numerical Experiments

For a multiobjective operation planning problem for 24 hours ($K = 24$) of an actual DHC plant, after formulating it as a multiobjective extended problem $P'(t, 24)$ which includes 334 0-1 variables, we apply the proposed interactive fuzzy satisficing method through the genetic algorithm. The problem involves three objective functions: the total running cost J_1, the surplus of the produced cold water over the predicted demand J_2 and the surplus of the produced steam over the predicted demand J_3.

In the following experiment, the proposed genetic algorithm is applied 10 times to each of minimax problems solved in the interactive method, where parameters in the genetic algorithm are set as: the population size (the number of individuals) $N = 70$, the crossover rate $p_c = 0.8$, the mutation rate $p_m = 0.005$, the maximal search generation $I_{\max} = 5000$. The numerical experiments are carried out on a personal computer (CPU: Intel PentiumII Processor 266MHz, C_Compiler: Microsoft Visual C++ 6.0).

First, according to Step 1 in the interactive fuzzy satisficing method, we calculate the (approximate) minimum $\underline{J}_i$ and the (approximate) maximum $\overline{J}_i$, $i = 1, 2, 3$. Table 1 shows these values.

Proceeding to Step 2, the hypothetical decision maker specifies membership functions $\mu_i(J_i(\boldsymbol{\lambda}(t,K)))$, $i = 1,2,3$ based on the individual minima and maxima of objective functions, and sets the initial reference membership levels $(\bar{\mu}_1, \bar{\mu}_2, \bar{\mu}_3) = (1.0, 1.0, 1.0)$.

Then, according to Step 3, the augmented minimax problem 20 for the current reference membership levels is solved by the proposed genetic algorithm. The results are shown in the second column of Table 2, where values in the parentheses in J_1 row present the number of switching of instruments.

Table 1. Individual minimum $\underline{J}_i$ and maximum $\overline{J}_i$, $i = 1, 2, 3$

Objective	Minimum $\underline{J}_i$	Maximum $\overline{J}_i$
J_1	1463317.37	2185775.47
J_2	3.80	660.04
J_3	62.73	787.89

Table 2. Interaction process

	1st	2nd	3rd
$\bar{\mu}_1$	1.0	1.0	1.0
$\bar{\mu}_2$	1.0	0.8	0.8
$\bar{\mu}_3$	1.0	0.8	0.5
μ_1	0.56	0.71	0.83
μ_2	0.56	0.51	0.63
μ_3	0.56	0.50	0.41
J_1	1675621(19)	1623197(14)	1597879(13)
J_2	40.98	35.95	25.33
J_3	173.74	220.13	239.75

In Step 4, since the decision maker cannot be satisfied with this result and he/she feels that J_1 should be improved even if J_2 and J_3 become worse, the reference membership levels are updated from $(1.0, 1.0, 1.0)$ to $(1.0, 0.8, 0.8)$, and return to Step 3. Again, the minimax problem (20) is solved for the current reference membership levels, and the results are shown in the third column of Table 2.

Furthermore, the decision maker hopes that J_1 becomes better at the expense of J_3 and updates the reference membership levels from $(1.0, 0.8, 0.8)$ to $(1.0, 0.8, 0.5)$. The results are shown in the fourth column of Table 2.

In this experiment, since the decision maker is satisfied with this result, this solution is the satisficing solution for the decision maker, and the interaction process stops.

6 Conclusion

In this paper, focusing on operation planning of district heating and cooling (DHC) plants, we formulated it as multiobjective nonlinear 0-1 programming problems and proposed an interactive fuzzy satisficing method through genetic algorithms. Furthermore, by applying the proposed method to numerical examples using actual plant data, the feasibility of the proposed method were demonstrated.

References

1. Sakawa M., Ushiro S., Kato K., Inoue T. (1999) Cooling load prediction through radial basis function network using a hybrid structural learning and simplified robust filter. Transaction of the Institute of Electronics, Information and Communication Engineers A **J82-A**, 31–39 (in Japanese)
2. Sakawa M., Kato K., Ushiro S. (2001) Cooling load prediction in a district heating and cooling system through simplified robust filter and multi-layered neural network. Applied Artificial Intelligence **15**, 633-643
3. Ito K., Yokoyama R. (1990) Optimal Planning of Co-Generation Systems. Sangyo Tosho, Tokyo (in Japanese)
4. Yokoyama R., Ito K. (1996) A revised decomposition method for MILP problems and its application to operational planning of thermal storage systems. Journal of Energy Resources Technology **118**, 277–284
5. Sakawa M., Yano H. (1985) An interactive fuzzy satisficing method using augmented minimax problems and its application to environmental systems. IEEE Transactions on Systems, Man, and Cybernetics **SMC-15**, 720–729
6. Sakawa M., Yano H., Yumine T. (1987) An interactive fuzzy satisficing method for multiobjective linear-programming problems and its application. IEEE Transactions on Systems, Man, and Cybernetics **SMC-17**, 654–661
7. Goldberg D. E. (1989) Genetic Algorithms in Search, Optimization, and Machine Learning. Addison Wesley, Reading, Massachusetts
8. Michalewicz Z. (1996) Genetic Algorithms + Data Structures = Evolution Programs. Springer-Verlag, 1992, Second, extended edition, 1994, Third, revised and extended edition, Berlin
9. Sakawa M. (2001) Genetic Algorithms and Fuzzy Multiobjective Optimization. Kluwer Academic Publishers, Boston
10. Gen M., Cheng R. (1996) Genetic Algorithms and Engineering Design. John Wiley & Sons, New York
11. Sakawa M., Kato K., Sunada H., Shibano T. (1997) Fuzzy programming for multiobjective 0-1 programming problems through revised genetic algorithms. European Journal of Operational Research **97**, 149–158
12. Sakawa M., Shibano T. (1996) Interactive fuzzy programming for multiobjective 0-1 programming problems through genetic algorithms with double strings. In: Da Ruan (Ed.) Fuzzy Logic Foundations and Industrial Applications. Kluwer Academic Publishers, Boston, 111–128
13. Sakawa M., Shibano T. (1998) An interactive fuzzy satisficing method for multiobjective 0-1 programming problems with fuzzy numbers through genetic algorithms with double strings. European Journal of Operational Research **107**, 564–574
14. Sakawa M., Kato K., Ushiro S., Ooura K. (1999) Fuzzy programming for general multiobjective 0-1 programming problems through genetic algorithms with double strings. 1999 IEEE International Fuzzy Systems Conference Proceedings **3**, 1522–1527

Adaptive Hybrid Genetic Algorithm with Fuzzy Logic Controller

YoungSu Yun[1] and Mitsuo Gen[2]

[1] School of Automotive, Industrial & Mechanical Engineering
 Daegu University, Kyungbook 712-714, Korea
 Phone: +82(53)850-4431, Fax: +82(53)850-6549
[2] Department of Industrial & Information Systems Engineering
 Ashikaga Institute of Technology, Ashikaga 326-8558, Japan
 Phone: +81(284) 62-0605 ext. 376, Fax: +81(284) 64-1071

Abstract. This paper proposes adaptive genetic operators in the hybrid genetic algorithm with a fuzzy logic controller. For the hybrid genetic algorithm (HGA), a rough search technique and an iterative hill climbing technique are employed to a genetic algorithm. The rough search technique is used to initialize the population of the genetic algorithm; its strategy is to make large jumps in the search space in order to avoid being trapped in local optima and the iterative hill climbing technique is also applied to find a better solution in the convergence region within the genetic algorithm loop. A crossover operator and a mutation operator used in the HGA are automatically adjusted for the adaptive ability during the search process of the HGA, and the fuzzy logic controller (FLC) regulates the applying ratios of these operators. For the comparison of the adaptive ability in the HGA, a conventional heuristic method and the proposed FLC are analyzed and compared in a numerical example. Finally, the best hybrid genetic algorithm with an adaptive ability is recommended.

1 Introduction

Genetic algorithms (GAs) are known to offer significant advantages over conventional methods by using simultaneously several search principles and heuristics. The most important ones include a population-wide search, a continuous balance between convergence and diversity, and the principle of building-block combination. Nevertheless, GAs can suffer from excessively slow convergence and premature convergence before providing an accurate solution because of their fundamental requirements; namely, not using a priori knowledge and not exploiting local search information. Also GAs may have a weakness in taking too much time to adjust fine-tuning structures of the GA parameters (crossover ratio, mutation ratio, and others). This kind of "blindness" may prevent them from being really of practical interest for a lot of applications.

To improve these weaknesses of GAs, various methodologies using adaptive operators of GAs and conventional heuristics for the hybridization of GAs have been developed (Gen and Cheng,1997, 2000; Ishibuchi, et al, 1994;

Li and Jiang, 2000; Renders and Flasse, 1996; Lee, Yun, and Gen, 2002; Yun, Gen, and Seo, 2002) [1–7].

Adaptive methodologies to adjust the fine-tuning structure of the parameters in GAs have two modes: one is to use fuzzy logic controllers (FLCs) and the other is heuristics. The parameters controlled by these methodologies are automatically regulated during the genetic search process. Thus, much time for the fine-tuning of these parameters can be saved and the GA search ability in finding the global optimum can be improved.

In the first mode, Gen and Cheng (2000), in their book, suggested various methods using FLCs[2]. Herrera and Lozano (2001) suggested and summarized adaptive genetic operators based on co-evolution with fuzzy behaviors[8]. The pioneering works in extending fuzzy logic technique to automatically regulation of the GA parameters were that of Lee and Takagi (1993), Xu and Vukovich (1994), Wang, Wang, and Hu (1997)[9–11]. Lee and Takagi focused on population size, Xu and Vukovich on generation number and population size, and Wang, Wang and Hu used two FLCs: one for the crossover ratio and the other for the mutation ratio. These parameters are considered as the input variables of GAs and are taken as the output variables of the FLC. For successfully applying a FLC to a GA, the key is to produce well-formed fuzzy sets and rules. Recently, Cheong and Lai (2000) suggested an optimization structure of GA using FLCs[12].

In the second mode, Mak, Wong, and Wang (2000) controlled the crossover and mutation ratios automatically according to the performance of GA operators in a manufacturing cell formulation[13]. Srinivas and Patnaik (1994), Wu, Cao, and Wen (1998) also controlled the ratios according to the fitness value of a GA population[14,15]. In the methodologies using conventional heuristics for the hybridization of GAs, several researchers have employed a method offering a priori knowledge before initializing the population of GAs, in order to improve the fundamental requirement of GAs on not using a priori knowledge. Davis (1991) suggested a theoretical algorithm for hybridization in order to offer a priori knowledge to GAs[16]. Li and Jiang (2000) proposed a rough search technique using simulated annealing for exploiting the knowledge[4].

To exploit local search information and find a better solution, various hybrid methods using conventional local search techniques have been developed. Gen and Cheng (1997) suggested a theoretical methodology using Lamarckian evolution and memetic algorithms for the hybridization of GAs[1]. Li and Jiang (2000) presented a new stochastic approach called the SA-GA-CA, which is based on the proper integration of a simulated annealing (SA), a GA, and a chemotaxis algorithm (CA) for solving complex optimization problems[4]. Yen et al. (1998) described two approaches for incorporating a simplex method into the GA loop as an additional operator[17]. These approaches for hybridization usually use the complementary properties of GAs and conventional heuristics.

By applying adaptive operators, a priori knowledge, and exploiting local search information, recently hybridized GAs are more effective and robustness than conventional GAs or conventional heuristics. However, there are always complementary properties between these various hybrid methods. Therefore, the aim of this paper is to develop an efficient hybrid algorithm and to compare the complementary properties of the proposed hybrid methods with an adaptive ability using a FLC and a heuristic method.

In Section 2, adaptive genetic operators using a FLC and a heuristic method are proposed. The hybrid concepts and logics underlying the proposed methods are provided in Section 3. In Section 4, four algorithms including one canonical GA, one hybrid GA, and two adaptive hybrid GAs utilizing the above concepts are proposed. Numerical example and test results used to compare the proposed algorithms are presented in Section 5, followed by the conclusion in Section 6.

2 Adaptive Genetic Operators (AGOs)

In this section, we proposed two adaptive methods using a FLC and a heuristic method in order to automatically regulate the ratios of the GA operators during the genetic search process.

2.1 AGO using a FLC

To automatically regulate the GA parameters using a FLC, we use the concept of Wang, Wang and Hu (1997)[11]. The main idea of this concept is that two FLCs: the crossover FLC and the mutation FLC are implemented independently to automatically regulate the crossover ratio and the mutation ratio during the genetic search process.

The heuristic updating strategy for the crossover ratio and mutation ratio is to consider the changes of the average fitness in the GA population of two continuous generations. For example, in a minimization problem, we can set the change in the average fitness at generation t-1, $\Delta \overline{eval}(V; t-1)$ and the change in the average fitness at generation t, $\Delta \overline{eval}(V; t)$. These values can be considered to regulate the crossover ratio p_C and the mutation ratio p_M as follows:

Procedure: regulation of p_C and p_M using the average fitness
begin
    if $\varepsilon \leq \Delta \overline{eval}(V; t-1) \leq \gamma$ and $\varepsilon \leq \Delta \overline{eval}(V; t) \leq \gamma$ **then**
        increase p_C and p_M for next generation;
    if $-\gamma \leq \Delta \overline{eval}(V; t-1) \leq -\varepsilon$ and $-\gamma \leq \Delta \overline{eval}(V; t) \leq -\varepsilon$
    then decrease p_C and p_M for next generation;
    if $-\varepsilon < \Delta \overline{eval}(V; t-1) < \varepsilon$ and $-\varepsilon < \Delta \overline{eval}(V; t) < \varepsilon$ **then**
        rapidly increase p_C and p_M for next generation;
end

where ε is a given real number in the proximity of zero, γ is a given maximum value of a fuzzy membership function, and $-\gamma$ is a given minimum value of a fuzzy membership function. The implementation strategy for the crossover FLC is as follows:

- **Input and output of the crossover FLC**
  The inputs to the crossover FLC are $\triangle \overline{eval}(V; t-1)$ and $\triangle \overline{eval}(V; t)$, and its output is the change in the crossover ratio $\triangle c(t)$.
- **Membership functions of $\overline{eval}(V; t-1)$, $\triangle \overline{eval}(V; t)$ and $\triangle c(t)$**
  The membership functions of the fuzzy input and output linguistic variables are used from the reference (Wang, Wang, and Hu, 1997).
- **Fuzzy decision table**
  Based on a number of experiments and the domain expert opinions, the fuzzy decision table is employed in Table 1.
- **Defuzzification table for control actions**
  For simplicity, the defuzzification table for determining the action of the crossover FLC was setup in (Wang, Wang, and Hu, 1997)

Table 1. Fuzzy decision table for crossover

$\triangle c(t)$		\multicolumn{8}{c}{$\triangle \overline{eval}(V; t-1)$}							
		NR	NL	NM	NS	ZE	PS	PM	PL	PR
$\triangle \overline{eval}(V;t)$	NR	NR	NL	NL	NM	NM	NS	NS	ZE	ZE
	NL	NL	NL	NM	NM	NS	NS	ZE	ZE	PS
	NM	NL	NM	NM	NS	NS	ZE	ZE	PS	PS
	NS	NM	NM	NS	NS	ZE	ZE	PS	PS	PM
	ZE	NM	NS	NS	ZE	PM	PS	PS	PM	PM
	PS	NS	NS	ZE	ZE	PS	PS	PM	PM	PL
	PM	NS	ZE	ZE	PS	PS	PM	PM	PL	PL
	PL	ZE	ZE	PS	PS	PM	PM	PL	PL	PR
	PR	ZE	PS	PS	PM	PM	PL	PL	PR	PR

NR - Negative larger, NL - Negative large, NM - Negative medium
NS - Negative small, ZE - Zero, PS - Positive small
PM - Positive medium, PL - Positive large, PR - Positive larger

The inputs of the mutation FLC are the same as for the crossover FLC and the output of which is the change in the mutation ratio $\triangle m(t)$. The procedure for its applying is detailed as follows:

Step 1 The input variables of the FLC for regulating the GA parameters are the change in the average fitness of two continuous generations ($t-1$ and t) as follows:

$$\triangle \overline{eval}(V; t-1), \triangle \overline{eval}(V; t)$$

Step 2 After normalizing $\overline{eval}(V; t-1)$ and $\triangle \overline{eval}(V; t)$, assign these values to the indexes i and j corresponding to the control actions in the defuzzification table.

Step 3 Calculate the changes of the crossover ratio $\triangle c(t)$ and the mutation ratio $\triangle m(t)$ as follows:

$$\triangle c(t) = Z(i,j) \times \alpha, \triangle m(t) = Z(i,j) \times \beta$$

where the contents of $Z(i,j)$ are the corresponding values of $\overline{eval}(V; t-1)$ and $\triangle \overline{eval}(V; t)$ for the defuzzification. The parameters α and β are a given value to regulate the increasing and decreasing range for the crossover ratio and the mutation ratio.

Step 4 Update the changes of the crossover ratio and the mutation ratio by using the following equation:

$$p_C(t) = p_C(t-1) + \triangle c(t), p_M(t) = p_M(t-1) + \triangle m(t)$$

where $p_C(t)$ and $p_M(t)$ are the crossover ratio and the mutation ratio at generation t, respectively.

2.2 AGO using a Heuristic Method

For the AGO using a heuristic method, we use the concept of Mrk, Wong, and Wang (2000)[13]. Adaptive crossover and mutation operators used by them prevent the genetic search process from prematurely conversing to local optimal solutions by regulating the balance of exploration and exploitation in the solution space. This scheme involves some rules to adjust adaptively the crossover and mutation ratios according to the performance of the genetic operators. The detailed procedure is as follows:

Procedure: regulation of $p_C(t)$ and $p_M(t)$ using the heuristic method
    begin
        if $f_{off\_size}/f_{pop\_size} \geq 0.1$ **then**
            $p_C(t) = p_C(t-1) + 0.05, p_M(t) = p_M(t-1) + 0.005$;
        if $f_{off\_size}/f_{pop\_size} \leq -0.1$ **then**
            $p_C(t) = p_C(t-1) - 0.05, p_M(t) = p_M(t-1) - 0.005$;
        if $-0.1 < f_{off\_size}/f_{pop\_size} < 0.1$ **then**
            $p_C(t) = p_C(t-1), p_M(t) = p_M(t-1)$;
    end

where f_{off_size} and f_{pop_size} are the fitness values of the offspring and parents, respectively. $p_C(t)$, $p_M(t)$, $p_C(t-1)$, and $p_M(t-1)$ are the crossover ratio and the mutation ratio in a current generation and the crossover ratio and the mutation ratio in a previous generation, respectively. In the cases of the $f_{off\_size}/f_{pop\_size} \geq 0.1$ and $f_{off\_size}/f_{pop\_size} \leq -0.1$, the adjusted ratio should not exceed the domain of 0.5 to 1.0 for the $p_C(t)$ and the domain of 0.05 to 0.10 for the $p_M(t)$. The scheme of the above procedure is evaluated in the every generation of genetic search process and changes the ratios of the crossover and mutation operators. Therefore, this mechanism is based on the fact that it encourages the well-performing operators to produce more offspring, while also reducing the chance for poorly performing operators to destroy the potential individuals during the recombination process.

3 Proposed Hybrid Concepts and Logics

In this section we propose the basic concepts and logics for constructing the searching procedures of the proposed hybrid methods. First, we apply a rough search technique for initializing the GA population. Secondly, a GA procedure is applied. In the last step, the local search technique to find a better solution is proposed.

3.1 Rough Search Technique

This technique can make large jumps in search space and is able to "jump out" of the local optima, also converges rapidly (Li and Jiang, 2000)[4]. But it is not the most effective method in searching optimum when it is used alone (Davis, 1991)[16]. However, it can be hybridized with other techniques. For example, if it will be used to initialize the population of a GA, this hybrid strategy can generate a better solution than using a conventional GA or a rough search technique alone. According to this concept, we develop a rough search technique as a new concept of the proposed hybrid methods. The applied technique is to initialize the population of a GA and the detailed heuristic procedure for its applying is as follows:

Step 1 Randomly generate as many strings as the population size within the allowable limits of the variables.

Step 2 Check the constraints. If they are satisfied, evaluate the fitness $f(x)$ of each string and store the one with the best fitness $f(x)$ as a string for the GA initial population.

Step 3 If the iteration does not run out, go to Step 1. If we find a fitness value better than the previous fitness $f(x)$ stored in Step 2, store the string as the next string for the GA initial population. This iterative procedure is repeated as many times as the generation number of the GA.

Step 4 If we do not have as many strings as the population size after applying Step 3, the remainders are randomly generated within the allowable limits of the variables.

3.2 Genetic Algorithm

For the representation of GAs, we use real-number representation instead of bit-string representation. Real-number representation has several advantages of (i) being better adapted to numerical optimization for continuous problems, (ii) speeding up the search over the bit-string representation, and (iii) easing the development of approaches for hybridizing with other conventional methods (Davis, 1991)[16]. The proposed GA is used as the main algorithm of the proposed hybrid methods and the detailed heuristic procedure for its applying is as follows:

Step 1: Initial population To get the initial population of a GA, we use the population obtained by the rough search technique proposed in section 3.1

Step 2: Genetic operators
- Selection: elitist selection strategy in enlarged sampling space (Gen and Cheng, 2000)
- Crossover: non-uniform arithmetic crossover operator
- Mutation: uniform mutation operator

Step 3: Stop condition If the maximum generation number is reached, then stop; otherwise, go to Step 2.

3.3 Local Search Technique

In this section, we suggest a method to hybridize a GA and a local search technique. This method is to incorporate a local search technique in the conventional GA loop (Gen and Cheng, 1997; Yen. et al., 1998)[1,17]. With this hybrid approach, the local search technique is applied to each newly generated offspring to move it to a local optimum before injecting it into the new population. For the proposed hybrid methods, we employ the iterative hill climbing method suggested by Michalewicz (1994)[18]. This technique can guarantee the desired properties of a local search technique for hybridization. The detailed procedure of the iterative hill climbing method, for minimization problem, is given as follows:

> **Procedure: Iterative hill climbing technique in the GA loop**
> begin
>   select an optimum string v_c in current GA loop;
>   randomly generate as many strings as the population size
>   in the neighborhood of v_c ;
>   select the string v_n with the optimal value of the objective
>   function f among the set of new strings;

```
    if f(v_c) < f(v_n) then
        v_c ← v_n;
    end
end
```


4 Proposed Algorithms for Experimental Comparison

In this section, we present four algorithms for experimental comparison, including one canonical genetic algorithm, one hybrid genetic algorithm without an adaptive ability, and two hybrid genetic algorithms with adaptive ones. The detailed procedures for the proposed algorithms are given as follows:

4.1 Canonical Genetic Algorithm (CGA)

This algorithm is to implement a GA with real-number representation, called CGA. The initial population of the CGA is randomly generated and the detailed procedure is the same as those in the Step 2 and 3 of Section 3.2, which are summarized in Figure 1 (a).

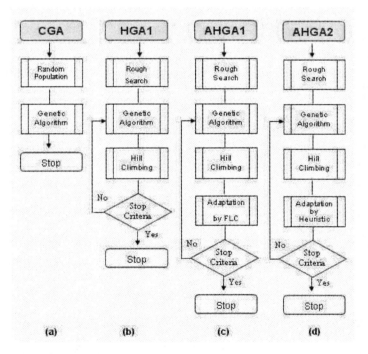

Fig. 1. Various hybrid genetic algorithms

4.2 Hybrid Genetic Algorithm (HGA)

This hybrid genetic algorithm is to apply the rough search technique for initializing the GA population and to apply the GA with the iterative hill climbing technique. The heuristic procedure is as follows:

Step 1 apply the rough search technique for making the initial population of the GA.
Step 2 apply genetic operators (i.e., the selection, crossover, and mutation operators) using the population obtained by Step 1.
Step 3 select an optimal solution among the feasible solutions obtained after Step 2.
Step 4 apply the iterative hill climbing technique using the optimal solution obtained in Step 3.
Step 5 go to Step 2 until the maximal generation number of the proposed algorithm is reached.

This procedure is also summarized in Figure 1 (b).

4.3 Adaptive Hybrid Genetic Algorithm 1 (AHGA1)

The first adaptive hybrid genetic algorithm is to apply the rough search technique for initializing the GA population and to apply the GA with the iterative hill climbing technique and the FLC suggested in Section 2.1. The heuristic procedure is as follows:

Step 1 apply the rough search technique for making the initial population of the GA.
Step 2 apply genetic operators (i.e., the selection, crossover, and mutation operators) using the population obtained by Step 1.
Step 3 select an optimal solution among the feasible solutions obtained after Step 2.
Step 4 apply the iterative hill climbing technique using the optimal solution obtained in Step 3.
Step 5 apply the fuzzy logic procedure for automatically regulating the GA parameters (i.e., the crossover and the mutation ratios).
Step 6 go to Step 2 until the maximal generation number of the proposed algorithm is reached.

This procedure is also summarized in Figure 1 (c).

4.4 Adaptive Hybrid Genetic Algorithm 2 (AHGA2)

The second adaptive hybrid genetic algorithm is the same algorithm as for the proposed AHGA1 except using the heuristic method suggested in section 2.2 instead of the FLC in section 2.1. This procedure is also summarized in Figure 1 (d).

5 Numerical Example

In this section, a numerical example is tested in order to compare the relative performance of the proposed algorithms. The parameters for the proposed algorithms are as follows: population size =10, initial crossover ratio = 0.7 initial mutation ratio = 0.01, generation number= 4000, and the search range of the iterative hill climbing technique = 2, The procedures of the algorithms were executed ten times and implemented in the Visual Basic language on IBM-PC NT P400 computer.

5.1 Test Problem

Test problem is the interesting nonlinear optimization problem provided by Himmealblau (1972)[19]. The results obtained are shown in Table 2.

Table 2. The computational results for test problem

	CGA	HGA	AHGA1	AHGA2
x_1	78.014	78.067	78.006	78.001
x_2	33.380	33.100	33.088	33.010
x_3	27.716	27.174	27.146	27.183
x_4	44.825	44.554	44.877	44.921
x_5	43.088	44.814	44.767	44.656
Best fitness $f(x)$	-30957.783	-31000.730	-31015.340	-31013.466
Average fitness	-30811.120	-30988.365	-30995.075	-30992.234
Average CPU Time (sec.)	2.2	2.6	2.4	5.3

In Table 2, the proposed HGAs (HGA, AHGA1, and AHGA2) show better results than the canonical GA (CGA) in the best fitness and the average fitness. Among the proposed HGAs, AHGA1 shows the best result in best fitness, average fitness, and average CPU time. This means that the proposed rough search technique, the iterative hill climbing technique, and the FLC enable the AHGA1 to search more precisely than the CGA, HGA, and AHGA2.

In terms of the average CPU time, the CGA, HGA, and AHGA1 show almost same results and these results are about two times faster than that of AHGA2. This implies that the proposed FLC used in AHGA1 is well regulating the GA operators and also can make the ratios of the operators having lower values than those of the AHGA2.

For the detailed comparison of the adaptive ability between the proposed FLC and the heuristic method, the behaviors of crossover ratio and the evolutionary behaviors of average fitness in the AHGA1 and AHGA2 are shown in Figure 2 and 3, respectively.

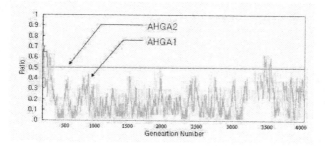

Fig. 2. Behaviors of crossover ratio

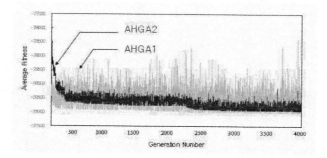

Fig. 3. Evolutionary behaviors of the average fitness

In Figure 2, the graph of the AHGA2 shows slightly the convergence process within about 100 generations and has the same value from the generation to the last generation. That of the AHGA1, however, has more various ratios and lower ratios than that of the AHGA2 during the genetic search process. This can prove that the average CPU time of the AHGA1 is faster than that of the AHGA2 in Table 2. Figure 3 shows the evolutionary behaviors of the average fitness for the proposed algorithms (i.e., the AHGA1 and AHGA2). This graph means the average fitness of all strings satisfying the constraints in each generation. The average fitness of the AHGA1 has various variations and converges more rapidly than that of the AHGA2. This implies that applying the FLC in the AHGA1 can increase the diversification of the population of the GA and makes the convergence of the GA to be more rapid than that of the AHGA2.

Based on these result analyses in the test problem, we can conclude that i) applying the rough search technique and the iterative hill climbing technique for the hybridization of the GA can be a good way to improve the performance of the GA, and ii) for the adaptive ability of the GA operators (i.e., the crossover and mutation operators), applying the FLC is more efficient than applying the heuristic method.

6 Conclusion

In this paper we have proposed several hybrid genetic algorithms and compared various hybrid abilities and adaptive ones. For constructing the hybrid genetic algorithms, we have combined a rough search technique and an iterative hill climbing technique into a genetic algorithm. A FLC and a heuristic method for comparing the adaptive ability of the hybrid genetic algorithms have been also employed. These methods have been tested in a numerical example and the results obtained are as follows:

i) The proposed rough search technique can make the hybrid algorithms converge more rapidly during the initial generations.

ii) The local search technique applied is intended to find a more accurate value after the convergence of the GA in each generation. This technique can save the CPU time of the proposed hybrid algorithms.

iii) By applying the proposed FLC to automatically regulate the GA parameters (i.e., the crossover and mutation operators), the search space of the hybrid algorithm becomes larger than those of the CGA, HGA without the FLC, and AHGA2 with the heuristic method. This in turn enables the hybrid algorithm with the FLC to find a better solution than the algorithm of CGA, HGA, and AHGA2.

References

1. Gen, M. and Cheng, R. (1997) Genetic Algorithms and Engineering Design. John-Wiley & Sons.
2. Gen, M. and Cheng, R. (2000) Genetic Algorithms and Engineering Optimization. John-Wiley & Sons.
3. Ishibuchi, H., Yamamoto, N., Murata, T. and Tanaka, H. (1994) Genetic algorithm and neighborhood search algorithms for fuzzy flow-shop scheduling problems. Fuzzy Sets and Systems, 67, 81-100.
4. Li, B. and Jiang, W. (2000) A novel stochastic optimization algorithm. IEEE Transactions on Systems, Man, and Cybernetics-Part B: Cybernetics, 30(1), 193-198.
5. Renders, J. M. and Flasse, S. P. (1996) Hybrid methods using genetic algorithms for global optimization. IEEE Transactions on Systems, Man, and Cybernetics-Part B: Cybernetics, 26(2), 243-258.

6. Lee, C. Y., Yun, Y. S. and Gen, M. (2002) Reliability optimization design for complex systems by hybrid GA with fuzzy logic control and local search. IEICE Transaction on Fundamentals of Electronics Communications and Computer Sciences, E85-A (4), 880-891.
7. Yun, Y. S., Gen. M. and Seo, S. L. (2002) Various hybrid methods based on genetic algorithm with fuzzy logic controller. to appear in Journal of Intelligent Manufacturing.
8. Herrera, H. and Lozano, M. (2001) Adaptive genetic operators based on co-evolution with fuzzy behaviors. IEEE Transactions on Evolutionary Computation, 5(2), 149-165.
9. Lee, M. and Takagi, H. (1993) Dynamic control of genetic algorithm using fuzzy logic techniques. Proceedings of the 5th International Conference on Genetic Algorithms, San Francisco, 76-83.
10. Xu, H. and Vukovich, G. (1994) Fuzzy evolutionary algorithm and automatic robot trajectory generation. Proceedings of the First IEEE Conference on Evolutionary Computation, IEEE Press, Piscataway, NJ, 595-600.
11. Wang, P. T., Wang, G. S. and Hu, Z. G. (1997) Speeding up the search process of genetic algorithm by fuzzy logic. Proceeding of the 5th European Congress on Intelligent Techniques and Soft Computing, 665-671.
12. Cheong, F. and Lai, R. (2000) Constraining the optimization of a fuzzy logic controller using an enhanced genetic algorithm. IEEE Transactions on Systems, Man, and Cybernetics-Part B: Cybernetics, **30(1)**, 31-46.
13. Mak, K. L., Wong, Y. S. and Wang, X. X. (2000) An adaptive genetic algorithm for manufacturing cell formulation. International Journal of Advanced Manufacturing Technology, 16, 491-497.
14. Srinivas, M. and Patnaik, M. (1994) Adaptive probabilities of crossover and mutation in genetic algorithms. IEEE Transactions on Systems, Man and Cybernatics, 24(4), 656-667.
15. Wu, Q. H., Cao, Y. J. and Wen, J. Y. (1998) Optimal reactive power dispatch using an adaptive genetic algorithm. Electrical Power and Energy Systems, 20(8), 563-569.
16. Davis, L. (1991) Handbook of Genetic Algorithms, Van Nostrand Reinhold.
17. Yen, J., Liao, J. C., Lee, B. J. and Randolph, D. (1998) A hybrid approach to modeling metabolic systems using a genetic algorithm and simplex method. IEEE Transactions on Systems, Man, and Cybernetics-Part B: Cybernetics, 28(2), 173-191.
18. Michalewicz, Z. (1994) Genetic Algorithms + Data Structures = Evolution Program, Second Extended Edition, Spring- Verlag.
19. Himmelblau, M. (1972) Applied Nonlinear Programming. McGraw-Hill, New York.

Finding Satisfactory Near-Optimal Solutions in Location Problems

María J. Canós, Carlos Ivorra, and Vicente Liern

Universitat de València, Departamento de Matemática Económico-Empresarial, Avda. Tarongers, s/n. E-46071 Valencia, Spain

Abstract. We develope and analyze a heuristic procedure to solve a fuzzy version of the p-median problem in which we allow part of the demand not to be covered in order to reduce the transport cost. This can be used to improve a given solution of the crisp p-median problem as well as to give to the decision-maker a range of alternative locations that can be adequate according to his or her own criteria.

1 Introduction

The (crisp) p-median problem is a location model formulated by Hakimi [6], [7] in the mid-1960s. It can be stated as follows: Let N be a connected, non-directed network with n nodes $v_1, \ldots, v_n$. Each node v_i has an associated positive weight w_i, usually called the demand at v_i. Each edge (v_i, v_k) has an associated positive length l_{ik}. We define the distance d_{ij} from v_i to v_j as the length of the shortest path joining v_i with v_j. The p-median problem consists in locating p facilities to cover the given demands in such a way that total transport costs are minimized. It is assumed that these costs are directly proportional to the travelled distances and to the quantities of product to be transported. A usual linear programming formulation is as follows:

$$\text{Min} \sum_{i=1}^{n} \sum_{j=1}^{n} d_{ij} x_{ij}$$
$$\text{s.t.} \sum_{i=1}^{n} x_{ij} = w_j \quad 1 \leq j \leq n$$
$$0 \leq x_{ij} \leq w_j y_i \quad 1 \leq i, j \leq n$$
$$\sum_{i=1}^{n} y_i = p$$
$$y_i \in \{0, 1\} \quad 1 \leq i \leq n$$

where x_{ij} is the part of the demand of the node v_j covered from the node v_i, and y_i is a binary variable taking the value 1 if there exists a facility in the node v_i.

This problem has been extensively studied (see, for instance, [5, 9]) and has been proved to be NP-hard [8]. Consequently, algorithms which are suitable for medium-sized instances, such as enumerative algorithms, cannot be used in larger instances.

In [2] and [3], we study a fuzzy version of this problem in which we allow part of the demand to remain uncovered when this generates a substantial reduction in costs. There, we define a fuzzy degree of satisfaction which combines the decision-makers interest in covering as much demand as possible with the benefit offered by a significant reduction in costs. We handle this fuzzy model in two ways: a mathematical programming model, and an explicit enumeration algorithm. However, as in the crisp problem, our proposals are only adequate for small and medium-sized instances. In the present paper we propose a heuristic method to find near-optimal fuzzy solutions. This can be used to improve a given solution of the crisp p-median problem as well as to give to the decision-maker a range of alternative locations that can be adequate according to his or her own criteria.

2 The fuzzy p-median problem

In this section, we recall the fuzzy p-median problem presented in [2]. As we have said, our fuzzy problem tries to improve a crisp p-median by considering the possibility of reducing the covered demand. So, we have two associated problems: the crisp problem, in which all the demand must be covered; and the fuzzy problem, which starts from the crisp p-median and searches for a more satisfactory partially feasible solution. This starting point can be the optimum, or a near optimum obtained with a heuristic procedure. In any case, we will refer to this solution as the crisp optimum.

The fuzzy problem has a flexible constraint, namely not leaving too much demand uncovered and a flexible goal, namely achieving a substantially lower cost. To express the constraint by means of a fuzzy set, we take as universe E the set of all $(V_0, w_1, \ldots, w_n)$, where V_0 is a collection of p nodes, and w_i are numbers satisfying $0 \leq w_i \leq w_i$ that represent the actually covered demand. The fuzzy constraint set will be a fuzzy subset of E given a membership function μ_f. A reasonable μ_f should take the value 1 for a solution that covers all the demand, and 0 if the covered demand is unacceptable to the decision-maker. The set of solutions with a substantially lower cost is defined by a membership function μ_g which should be 0 in the solutions whose cost is greater than, or equal to, the optimal crisp cost, and should be 1 in those solutions whose cost is less than the best expectations of the decision-maker. Our fuzzy decision set will be the intersection of these two fuzzy sets. Its membership function will be given by

$$\mu_D = \min\{\mu_f, \mu_g\}. \qquad (1)$$

We look for the solution where μ_D attains its maximum.

Here, we assume that the decision-maker has no preferences for reducing the demand in one vertex or another. Then, the function μ_f only depends on the total uncovered demand

$$x = \sum_{i=1}^{n} w_i - \sum_{i=1}^{n} w'_i.$$

We choose a piecewise linear membership function

$$\mu_f(V_0, w'_1, \ldots, w'_n) = f(x) = \begin{cases} 1 & if \quad x < 0 \\ 1 - \frac{x}{p_f} & if \quad 0 \leq x \leq p_f \\ 0 & if \quad x > p_f \end{cases} \quad (2)$$

where p_f is the maximum demand that the decision-maker will admit to leave uncovered. In the same way, we take μ_g as a piecewise linear function of the reduction of cost provided by the solution. So, for a given solution $(V_0, w_1, \ldots, w_n)$ we define:

$\delta(i) = \min_{j \in V_o} d_{ij}$, the distance from v_i to the nearest facility,

$z = \sum_{i=1}^{n} \delta(i) w'_i$, the cost of the solution,

$y = z^* - z,$ where z^* is the optimal crisp cost,

and

$$\mu_g(V_0, w'_1, \ldots, w'_n) = g(y) = \begin{cases} 0 & if \quad y < 0 \\ \frac{y}{p_g} & if \quad 0 \leq y \leq p_g \\ 1 & if \quad y > p_g \end{cases} \quad (3)$$

where p_g is the greatest cost reduction that the decision-maker expects to obtain.

With these choices, the fuzzy p-median problem is equivalent to the following mathematical programming problem [2, 3]:

$$\begin{aligned}
& \text{Max } \lambda \\
& \text{s.t. } \lambda + \sum_{i=1}^{n} \sum_{j=1}^{n} \frac{d_{ij}}{p_g} x_{ij} \leq \frac{z^*}{p_g} \\
& \lambda - \sum_{i=1}^{n} \sum_{j=1}^{n} \frac{1}{p_f} x_{ij} \leq 1 - \frac{\sum_{j=1}^{n} w_j}{p_f} \\
& 0 \leq x_{ij} \leq w_j y_i \qquad 1 \leq i,j \leq n \\
& \sum_{i=1}^{n} x_{ij} \leq w_j \qquad 1 \leq j \leq n \\
& \sum_{i=1}^{n} y_i = p \\
& y_i \in \{0,1\} \qquad 1 \leq i \leq n
\end{aligned} \quad (4)$$

In the following sections we develop a specific heuristic procedure for the fuzzy p-median problem. Initially, we show that in practice we can consider μ_g as a function of V_0 and x.

Theorem 1. *Among the solutions $(V_0, w_1, \ldots, w_n)$ with a fixed set of nodes V_0 and a fixed total uncovered demand x,*

1) The function μ_D attains its maximum in the solutions for which μ_g is maximum.
2) These solutions are those leaving uncovered the demand corresponding to the farthest nodes from their respective facilities.

Let us call $\lambda(V_0, x)$ the maximum value of μ_D among the solutions with a set of nodes V_0 and total uncovered demand x. A solution of the fuzzy p-median problem is an $(n+1)$-tuple $(V_0, w_1, \ldots, w_n)$ for which the membership degree $\mu_D(V_0, w_1, \ldots, w_n)$ is maximum or equivalently a pair (V_0, x) for which $\lambda(V_0, x)$ is maximum, since from x we can calculate the demands $w_1, \ldots, w_n$ providing the best μ_D.

In Section 3 we give an algorithm to calculate $\lambda(V_0) = \max_x \lambda(V_0, x)$ for a given V_0. In Section 4 we propose an interchange heuristic procedure to seek a location V_0 with maximum $\lambda(V_0)$.

3 Calculating the satisfaction level

Let us describe a subroutine to calculate $\lambda(V_0)$ for a given location V_0. Firstly, we calculate $\delta(i) = \min_{v_j \in V_o} d_{ij}$ for each node, and the cost $z = \sum_{i=1}^{n} \delta(i) w_i$, of the location V_0. If we are working with a near optimal crisp z^* and $z < z^*$, then we have found a better optimum and we should actualize $z^* = z$. So assume that $z \geq z^*$.

We order $V = \{v_{i_1}, \ldots, v_{i_n}\}$ in such a way that

$$\delta(i_1) \geq \delta(i_2) \geq \ldots \geq \delta(i_n).$$

Note that the nodes $v_i \in V_0$ will be in the final positions with $\delta(i) = 0$. Now, take the maximum r such that

$$\sum_{k=1}^{r} \delta(i_k) w_{i_k} \leq z - z^*.$$

We are looking for the demand we need to reduce in order to make that the cost of location V_0 with partially covered demand equals the optimal crisp cost z^*. At this time, we must leave uncovered all the demand of nodes $v_{i_1}, \ldots, v_{i_r}$ and possibly part of the demand of $v_{i_{r+1}}$. The minimum reduction a of demand which makes the cost equal to z^* is determined by

$$\sum_{k=1}^{r} \delta(i_k) w_{i_k} + \delta(i_{r+1}) \left[a - \sum_{k=1}^{r} w_{i_k} \right] = z - z^*.$$

Whatever solution (V_0, x) satisfies $\mu_g(V_0, x) = 0$ if $x \leq a$ and $\mu_g(V_0, x) > 0$ if $x > a$ (see Figure 1). If $a \geq p_f$ then we have exceeded the tolerance level for reducing the demand with $\mu_g(V_0, a) = 0$, thus $\lambda(V_0) = 0$, and we have finished. Otherwise, calculate

$$s = \sum_{k=1}^{r+1} w_{i_k},$$

the maximum demand we can reduce before starting with node $v_{i_{r+2}}$, and $z_a = z*$, the cost of the fuzzy solution (V_0, a).

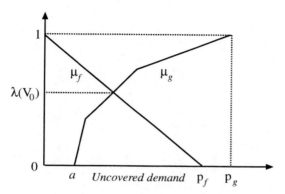

Fig. 1. First case

In the general case, we have two possibilities which are illustrated in Figure 2. We are dealing with node $v_{i_{r+1}}$. The current reduced demand is a, and s is the reduction of demand that would be attained by leaving node $v_{i_{r+1}}$ completely uncovered. The possibilities are (i) $\mu_g(s) < \mu_f(s)$ or (ii) $\mu_g(s) \geq \mu_f(s)$.

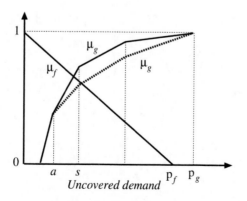

Fig. 2. General case

In case (i) all the demand of $v_{i_{r+1}}$ must be left uncovered, so we recalculate z_a, a, s and r according to the following rules:

$$z_a = z_a - \delta(i_{r+1})(s-a); \quad a = s; \quad s = s + w_{r+2}; \quad r = r+1.$$

In case (ii) $\lambda(V_0)$ is attained with a reduction of demand x such that $a \leq x \leq s$. To decide which is the case and calculate x in the second instance, we note that

$$\mu_f(x) = 1 - \frac{x}{p_f}; \quad \mu_g(x) = \frac{z^* - z_a + (x-a)\delta(i_{r+1})}{p_g}.$$

In case (ii), we obtain that

$$x = \frac{p_f p_g - p_f(z^* - z_a) + p_f a \delta(i_{r+1})}{p_f \delta(i_{r+1}) + p_g}$$

and hence

$$\lambda(V_0) = \mu_f(x) = \frac{p_f \delta(i_{r+1}) + z^* - z_a - a\delta(i_{r+1})}{p_f \delta(i_{r+1}) + p_g}.$$

These considerations lead to the following algorithm:

Inputs: $D = [d_{ij}], w_j, z^*, V_0, p_f, p_g$.

STEP 0:

Calculate $\delta(i) = \min\limits_{v_j \in V_o} d_{ij}$, $z = \sum\limits_{i=1}^{n} \delta(i) w_i$. If $z < z^*$, then make $z^* = z$.
Calculate the permutation $i_1, \ldots, i_n$ such that $\delta(i_1) \geq \delta(i_2) \geq \ldots \geq \delta(i_n)$.
Calculate the maximum r such that $z = \sum\limits_{k=1}^{r} \delta(i_k) w_{i_k} \leq z - z^*$ and a.

STEP 1: If $a \geq p_f$ then $\lambda(V_0) = 0$. Stop. Otherwise, go to step 2.

STEP 2: While

$$\frac{z^* - z_a + (s-a)\delta(i_{r+1})}{p_g} < 1 - \frac{s}{p_f},$$

assign $z_a = z_a - \delta(i_{r+1})(s-a); \quad a = s; \quad s = s + w_{r+2}; \quad r = r+1.$

STEP 3: Compute

$$\lambda(V_0) = \frac{p_f \delta(i_{r+1}) + z^* - z_a - a\delta(i_{r+1})}{p_f \delta(i_{r+1}) + p_g}.$$

Outputs: $\lambda(V_0)$.

4 An interchange heuristic procedure

In this section we propose a descend method of local optimization. It is well known that the efficiency of these methods depends on both the neighbourhood structure and the search strategy [11]. As we have seen, we are looking for a set of p nodes V_0 with maximum $\lambda(V_0)$. We define the neighbourhood of V_0 as the family of all sets of nodes obtained from V_0 with just one swap. The basic search scheme is the same as Teitz and Barts algorithm [13], i. e. we jump from a location V_0 to the location V_1 in its neighbourhood with the greatest $\lambda(V_1)$. However, we tested a variant search strategy in which we jump to the first location V_1 that we find in the neighbourhood of V_0 with a greater satisfaction level $\lambda(V_1)$. To distinguish them, we call one strategy 1 and the other, strategy 2, respectively.

At this point, we face a difficulty which has no equivalent in crisp combinatorial optimization problems. In this kind of problem, it is very unusual that the objective function takes the same value for every element in a given neighbourhood. On the contrary, if we are considering a location V_0 whose total transport cost z is much greater than z^*, it can happen that λ takes the value 0 at V_0 and at all nearby locations. In this case, the heuristic would stop. To handle this case, we stipulate that while λ remains null, the interchange rule must be to jump to the location with lowest cost z instead of the above-mentioned locations.

At last, we note that when comparing the satisfaction of a given location V_0 with the satisfaction of its neighbours, the crisp optimum z^* must be actualized as mentioned in Section 3. In this case, $\lambda(V_0)$ must be recalculated and we must start the comparison process again.

Therefore, the procedure is as follows:

Inputs: $D = [d_{ij}], w_j, z^*, p_f, p_g$.

STEP 0: Select a starting location V_0 and calculate $\lambda(V_0)$.

STEP 1: For each location V in the neighbourhood of V_0 calculate $\lambda(V)$ and the cost $z(V)$. Call V_1, V_2 the locations with greatest $\lambda(V_1)$ and lowest $z(V_2)$ respectively.

STEP 2: If z^* has been actualized in step 1, then recalculate $\lambda(V_0)$ and go to step 1.

STEP 3: If $\lambda(V_1) = 0$, then assign $V_0 = V_2$ and go to step 1.

STEP 4: If $\lambda(V_1) \leq \lambda(V_0)$ stop.

STEP 5: If $\lambda(V_1) > \lambda(V_0)$, then assign $V_0 = V_1$ and go to step 1.

Outputs: V_0, $\lambda^* = \lambda(V_0)$.

Remark 1. From λ^* we can calculate the uncovered demand $x = p_f(1 - \lambda^*)$ and the reduction of cost $p_g\lambda^*$. The demand must be reduced from the nodes v_i with greatest $\delta(i)$ (according to the notation of Section 3).

Remark 2. The procedure can cycle in step 3 if there exist two or more neighbouring locations with the same minimal cost in a zone with null satisfaction level. We can avoid this by introducing random jumps when a cycle is detected. However, this is a very unlikely situation and so it is easier to start again with another location.

Remark 3. The procedure we have described corresponds to what we have called strategy 1. Strategy 2 can be implemented by leaving step 1 as soon as a location V_1 is found with $\lambda(V_1) > \lambda(V_0)$. Note that in this case we have $\lambda(V_1) \neq 0$ and hence location V_2 will be unnecessary.

5 Computational results

5.1 Testing the procedure

We have checked our algorithm with some of the test problems of the website [1] that has been used in many location papers. Specifically, we have chosen three different networks with 100 nodes and another three networks with 200 nodes. The rows in Table 1 contain (in this order) the data corresponding to problems called there *pmed1*, *pmed4*, *pmed5*, *pmed7*, *pmed9* and *pmed10* including the optimal crisp costs z^* (the demand of each node equals 1). To calculate satisfactions, we have taken 5% of the total demand for p_f and 10% of z^* for p_g. The third column contains the fuzzy optimum of each problem, which has been calculated with the package GAMS® (solver OSL).

We have run our procedure 100 times for each strategy and problem starting from a random location. This has been performed on a 600 MHz iMac®. The time of each execution has not exceeded 5 minutes in the worst case. The fourth column of Table 1 shows the number of times the optimum has been attained and the last column shows the average deviation of the global satisfaction λ obtained in each execution.

We see that the frequency does not depend on the size of the problem, nor on the chosen strategy. However Strategy 1 seems to be more reliable in the sense that its results are always acceptable while Strategy 2 is much more oscillatory. The mean number of interchanges of both strategies for each problem is similar.

On the other hand, the average deviation remains satisfactory in every case. Note that in the worst case, the deviation of 5.01% corresponds to a solution for which the reduction of total demand is the same as the optimal reduction (4 units) and whose cost reduction is just 6.28 less than the optimal one (81.55). For interpreting this value let us observe that the decision-makers best expectation is a reduction of 125.5, and so a reasonable reduction could be about 62.

Table 1. Computational results

Problem	Size		Crisp cost	Fuzzy optimum	Frequency		Average deviation %	
	n	p	z^*	λ^*	(1)	(2)	(1)	(2)
1	100	5	5819	52.70	64	92	0.30	0.06
2		20	3034	60.13	32	8	2.90	2.99
3		33	1355	66.67	38	54	0.80	0.39
4	200	10	5631	58.00	50	100	1.82	0
5		40	2734	65.89	14	0	0.97	1.31
6		67	1255	64.98	20	3	3.6	5.01

Finally, it is striking that Strategy 2 always obtains the optimal satisfaction in problem 4 and never in problem 5. This seems to be related with the number of local optima. Indeed, in problem 5 we have reached 21 near-optimal solutions.

5.2 Analyzing the outputs

In order to use our procedure as a DSS tool we should take into account not only the best near-optimum obtained but all the available near-optima or at least the better ones. Let us analyze in detail the outputs of problems 1 and 2. In Problem 1 we have found two fuzzy near-optimal locations:

Table 2. Near-optima for Problem 1

	Location	λ	z	Reduction of cost	Reduction of demand
L_1	$v_7, v_{37}, v_{42}, v_{91}, v_{99}$	52.70	6303	306.65	2
L_2	$v_7, v_{13}, v_{25}, v_{65}, v_{91}$	52.27	5821	304.16	2

L_1 is the fuzzy optimum. Since the crisp optimal cost is 5819 m.u., we can see that neither of them is a crisp optimum. In fact, the nearest fuzzy optimum corresponds to the location $v_7, v_{37}, v_{42}, v_{91}, v_{99}$, and by means of the Teitz and Barts algorithm for the crisp p-median problem we have found 5 intermediate locations, i. e. if we order the possible locations in decreasing order of crisp cost, the fuzzy optimum is at least in the seventh place.

On the other hand, L_1 and L_2 have very similar characteristics, so the decision maker could be interested in any of them.

Problem 2 has a very different behaviour. We have found many near-optimal solutions. Table 3 shows the data associated to the solutions corresponding to the best four values of λ. For each of these values there are several alternative locations which happen to have the same crisp cost z (contained also in Table 3).

Table 3. Near-optima for problem 2

Location group	λ	z	Reduction of cost	Reduction of demand
A	60.13	3034	182.43	2
B	59.58	3038	180.76	2
C	58.47	3046	177.39	2
D	58.06	3049	176.15	2

In Figure 3 we have detailed the different locations corresponding to each group. The columns correspond to nodes v_1, v_3, v_5, v_6, v_7, v_8, v_9, v_{10}, v_{12}, v_{13}, v_{22}, v_{26}, v_{27}, v_{34}, v_{38}, v_{48}, v_{50}, v_{51}, v_{55}, v_{58}, v_{60}, v_{62}, v_{63}, v_{66}, v_{71}, v_{72}, v_{77}, v_{83}, v_{87}, v_{91}, v_{93}, v_{96}, v_{100}. Each shaded square corresponds to a node in the location.

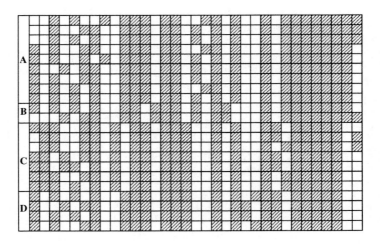

Fig. 3. Location groups of problem 2

We see that each location has 10 fixed nodes and another 10 ranging in a set of 23 possibilities. This provides a considerable margin to the decision-maker. We think that this kind of diagram can also give a global insight about the fuzzy structure of the solution.

6 Conclusions

Our algorithm offers a good behaviour on the problems we have tested. It uses to end with a good near optimal solution, but the final location seems to be very sensitive to the starting location. We think this can be significantly improved by considering fuzzy versions of simulated annealing procedures. However, we must point out that finding the exact optimal solution is not the main goal of an algorithm of this kind. A better use is to give the decision-maker not just the best location obtained, but many near optimal solutions corresponding (in many cases) to various alternative locations. In this way, we are selecting among many uninteresting possibilities, some locations that the decision-maker can consider with regard to other factors not formalized in the model. Indeed, the satisfaction degree λ can be easily modified to reflect any desirable property of a possible location.

When applied to a very large problem it can be useful to stipulate a lower bound for the desirable satisfaction degree of a solution, so we can stop as soon as a location is found with a satisfaction of, say, 50%. Our calculations indicate that when the starting location has a reasonable cost (for instance, if we take a crisp near-optimal) then solutions with acceptable satisfaction are very quickly found.

Finally, the empirical results shown in Subsection 5.2 suggest that the Heuristic Concentration method could be efficiently applied to refine the output as well as the analysis of the solutions.

References

1. Beasley, J. E. OR-Library <mscmga.ms.ic.ac.uk/jeb/orlib/pmedinfo.html>.
2. Canós M.J., Ivorra C., Liern V. (1999), An exact algorithm for the fuzzy p-median problem, European Journal of Operational Research **116**, 80–86.
3. Canós M.J., Ivorra C., Liern V. (2001), The fuzzy p-median problem: A global analysis of the solutions, European Journal of Operational Research **130**, 430–436.
4. Chanas, S. and Kuchta, D. (1994), Discrete Fuzzy Optimization, in Delgado, M., Verdegay, J.L. and Vila, M.A. Fuzzy Linear Programming: From classical methods to new applications, Fuzzy optimization: recent advances, Physica-Verlag, Heidelberg.
5. M.S. Daskin (1995), Network and Discrete Location: Models, Algorithms and Applications, Wiley, Nueva York.
6. S.L. Hakimi (1964), Optimum Locations of Switching Centers and the Absolute Centers and Medians of a Graph, Operations Research, **12**, 450–459.
7. S.L. Hakimi (1965), Optimum Distribution of Switching Centers in a Communication Network and Some Related Graph Theoretic Problems, Operations Research, **13**, 462–475.

8. O. Kariv y S.L. Hakimi (1979), An Algorithmic Approach to Network Location Problems. II: The p-Medians, SIAM Journal on Applied Mathematics, **37**, 539–560.
9. M. Labb, D. Peeters y J.F. Thisse (1995) Location on Networks in: M.O. Ball, T.L. Magnanti, C.L. Monma y G.L. Nemhauser (eds.) Network Routing, Chapter 7, Elsevier Science, Amsterdam.
10. Lai, Y. J. and Hwang, C. L., (1992), Fuzzy Mathematical Programming: Methods and Applications, Springer-Verlag, Berlin.
11. Reeves, C. R. (ed.) (1995), Modern Heuristic Techniques for Combinatorial Problems, McGraw Hill Book Company, London.
12. Slowinski, R. (1998), Fuzzy Sets in Decision Analysis, Operations Research and Statistics, Kluwer Academic Publishers.
13. M.B. Teitz y P. Bart (1968), Heuristic Methods for Estimating the Generalized Vertex Median of a Weighted Graph, Operations Research, **16**, 955–951.
14. Zimmermann, H.J. (1997), Fuzzy Mathematical Programming, in T.Gal, H.J. Greenberg (eds.), in Advances in Sensitivity Analysis and Parametric Programming, Kluwer Academic Publishers, Boston.

Route Choice Making Under Uncertainty: a Fuzzy Logic Based Approach

Vincent Henn

Traffic Engineering Laboratory (LICIT), INRETS–ENTPE, rue Maurice Audin, 69 518 Vaulx-en-Velin cedex, France

Abstract. This paper deals with the representation of uncertainty in the drivers' route choice process. The objective is not to find the *best* route, but rather to model the way drivers actually choose, without knowing perfectly the traveling conditions.

We propose two different models based on fuzzy sets in order to represent this lack of knowledge, but also the non-willingness of drivers of an exact optimization.

The first one is based on the comparison of fuzzy generalized costs gathering both imprecision and uncertainty and it is proved to have inconsistencies, so that it is not suitable for uncertainty representation.

The second one is based on an individual behaviour which is justified by decision theoretic arguments. It enables to represent the effect of uncertainty on individual choices as well as the complexity of the interactions between those choices when they are made simultaneously on a road.

1 Introduction

Traffic managers are willing to optimize the functioning of their roads network, by improving the traffic conditions, reducing the congestion as well as the noise and pollutants emissions. The problem is that traffic flows are made of independent drivers who seek their own individual optimum.

Thus, before controlling the traffic (towards a "social" optimum), traffic managers have to understand how drivers behave. For this, they use models or representations of this behaviour in order to predict how traffic flows spread out over the roads network. In this paper, we will focus our attention on the drivers' route choice process and on the resulting models which are referred to as *traffic assignment models*.

Up to a few years ago, such models used to suppose that drivers knew everything and were able to optimize their travel by making the best route choice. Such an hypothesis is of course not exactly correct because drivers are not always aware of the traffic conditions they will encounter during their trip, for example. But, anyway, it makes it possible to formalize properly the traffic assignment problem.

However, for a few years, the emergence of Advanced Travellers Information Systems (ATIS) has made this assumption to be revised. Indeed, those systems, giving information on the traveling conditions to drivers, are supposed to enhance the level of service of roads and reduce congestion by making some people divert from congested areas. It is thus clear that the assumption of drivers' perfect knowledge cannot be maintained in order to study the effects of some ATIS. Indeed, what would be the interest of any additional information?

In this paper, we propose to explore the non-optimality of the drivers' route choice and we will present a model based on the new assumption that drivers not only *don't know* the precise traffic conditions, but they also *don't care* about an exact optimization of their travel costs. They are then supposed to use a sort of heuristic leading them to a *nearly* optimal choice.

In a first section, we will begin by formulating the route choice problem and exploring the classical models. In the following, we will explain how fuzzy logic is a good framework for representing imperfections in the route choice process and we will propose some models of route choice.

2 Modeling Traffic Assignment

2.1 Definition of the Problem

The aim of a traffic assignment model is to represent the way drivers choose their route at each intersection.

In this paper, we will simplify this problem by considering a simple network with only one origin O for travels, one destination D and n isolated routes between those two points. Of course networks generally have more than one single origin-destination pair and routes are not independent because they intersect or share some portions of roads. Meanwhile, such an assumption will enable us to focus on the route choice problem without considering the interactions between routes.

Let us note q the total flow (that is: the number of vehicles passing through one point, by time unit) from O to D and q_i the flow on each route i (of course, $\sum_i q_i = q$).

Our objective is to determine the proportion $\gamma_i = q_i/q$ of traffic using each route i (with $\sum_i \gamma_i = 1$).

2.2 Individual Rational Behaviour

Individual drivers can be considered to be *free* and choose their route independently, without cooperating one with each other, but also to be *rationale* and try to minimize the cost of their trip.

The cost of a trip can include a lot of factors such as the travel time, the monetary costs (petrol, tolls, etc.), the travel distance, the complexity of a path, and so on. But a very large number of drivers are to be considered at the same time (several hundreds of thousands for the morning peak hour in a medium-sized city) so that a multi-criteria analysis would be very complex.

Finally, we have chosen to consider that the only criterion taken into account by drivers is a single cost, gathering all those factors.

We are not interested in the way those factors are combined and, in a general way, we only consider that the cost C_i of a route i is an increasing function of the traffic q_i using this route. This represents the fact that drivers are embarrassed by each others.

For illustrative reasons, we will furthermore suppose that this cost is only made of the travel time it takes to reach the destination starting from the origin.

Drivers' rationality implies that the vector γ of proportions of traffic using each route is a function of the vector of costs C of the different possible routes. Those latter costs are also functions of the traffic $q_i = \gamma_i.q$ on each route. So, finally, the traffic assignment problem can be formulated by the following fixed-point problem:

$$\gamma = \mathcal{F}\big(C(\gamma)\big) \tag{1}$$

with the function $\mathcal{F}$ resulting from the aggregation of individual route choices.

2.3 Deterministic Equilibrium

Let us consider a stationary situation where traffic conditions repeat exactly the same everyday (without incident, with a constant demand for travel, etc.). We can suppose that drivers are perfectly aware of those conditions and, thus, we can further suppose that an equilibrium is reached where each driver achieves to minimize his/her travel cost.

Indeed, if one day a driver used a route i with a cost C_i higher than the cost C_j of another route j, (s)he would likely change his/her route the day after and rather choose route j.

Therefore, in such an equilibrium situation, the cost on all actually used route is equal and minimum. This principle is referred to as the Wardrop principle [1] in the traffic literature (it is nothing but a special case of a Nash equilibrium).

This equilibrium state has been extensively studied in the literature and it is an important basis for traffic assignment, but it is not realistic. Indeed, we know that drivers are not aware of everything and this is exactly the reason why giving information to them (for example by ATIS) is worthwhile.

2.4 Situations with Imperfect Drivers' Knowledge

Zero vs Full Information If we wish to study the advantage that drivers can take of any supplementary information, then it is clear that we have to consider that they do not know everything.

But on the other hand we cannot consider that drivers do not know anything before they get informed. Indeed, without ATIS, drivers are able to choose their route and they make some nearly optimal choice rather than making some completely random choice.

So, finally, we have to consider some intermediate situations, between the extremes of a full-information case (corresponding to the Wardrop equilibrium) and a zero-information one where drivers would be blind and would choose their route on a pure random basis.

Recurrent Situations We will focus our attention on "recurrent" situations where traffic conditions repeat every day *almost* the same, so that drivers are familiar with those conditions, but are not completely aware of the current state of the traffic, because of some random effect.

In such situation, we can suppose that an equilibrium is reached, thanks to the familiarity of divers, but this state is not necessarily optimal, due to randomness.

Recurrent situations are defined by several possible states ω of traffic. Each one is actually a vector of traffic conditions giving the demand for travel between origin and destination, the capacity of the routes, the incidents, etc.

The set Ω of the possible states of traffic is supposed to be not too big so that drivers have experienced each state ω several times and know them all.

In such situations, we can assume that:

- Drivers know the set Ω of possible traffic conditions.
- For each vector ω of conditions, they know the cost $C_i(\omega)$ they would incur on each route i.
- They know the "chance" that each vector of conditions occurs. For example they may have drawn some probability or possibility measures from the previous days experiences.
- But they do not know the current conditions ω^* which they will actually encounter.

With such assumptions and given the rationality principle, traffic will be in an equilibrium state where no driver will be surprised by the *a posteriori* incurred cost, even if his/her choice is not optimal in regard to the current situation.

3 Fuzzy Sets as a Basis for Representing Imperfections

So, finally, we are interested in considering the imperfections which lie in the drivers' route choice process, but only in some situations where those imperfections are limited and known.

Fuzzy logic gives a worthwhile framework for representing such imperfections. We will consider three of them, covering the different semantics of fuzzy sets [2]: Imprecision in the costs perception; Uncertainty on the traveling conditions and Fuzzy preference between costs.

3.1 Imprecision of Perception

Two values of a cost are difficult to distinguish from the drivers' point of view when they are too close. Actually, drivers are not interested in this distinction and do not care about the difference between close values, just because "13 minutes" is nearly the same as "13 minutes and 3 seconds".

We can thus suppose that drivers handle *imprecise* costs represented by fuzzy numbers rather than crisp ones. It is important to understood that this imprecision is due to the non-capacity or non-willingness of drivers to distinguish between different values for the same cost.

The membership function of the *perceived* cost $\widetilde{C}_i^P$ of route i is linked to the degree of similarity with a typical element which is the *real* cost C_i. It can be, for example, constructed with a similarity (or proximity) measure S, a function from $\Re \times \Re$ into $[0;1]$ verifying (see [3]):

$$\forall x \in \Re, \quad S(x,x) = 1$$
$$\forall x, y \in \Re, \quad S(x,y) = S(y,x)$$
$$\forall x, y, z \in \Re, \quad S(x,z) \geq \min\{S(x,y); S(y,z)\}$$

The membership function of the fuzzy costs is then defined by:

$$\forall x \in \Re, \quad \mu_{\widetilde{C}_i^P} = S(x, C_i) \qquad (2)$$

An example of a similarity function that can be used is the following function with parameter ϵ:

$$S(x,y) = \begin{cases} 1 - \dfrac{|x-y|}{\epsilon} & \text{if } |x-y| \leq \epsilon \\ 0 & \text{otherwise} \end{cases}$$

3.2 Uncertainty on the Traveling Conditions

Fuzzy logic and, more specifically, possibility theory can represent the randomness of traffic conditions which is the source of ignorance of drivers: they do not know the current conditions, but they know the set Ω of *possible* conditions.

This set is rather fuzzy but the semantic is here different from that of the perceived cost $\widetilde{C}_i^P$. Indeed, the grade of membership of a vector of condition ω does not represent any similarity degree, but it represents a possibility level $\pi(\omega)$ that this event might occur.

The membership function (or, equivalently, the possibility distribution) can be constructed based on two extreme sets of conditions.

Let us introduce Ω^+ the (classical) set of conditions which are *fully possible*, because for example they have been observed with a frequency higher than a level α^+. This set is the kernel of Ω.

In a similar way, let us define the set Ω^- of conditions which are *fully impossible*, because they have occurred with a frequency lower than a level α^-. The complementary of this set is the support of Ω.

Let us define Ω^0 the set of intermediate conditions, not belonging to Ω^+, nor to Ω^-:

$$\Omega^0 = \{\omega : \omega \notin \Omega^+; \omega \notin \Omega^-\}$$

For simplicity, we will construct the membership function of Ω by a linear interpolation between Ω^- (where $\mu_\Omega = \underline{0}$) and Ω^+ (where $\mu_\Omega = \underline{1}$). We will see later that such a simple shape has no important effect on the resulting assignment.

3.3 Grade of Preference Between Costs

A third type of fuzziness that is introduced in the route choice model is the notion of grade of preference.

It is clear that a driver's preference for spending 13 minutes on route 1 is higher than his/her preference for 14 minutes on route 2. But this latter preference is also much higher than the preference for spending 30 minutes on route 3. So we have to introduce the notion of degree of preference, with values into $[0;1]$.

Let us define a preference measure ϕ between costs. This measure $\phi(x|X)$ gives the preference for paying the cost x among the set of possible costs X. This measure must verify:

$$\forall x, y \in X, \quad x < y \implies \phi(x|X) \geq \phi(y|X) \quad (3)$$

$$\max_{x \in X} \phi(x|X) = 1 \quad (4)$$

For simplicity of notation, we will omit the set X of possible costs and the preference for the costs on route i will be simply denoted $\phi(C_i)$.

Since we suppose that drivers handle fuzzy costs rather than crisp ones, we have to extend the preceding measure $\phi(C_i)$ into a measure of preference $\varphi(\widetilde{C}_i^P)$ between fuzzy perceived costs. The problem is then to define the notion of ranking (the "<" relation). This issue has been widely addressed in the fuzzy literature and many measures have been proposed (see for example, among others, [4–7]).

4 Fuzzy Sets Based Heuristics for Traffic Assignment

We will present in this section two models for representing the heuristic used by drivers to find their optimal route. The first one is based on the comparison of the predicted costs of the routes, whereas the second one is based on some decision-theoretic basis.

4.1 Comparison of fuzzy costs

The first idea to represent drivers' rationale in route choice making is to combine both imprecision and uncertainty into single generalized costs and compare those costs in order to find the best route to choose.

Even if such an idea is appealing and has been used by many authors (see, for example [8–13]), we will show that it can pose some problems.

Fuzzy generalized costs The basis of this approach is to combine different semantics of fuzzy sets. As we have already seen, the membership function of a fuzzy set can both represent a similarity degree with a typical element of the set or a possibility level that an event occurs [2].

We have already mentioned that in recurrent situation, for each vector ω of conditions, drivers are supposed to know the incurred costs $C_i(\omega)$ on each route i. Such costs are not handled directly, but by the way of imprecise fuzzy perceived costs $\widetilde{C}_i^P(\omega)$. Drivers also know the possibility level $\pi(\omega) = \mu_\Omega(\omega)$ that each state ω occurs.

From this stuff, a fuzzy generalized cost $\widetilde{C}_i$ can be devised gathering both *imprecision* due to non-willingness of drivers to handle costs to the nearest second and *uncertainty* due to their incapability of predicting the state of traffic. Such a cost is the union of the possible perceived costs $\widetilde{C}_i^P(\omega)$ in each state ω. Its membership function is given by:

$$\forall x \in \Re, \ \mu_{\widetilde{C}_i}(x) = \max_{\omega \in \Omega} \min \left\{ \mu_{\widetilde{C}_i^P(\omega)}(x); \pi(\omega) \right\}$$

$$= \max_{\omega \in \Omega} \min \left\{ S\left(x, \widetilde{C}_i^P(\omega)\right); \pi(\omega) \right\} \quad (5)$$

This generalized cost can be seen as a *predicted* cost of route i and the membership level of an element x represents something like the possibility level that the value x is *perceived* by a driver.

It is interesting to note that even if the two natures of fuzziness have been considered separately, they cannot be extracted anymore from the final cost. Indeed, given the membership function of $\widetilde{C}_i^P$, it is impossible to come back to the initial cost C_i, neither to the possibility $\pi(\omega)$ of a situation or to the perceived cost $\widetilde{C}_i^P$.

Route Choice model Once the different routes are evaluated with a predicted cost, they can be ranked by comparing those costs, by mean of a preference relation.

The basic idea is to link the probability p_i of choosing route i with the preference degree $\varphi(\widetilde{C}_i)$ for the predicted cost on this route. This means: the more preferred is a route, the more probably it will be chosen.

The simplest model assumes the proportionallity between probability p_i and preference $\varphi(\widetilde{C}_i)$ and verifies the normality condition of probability ($\sum_i p_i = 1$). It yields:

$$p_i = \frac{\varphi(\widetilde{C}_i)}{\sum_j \varphi(\widetilde{C}_j)} \qquad (6)$$

The proportion γ_i of traffic on route i is then equal to this probability, due to the law of large numbers.

We have studied such a model in [11,14] in the particular case where the preference measure is based on the Dubois and Prade comparison indices [4]. With its light formulation it has been proved to be equivalent to the classical discrete choice model (such as logit or probit) when using specific membership functions.

Limitations In spite of its interesting aspects and its quite widespread use, such a model can be criticized and the following reproaches can be addressed.

A first drawback is that the model does not take into account the correlation of uncertainty between costs.[1] Indeed, even if the routes are independent, the uncertainty on their costs can be linked to the same external conditions. For example some rainy weather can induce possible increases in the costs on all the routes, and those increases are not independent.

We have also exhibited in [17] some paradoxical behaviour when uncertainty was taken into account. This problem was not linked with the simple model (eq. 6) briefly presented here and could be observed even with a more valid transformation from possibility to probability (we used, for example, an information-invariant transformation, as proposed by Klir [18]).

Finally, we can question on the validity of the underlying individual model. Actually, it has not been formalized, it has no axiomatic basis and we are not sure it is valid. Indeed, the three semantics of fuzzy sets (imprecision, uncertainty and preference) have been combined without any special care and without preserving their particularity.

Perhaps this can be a source of problem. For example, what is the *meaning* of summing up preference degrees in eq 6? This question can be linked to the more general problematic of probability-possibility transformations (see [19] for a review on the links between those two theories).

Our own conclusion is that such an approach cannot be considered any longer and we propose to develop another one, based on some decision theory fundamentals.

---

[1] This problem is quite similar to a classical one in traffic assignment literature referred to as the problem of *independence of irrelevant alternatives* (see for example [15, p. 294]).

4.2 Decision-Theoretic Basis

Individual choice model For given conditions ω, the cost on route i is $C_i(\omega)$ and it is supposed to be perceived by drivers as a fuzzy cost $\widetilde{C}_i^P(\omega)$ as defined by eq. 2.

Without uncertainty (when the set of possible conditions is reduced to a singleton), drivers are supposed to try to minimize their travel cost so that they will choose route i^* with the highest preferred cost. Since their perception is fuzzy, they will choose route i^* such that $\varphi(\widetilde{C}_{i^*}^P) = 1$.

The problem is to consider "recurrent situations" where drivers do not know the conditions they will encounter during their trip. So they have to take into account the set of all the possible costs $\{\widetilde{C}_i^P(\omega)\}_{\omega \in \Omega}$. If one route is always preferred to the others, that is if $\forall \omega : \varphi(\widetilde{C}_i^P(\omega)) = 1$, it means that this route is *always* the best one, whatever are the traveling conditions, and it can be chosen by a driver. But, generally, it is not the case and a decision model has to be defined, combining both preference and possibility levels. We use the following preference index for each route, considering all the possible conditions [20]:

$$\nu_i = \inf_{\omega \in \Omega} \max\left\{1 - \pi(\omega) \; ; \varphi\left(\widetilde{C}_i^P(\omega)\right)\right\} \quad (7)$$

This is a *pessimistic* index in the sense that it mainly considers the worst (the less preferred) possible situations. It implicitly assumes that possibility and preference measures are commensurable and can be combined.

A symmetrical *optimistic* index could as well be used, depending on the type of drivers' behaviours [21]:

$$\nu_i' = \sup_{\omega \in \Omega} \min\left\{\pi(\omega) \; ; \varphi\left(\widetilde{C}_i^P(\omega)\right)\right\} \quad (8)$$

Every driver facing a set of routes with a vector ν of preference indices is then supposed to choose one route i^* such that

$$\nu_{i^*} = \max_i \nu_i \quad (9)$$

Such a decision model has been justified by [22,20] in the frame of decision theory under uncertainty by an axiomatic similar to the ones given by Savage [23] or by von Neuman and Morgenstern [24] to justify the expected utility criterion.

Traffic Assignment Model The proportion γ_i of drivers choosing route i is calculated by aggregating the individual choices based on the preceding individual model. So, if we consider a homogeneous population with identical drivers, eq. 9 implies that no driver can choose a route with non-maximal preference. This yields to:

$$\nu_i < \max_j \nu_j \implies \gamma_i = 0 \qquad (10)$$

It can be seen as a necessary condition for traffic to be in equilibrium when drivers follow the preceding individual model. Indeed, in "recurrent" situations, we can suppose that an equilibrium is observed where the *ex post* realized costs are coherent with the *ex ante* predicted possible costs.

To further calculate the traffic proportions on each route, we have to consider the fact that costs depend on the traffic which is a consequence of individual choices based on the same costs.

The real cost on each route i can be represented by a traffic flow model, depending on the flow q_i on this route and on some parameters such as, for example, the capacity Q_i of the road and the free-flow cost C_i^0 which would be incurred if the route was empty. For example, a simple cost function can be used, such as the one proposed by the American Bureau of Public Roads [15, p. 358]:

$$C_i(q_i) = C_i^0 \left(1 + \alpha \left(\frac{q_i}{Q_i}\right)^\beta\right) \qquad (11)$$

(Typical values for the parameters are $\alpha = 0.15$ and $\beta = 4$.)

The problem is then to find the traffic assignment verifying the equilibrium relation. For this, we will consider the set of all possible traffic assignment ($\gamma \in [0,1]^n$) and we will extract an assignment vector γ^* verifying eq. 10.

The capacity, the free-flow cost and the total demand q may be function of the vector of conditions ω. But, since we assume that drivers do not know the current traffic conditions, traffic assignment cannot be a function of ω.

The flow on each route i depends on the traffic assignment γ_i and on the total demand: $q_i = \gamma_i.q$. So, the cost on route i depends on the assignment coefficient γ and on the vector of conditions ω:

$$C_i(\gamma_i, \omega) = C_i^0(\omega) \left(1 + \alpha(\omega) \left(\frac{\gamma_i.q(\omega)}{Q_i(\omega)}\right)^{\beta(\omega)}\right) \qquad (12)$$

The perceived fuzzy cost $\tilde{C}_i^P(\gamma_i, \omega)$ can be calculated using eq. 2, as a function of the assignment and the state of traffic.

The preference index can then be calculated, using eq. 7–8, as a function of γ:

$$\nu_i(\gamma) = \inf_{\omega \in \Omega} \max\left\{1 - \pi(\omega) \,;\, \phi\left(\tilde{C}_i^P(\gamma_i, \omega)\right)\right\} \qquad (13)$$

An equilibrium is reached if no driver is surprised by the incurred cost. That is, an equilibrium is reached if the preference indices $\nu(\gamma)$ for the routes, given by eq. 13 are compatible with the rationality equation in eq. 10.

Equilibrium heuristic A simple trial-and-error heuristic can be devised to find such an equilibrium, by progressively correcting the traffic assignment in regards to the rationality principle:

Initialization: Take an arbitrary assignment γ^0. Without any further information, the traffic can be homogeneously assigned on all the n routes: $\forall i,\ \gamma_i^0 = 1/n$.

Step k: Correction of the current assignment γ^k:
- With the current traffic assignment vector γ^k, calculate the possible real costs (eq. 12) and the possible perceived costs on each route (eq. 2). Then calculate the preference vector ν^k by using eq. 13.
- Define IN^k the set of routes that *can* be used in regards to the preference indices, and OUT^k, the set of routes that could be used at step $k-1$ and that cannot at step k:

$$\text{IN}^k = \left\{ i : \nu_i^k = \max \nu_j^k \right\} \quad (14)$$

$$\text{OUT}^k = \left\{ i \in \text{IN}^{k-1} : \nu_i^k < \max_j \nu_j^k \right\} \quad (15)$$

- A proportion Γ^k of traffic has been assigned in excess at step k and is to be re-assigned at the next step to routes belonging to IN^k:

$$\Gamma^k = \sum_{i \in \text{OUT}^k} \gamma_i^k \quad (16)$$

- Only a portion λ of Γ^k is actually re-assigned, in order to avoid overreaction. Traffic assignment at the next step is calculated as follows:

$$\forall i \in \text{IN}^k : \gamma^{k+1} = \gamma^k + \lambda.\Gamma^k/|\text{IN}^k| \quad (17)$$
$$\forall i \in \text{OUT}^k : \gamma^{k+1} = (1-\lambda).\gamma^k \quad (18)$$

Exit condition If the wrongly assigned traffic proportion is still important ($\|\Gamma^k\| > \epsilon$), go to the next step: $k \leftarrow (k+1)$.
Else, an equilibrium is reached: end of the heuristic.

5 Analysis of the Model

In this section, we will study the behaviour of the proposed model and particularly the effects of introducing fuzziness in the route choice onto the traffic assignment.

Actually, we will only consider the pessimistic behaviour (eq. 7) since the optimistic one (eq. 8) seems to be too much optimistic and routes are often rated at the highest level ($\nu_i = 1$), so that the model is enable to discriminate between routes.

5.1 Imprecision and Grade of Preference

Even though the two concepts of imprecision and grade of preference had been introduced separately, they will be gathered in the following.

Indeed, the fuzzy perceived costs represent an interesting intermediate formal step between real costs and the grade of preference for routes. But nevertheless, it is difficult to validate such a fuzzy perception because it cannot be observed anywhere since it only exists in drivers' mind, whereas travel costs as well as preferences can be observed and measured in reality.

Furthermore, a fuzzy preference measure is more difficult to be built between fuzzy costs than between real costs (as we mentioned earlier) and, of course, such a latter measure can also easily be supposed to encompass the fuzziness of drivers' perception

So, finally, when calculating the preference for a route in eq. 13, we will not consider the intermediate step of fuzzy perception but only the fuzzy preference between real costs $\varphi(C_i(\gamma_i, \omega))$.

We have applied the model on some theoretical networks and we have found that the effect of this preference measure was very weak and traffic assignment was very little sensitive to those parameters. In particular, this means that the shape of membership functions or preference measures is not critical and can be designed easily, for example using arbitrarily some simple trapezoidal shapes, as we proposed.

5.2 Influence of Uncertainty

On the contrary, the effects of uncertainty are much more important. Actually, it is a good thing since the model was designed in order to represent those effects.

For example, let us consider a simple 2-routes network with similar routes ($C^0 = 10$ min., $Q = 20$ veh./h, $q = 30$ veh./h), but with uncertainty on the capacity of route 1. Every state of traffic ω corresponds to a given capacity $Q_1 \in [(1-\eta).20; (1+\eta).20]$. Figure 1 depicts the effects of the amount of uncertainty $\eta \in [0; 50\%]$ on the proportion of traffic on route 1 at equilibrium.

It can be observed that, when the uncertainty increases on route 1, the proportion on route 2 increases, which means that drivers prefer route 2. This is not surprising since they are risk-averse.

But such an intuitive reasoning is not always valid and an individual pessimistic behaviour does not always lead to a higher proportion of traffic on the "safer" route, that is on the route with the minimum uncertainty. Hence, given the preceding network, another type of uncertainty on the route 1 may have lead to an opposite behaviour, with a higher proportion of traffic on route 1 ($\gamma_1 > 0.5$).

To illustrate this quite paradoxical behaviour and in order to clarify the results, we have considered a possibility distribution with values only in $\{0; 1\}$.

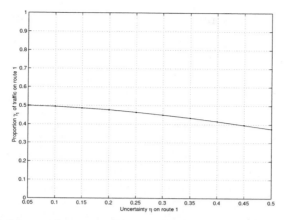

Fig. 1. Influence of uncertainty on the proportion of traffic on route 1

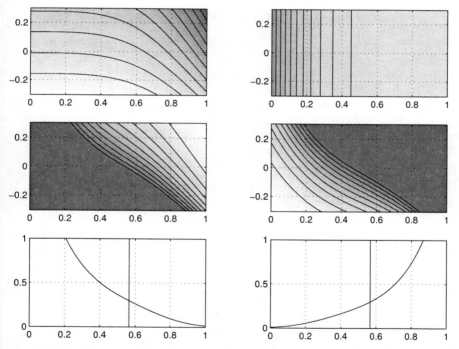

Fig. 2. Calculation of the traffic assignment. Left column is for route 1 and right one is for route 2; darker colors are for higher values; first row depicts the costs C_i depending on γ_1 (x-axis) and ω (y-axis); second row depicts the preference $\phi(C_i)$ for a cost, depending on γ_1 and ω; third row depicts the preference $\nu_i(\gamma)$ for a route, depending on γ_1; the equilibrium assignment such that $\nu_1 = \nu_2$ is pointed out.

The set Ω of possible conditions is then a classical (non-fuzzy) set. In such a situation, the calculation of the equilibrium can be made directly, without using the presented trial-and-error heuristic.

Figure 2 depicts, for example, the calculation of the assignment in the case when the capacity is certain and fixed ($Q_1 = Q_2 = 20$ veh./h) and the free-flow cost on route 1 is uncertain: $C_1^0 \in [(1-\eta).10; (1+\eta).10]$ (with $\eta = 0.3$). As can be seen on the first row, the cost on route 2 does not depend on the state of traffic ω, but only on the traffic assignment. Nevertheless, the preference for each route depends (in both cases) on the state of traffic and on the assignment. And, finally, there is no reason why the equilibrium assignment γ^* (such that the preference for each route is equal) would be higher or lower than the 0.5 medium value (here, equilibrium is reached for $\gamma_1 = 0.56$).

It must be understood that the collective flow behaviour is more complicated than the individual one, due to interactions between individual choices.

6 Conclusions and Perspectives

We have presented in this paper a general framework for modeling the fact that drivers do not make their route choice with an exact and optimal procedure, but rather based on an heuristic considering explicitly the uncertainty lying in the traveling conditions.

The presented model gives interesting results and demonstrates the complexity of a collective behaviour (even with some simple individual assumptions) and thus it shows the necessity to take into account uncertainty at the individual stage.

One of the major goals of the model was to be able to represent the modification of uncertainty because of the reception of new information by drivers.

Given the proposed framework, it is easy to see that the introduction of information is quite straightforward by combining the initial costs predictions with the new ones, given by information. Such a combination can be modulated by a drivers' confidence into the information medium for example. Since all those costs are represented by fuzzy sets, they can easily be handled and merged. This issue has already been addressed elsewhere (see [11,14]) and was beyond the scope of the present paper.

As for validation of the proposed model, a lot of things remain to be done, but several ideas can already be noted.

Before validating the collective behaviour of the model (which requires heavy experiences), the individual (pessimistic) model will be explored in a psychological point of view, in order to be validated, since the given decision-theory justification is not sufficient.

It is interesting to note that this validation is rather difficult because we do not look for an heuristic finding the *best* route choice (based on some optimization criterion), but the *most representative* one, that is the closest to the drivers' behaviour. For this, we do not search for an uncertainty reduction (and a reduction of the non-optimality of the route choices), but, on the contrary, we look for the wrong and non-optimal heuristic which is the most exact, the most close to drivers' reality. This issue is to be addressed in the future.

References

1. Wardrop, J. G. (1952). Some theorical aspects of road traffic research. Proc. of the Institution of Civil Engineers, **II.1**, 325–378.
2. Dubois, D., and Prade, H. (1997). The three semantics of fuzzy sets. Fuzzy Sets and Systems, **90**, 141–150.
3. Zadeh, L. A. (1971). Similarity relations and fuzzy ordering. Information sciences, **3**, 177–200.
4. Dubois, D., and Prade, H. (1983). Ranking fuzzy numbers in the setting of possibility theory. Information sciences, **30**, 183–224.
5. Delgado, M., Verdegay, J. L., and Villa, M. A. (1988). A procedure for ranking fuzzy numbers using fuzzy relations. Fuzzy Sets and Systems, **26**, 49–62.
6. Cheng, C.-H. (1998). A new approach for ranking fuzzy numbers by distance method. Fuzzy Sets and Systems, **95**, 307–317.
7. Kundu, S. (1998). Preference relation on fuzzy utilities based on fuzzy leftness relation on intervals. Fuzzy Sets and Systems, **97**, 183–191.
8. Akiyama, T., and Kawahara, T. (1997). Traffic assignment model with fuzzy travel time information. In 9th mini EURO Conf. Fuzzy sets in Traffic and Transport Systems, Budva, Yougoslavia.
9. Bierlaire, M., Burton, D., and Lotan, T. (1993). On the behavioural aspects of modal choices. GRT Report 93/22, Facultés Universitaires ND de la Paix, Namur, Belgique.
10. Chen, H.-K., and Chang, M.-S. (1998). A fuzzy dynamic optimal route choice model. In Transportation Research Board annual meeting, Washington.
11. Henn, V. (2000). Fuzzy route choice model for traffic assignment. Fuzzy Sets and Systems, **116**, 77–101.
12. Okada, S., and Soper, T. (2000). A shortest path problem on a network with fuzzy arc lengths. Fuzzy Sets and Systems, **109**, 129–140.
13. Perincherry, V., Kikuchi, S., and Dell'Orco, M. (2000). Use of possibility theory when dealing with uncertaity in modelling choice. In Euro Working Group on Transportation meeting, Rome.
14. Henn, V. (2001). Traffic information and traffic assignment: towards a fuzzy model. PhD thesis, Saint-Étienne University, France. In french.
15. Sheffi, Y. (1985). Urban transportation networks: Equilibrium analysis with mathematical programming methods. Prentice Hall, New Jersey.
16. (2002). 13th mini-Euro conference "Handling Uncertainty in the Analysis of Traffic and Transportation Systems", Bari, Italy.
17. Henn, V. (2002). What is the meaning of fuzzy costs in fuzzy traffic assignment models? In [16].
18. Klir, G. J. (1990). A principle of uncertainty and information invariance. Int. Journal of General Systems, **17**, 249–275.

19. Dubois, D., Nguyen, H. T., and Prade, H. (1999). Possibility theory, probability and fuzzy sets: misunderstanding, bridges and gaps. In Dubois, D., and Prade, H., editors, Fundamental of fuzzy sets, The handbook of fuzzy sets, pages 343–438. Kluwer academic publisher.
20. Dubois, D., Prade, H., and Sabbadin, R. (2001). Decision theoretic foundations of qualitative possibility theory. European Journal of Operational Research, **128**, 459–478. Invited review.
21. Dubois, D., Prade, H., and Sabbadin, R. (1997). A possibilistic logic machinery for qualitative decision. In Proc. of the AAAI 1997 spring symposium series (Qualitative preferences in deliberation and practical reasonning), pages 47–54, Standford University, California.
22. Dubois, D., and Prade, H. (1995). Possibility theory as a basis for qualitative decision theory. In 14th Int. joint conference on artificial intelligence, pages 1924–1930, Montreal, Canada.
23. Savage, L. J. (1954). The foundations of statistics. Wiley, New York.
24. von Neumann, J., and Morgenstern, O. (1944). Theory of games and economic behavior. Princeton Univ. Press, Princeton, NJ.

A New Technique to Electrical Distribution System Load Flow Based on Fuzzy Sets

Carlos A. F. Murari, Marcelo A. Pereira, and Marcelo M. P. Lima

State University of Campinas - UNICAMP
DSEE / FEEC / UNICAMP
C.P. 6101, 13081-970, Campinas, SP, BRASIL

Abstract. The researchers in electrical energy systems are always developing new models for the electrical networks, trying to incorporate the technicians' and engineers' knowledge in computational algorithms to get results as near as of the reality. We applied fuzzy numbers (bell shape) to represent imprecise variables and we present a new technique to electrical distribution system load flow based on fuzzy sets. With this new computational algorithm is possible to deal with information like *the voltage in the node k is high* and the results can also be interpreted through linguistics terms.

1 Introduction

Actually, there is no enough real-time information on the electrical distribution systems because there are too many lines and components and it is not economically justifiable to put real-time monitoring and control devices everywhere along feeders and laterals. Normally the only information available is the total feeder current recorded at substation [11]. Even for systems that have others monitoring points, some imprecision always exists. In addition, human experts tend to use linguistic terms such as *heavy load* or *worst voltage level* to describe system condition. A special feature of the fuzzy set is to deal with these kind of imprecision and linguistic terms.

In the past, the technicians' and engineers' knowledge could not be used integrally in the modeling of the electric power systems. Human information like *the power flow in the line k-m is high* could not be quantified and inserted in a computational algorithm that implied in an immediate decision. A deterministic or probabilistic power flow was not capable to represent this kind of information integrally [12], because in the deterministic method the data are considered constants and the probabilistic method is based on repetitions of events or in experimental data. The probabilistic power flow has another disadvantage: it needs more complexes calculation routines [8]. In many cases, the uncertainties in the electrical networks were unknown.

The theory of fuzzy sets [1] and the theory of possibilities [3] made possible the development of fuzzy load flow [9–12]. A DC and AC Fuzzy Load Flow is proposed in [9]. The DC Load Flow is used as a first approach to obtain a fuzzy description of the angles and active power flows. The AC Load Flow is obtained using an incremental technique. The method combines deterministic and fuzzy values using the Jacobian matrix of Newton-Raphson method. The authors comment that their operations cannot convert directly fuzzy loads (P,Q) into (S,θ) form because fuzzy numbers operations present certain restrictions. They proposed load flow algorithms that use membership functions with trapezoidal shape (Fig.1) for the representation of the variables of the problem.

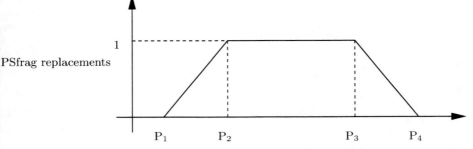

Fig. 1. Trapezoidal membership function

We present a new technique to electrical distribution system load flow based on fuzzy sets. The fuzzy numbers to active and reactive losses, voltage and current have an appropriate membership function that facilitates to overcome the fuzzy numbers' operations.

2 Fuzzy Number

A fuzzy number A is defined [4,14] as a fuzzy set on the space of real numbers $\Re$, whose membership function is a continuous function or continuous in parts, that satisfies the following conditions:

$A(x) : (-:,+:) \to [0,1], \ x \in \Re$

$\exists \ m \in \Re$ such that $A(m) = 1$

A is nondecreasing on (-:,m]

A is nonincreasing on [m,+:)

The problem's knowledge is fundamental to choose the membership function and also it must take in account the fuzzy numbers' operations to avoid problems with the membership function that result of operations such as multiplication and division [4; 14]. The extension principle of Zadeh [14] is the start point to obtain a relation between fuzzy sets.

The membership function chosen to deal with our problem is the same presented in [11] and the equation (1) and Fig. 2 describe it.

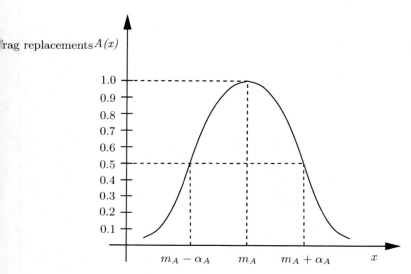

Fig. 2. Bell shape membership function

$$\tilde{A} = A(x) = \frac{1}{1 + (\frac{x - m_A}{\alpha_A})^2} \qquad (1)$$

Equation (1) can be also represented by $A(m_A, \alpha_A)$ where m_A corresponds to the value of larger degree $(A(m) = 1)$, and the parameter α_A, which is called the spread of the function, describes the degree of fuzziness with the fuzzy variable.

3 Fuzzy Numbers' Operators

Let $A(m_A, \alpha_A)$ and $B(m_B, \alpha_B)$ two fuzzy numbers, defined by equation (1). Their algebraic operations and membership functions are:

Sum

$$C(m_C, \alpha_C) = A \oplus B \qquad (2)$$

such that $m_C = m_A + m_B$ and $\alpha_C = \alpha_A + \alpha_B$

Subtraction

$$D(m_D, \alpha_D) = A \ominus B \qquad (3)$$

such that $m_D = m_A - m_B$ and $\alpha_D = |\alpha_A - \alpha_B|$

Multiplication

$$E(m_E, \alpha_E) = A \otimes B \qquad (4)$$

such that $m_E = m_A.m_B$ and $\alpha_E = min|m_E - \beta|, |\eta - m_E|$

$$\begin{aligned}\beta = min\ [&|(m_A - \alpha_A).(m_B - \alpha_B)|,\\ &|(m_A - \alpha_A).(m_B + \alpha_B)|,\\ &|(m_A + \alpha_A).(m_B - \alpha_B)|,\\ &|(m_A + \alpha_A).(m_B + \alpha_B)|]\end{aligned} \qquad (5)$$

$$\begin{aligned}\eta = max\ [&|(m_A - \alpha_A).(m_B - \alpha_B)|,\\ &|(m_A - \alpha_A).(m_B + \alpha_B)|,\\ &|(m_A + \alpha_A).(m_B - \alpha_B)|,\\ &|(m_A + \alpha_A).(m_B + \alpha_B)|]\end{aligned} \qquad (6)$$

$\oplus, \ominus$ and $\otimes$ represent sum, subtraction and multiplication fuzzy numbers' operators.

The definition of variables β and η originates from the interval analysis such as in [14].

Both subtraction and multiplication operators result in a narrower interval $[(m_A - \alpha_A, m_A + \alpha_A]$. The α_D (subtraction) and α_E (multiplication) values are calculated so that the resulting membership function can be represented by the equation (1). Thus, the fuzzy number properties and the main region around m_D and m_E are preserved.

The fuzzy DC power flow presented in [9] superimposes a deterministic DC load flow with a fuzzy possibility distribution to obtain the possibility distributions of bus angles and active power flows without operating with fuzzy numbers. By using the fuzzy numbers' operators, described in this section, a fuzzy DC power flow can also be implemented. If the loads are described as fuzzy numbers and the matrix B (see equation (10)) has crisp values, the bus angles are determined as fuzzy numbers by applying sum and multiplication operators.

For AC power flow, the problem concerned to a phenomenon of accumulation of fuzziness [14] can occur. It was necessary the definition of a new fuzzy multiplication operators to be used in the calculations of the flows. In fuzzy mathematics the choice of the best functions and operators depends on the type of application.

Multiplication (two options)

Option 1

$$E(m_E, \alpha_E) = A \otimes B \text{ such that } m_E = m_A.m_B \text{ and } \alpha_E = \alpha_{ord}(3)$$

$\alpha_{ord}(3)$ is the third element of the vector α_{ord}:

$$\alpha_{ord} = str \; \{|m_E - [(m_A - \alpha_A).(m_B - \alpha_B)]|, \\ |m_E - [(m_A - \alpha_A).(m_B + \alpha_B)]|, \\ |m_E - [(m_A + \alpha_A).(m_B - \alpha_B)]|, \\ |m_E - [(m_A + \alpha_A).(m_B + \alpha_B)]|\} \quad (7)$$

and *str* is a routine that places the elements in growing order.

Option 2

$$E(m_E, \alpha_E) = A \underline{\otimes} B \text{ such that } m_E = m_A.m_B \text{ and } \alpha_E = \alpha_{ord}(2)$$

$\alpha_{ord}(2)$ is the second element of the vector α_{ord}

For the functions sin and cos we developed a formulation to possibility the application of these functions in a fuzzy way (the following expressions are an approach for *sin* and *cos fuzzy*).

sin fuzzy ($\underline{sin}$)

$$\underline{sin}(m_\theta, \alpha_\theta) = (m_{sin}, \alpha_{sin})$$
$$m_{sin} = sin(m_\theta) \quad (8)$$
$$\alpha_{sin} = (|(|sin(m_\theta + \alpha_\theta)| - |m_{sin}|) + |(|m_{sin}| - |sin(m_\theta - \alpha_\theta)|)|)/2$$

cos fuzzy ($\underline{cos}$)

$$\underline{cos}(m_\theta, \alpha_\theta) = (m_{cos}, \alpha_{cos})$$
$$m_{cos} = cos(m_\theta) \quad (9)$$
$$\alpha_{cos} = (|(|cos(m_\theta + \alpha_\theta)| - |m_{cos}|) + |(|m_{cos}| - |cos(m_\theta - \alpha_\theta)|)|)/2$$

Starting from entrance data (loads, generations, etc.) with imprecision and using fuzzy mathematical operations, we propose a power load flow to supply results with fuzzy numbers, being used membership functions with bell shape (Fig. 2 and equation (1)).

With this kind of data representation, it was possible to develop algebraic operations that were easily implemented in our computational algorithm and the data and results require only just two parameters. The imprecision was attributed only to the active and reactive power whose values correspond to the parameter m and the parameter α defines a certain imprecision, fixed in 5% of m in this application.

4 Fuzzy Load Flow

The fuzzy load flows are based in the conventional load flows: DC Load Flow (DCLF) and Newton-Raphson Load Flow (NRLF), where some mathematical operations were substituted by fuzzy operations.

4.1 Fuzzy DC Load Flow - FDCLF

The FDCLF is simulated in two iterations to include the active power losses and to obtain a better estimate for the angles.

Procedure:
Starting from the active powers in the nodes $[\tilde{P}]$ solves the equation:

$$[\tilde{\theta}] = [B]^{-1} \otimes [\tilde{P}] \tag{10}$$

Having this first estimate for the angles, it is possible to calculate the losses in the connections:

$$\tilde{P}_{km}^{losses} = \tilde{\theta}_{km}^2 \otimes g_{km} \tag{11}$$

Add half of the values of the losses in the respective nodes and obtain a new vector $[\tilde{P}]$.
Resolve the equation (10) to obtain the final values for the angles.
Calculate the active flow in the lines:

$$\tilde{P}_{km} = \sum_{m \in \Omega_k} x_{km}^{-1} \otimes \tilde{\theta}_{km} \tag{12}$$

4.2 Fuzzy Newton-Raphson Load Flow - FNRLF

The FNRLF was developed in the following way:

With the point of larger degree (m) of the data, solve a conventional NRLF to obtain the angles and the magnitudes of the voltages that will be the values corresponding to the central points (m) of the possibility distributions for these variables.

Solve
$$[\alpha_{\theta v}] = [J]^{-1} \cdot [\alpha_{LIQPQ}] \tag{13}$$
to determine the deviations (α) of these variables.

J is the Jacobian matrix obtained in the last iteration of NRLF and the vector α_{LIQPQ} is formed by the deviations (α) of the active and reactive power data.

Calculate the active and reactive powers in the nodes:

$$\tilde{P}_k = \tilde{V}_k \sum_{m \in K} \tilde{V}_m \otimes [G_{km} \otimes \underline{cos}(\tilde{\theta}_{km} \oplus \varphi_{km}) \oplus B_{km} \otimes \underline{sin}(\tilde{\theta}_{km} \oplus \varphi_{km})]$$

$$\tilde{Q}_k = \tilde{V}_k \sum_{m \in K} \tilde{V}_m \otimes [G_{km} \otimes \underline{sin}(\tilde{\theta}_{km} \oplus \varphi_{km}) \ominus B_{km} \otimes \underline{cos}(\tilde{\theta}_{km} \oplus \varphi_{km})] \tag{14}$$

Calculate the active and reactive power flows:

$$\tilde{P}_{km} = g_{km} \otimes \tilde{V}_k^2 \ominus tap_{km} \otimes \tilde{V}_k \otimes \tilde{V}_m \otimes$$
$$[g_{km} \otimes \underline{cos}(\tilde{\theta}_{km} \oplus \varphi_{km}) \oplus b_{km} \otimes \underline{sin}(\tilde{\theta}_{km} \oplus \varphi_{km})]$$

$$\tilde{P}_{mk} = g_{km} \otimes (tap_{km} \otimes \tilde{V}_m)^2 \ominus tap_{km} \otimes \tilde{V}_k \otimes \tilde{V}_m \otimes$$
$$[g_{km} \otimes \underline{cos}(\tilde{\theta}_{km} \oplus \varphi_{km}) \ominus b_{km} \otimes \underline{sin}(\tilde{\theta}_{km} \oplus \varphi_{km})] \tag{15}$$

$$\tilde{Q}_{km} = -(b_{km} + b_k^{shunt}) \otimes \tilde{V}_k^2 \oplus tap_{km} \otimes \tilde{V}_k \otimes \tilde{V}_m \otimes$$
$$[b_{km} \otimes \underline{cos}(\tilde{\theta}_{km} \oplus \varphi_{km}) \ominus g_{km} \otimes \underline{sin}(\tilde{\theta}_{km} \oplus \varphi_{km})]$$

$$\tilde{Q}_{km} = -(b_{km} + b_k^{shunt}) \otimes (tap_{km} \otimes \tilde{V}_m^2) \oplus tap_{km} \otimes \tilde{V}_k \otimes \tilde{V}_m \otimes$$
$$[b_{km} \otimes \underline{cos}(\tilde{\theta}_{km} \oplus \varphi_{km}) \oplus g_{km} \otimes \underline{sin}(\tilde{\theta}_{km} \oplus \varphi_{km})] \tag{16}$$

Calculate the active power losses:

$$\tilde{P}_{km}^{Losses} = (g_{km} \otimes \tilde{V}_k^2) \oplus (g_{km} \otimes (tap_{km} \otimes \tilde{V}_m)^2 \ominus$$
$$2 \otimes tap_{km} \otimes \tilde{V}_k \otimes \tilde{V}_m \otimes g_{km} \otimes \underline{cos}(\tilde{\theta}_{km} \oplus \varphi_{km}) \tag{17}$$

5 Tests

The FDCLF and the FNRLF were applied to the system IEEE30 [2]. The tables present some of results obtained with FDCLF and FNRLF corresponding to the points ($m-\alpha$) and ($m+\alpha$) and they are compared with the results obtained with the conventional DCLF and NRLF applied in two data base: $[P_{min} = m_P - \alpha_P; Q_{min} = m_Q - \alpha_Q]$ and $[P_{max} = m_P + \alpha_P; Q_{max} = m_Q + \alpha_Q]$.

5.1 Results with DCLF and FDCLF

Table 1 presents some values of the angles, table 2 some active power flows, table 3 the active power in the slack node and table 4 some active power losses.

Table 1. Angles (degrees)

	DCLF		FDCLF		|ERROR|(%)	
Node	θ^{Pmin}	θ^{Pmax}	$\theta^{m-\alpha}$	$\theta^{m+\alpha}$	$\theta^{m-\alpha}$	$\theta^{m+\alpha}$
2	-5.50	-5.99	-5.46	-6.02	0.62	0.55
3	-8.23	-8.98	-8.20	-9.00	0.25	0.22
5	-14.26	-15.41	-14.23	-15.44	0.21	0.19
8	-12.16	-13.24	-12.13	-13.26	0.22	0.19
12	-15.66	-17.16	-15.63	-17.18	0.16	0.14
14	-16.71	-18.31	-16.68	-18.34	0.15	0.13
28	-16.98	-18.60	-16.95	-18.63	0.22	0.19

Table 2. Active power flow (MW)

	DCLF		FDCLF		|ERROR|(%)	
Line	P_{km}^{min}	P_{km}^{max}	$P_{km}^{m-\alpha}$	$P_{km}^{m+\alpha}$	$P_{km}^{m-\alpha}$	$P_{km}^{m+\alpha}$
1 - 2	166.857	181.776	165.824	182.781	0.62	0.55
1 - 3	77.151	82.922	77.181	82.888	0.25	0.22
2 - 5	-15.559	-15.636	-15.666	-15.529	0.04	0.04
14 - 15	5.292	5.822	5.291	5.823	0.07	0.05
15 - 18	0.909	0.971	0.908	0.973	0.03	0.02
23 - 24	-2.459	-2.741	-2.460	-2.740	0.19	0.16

5.2 Results with NRLF and FNRLF

Tables 5 and 6 present some values of the magnitudes and angles of voltages, respectively. The errors for the magnitudes are less than 0.01% and for the angles are less than 0.04%.

Table 3. Active power (MW)

	DCLF		FDCLF		|ERROR|(%)	
Slack Node	P_g^{min}	P_g^{max}	$P_g^{m-\alpha}$	$P_g^{m+\alpha}$	$P_g^{m-\alpha}$	$P_g^{m+\alpha}$
1	244.38	266.40	243.15	267.59	1.71	1.54

Table 4. Active power losses (MW)

	DCLF		FDCLF		|ERROR|(%)	
Line	P_{loss}^{min}	P_{loss}^{max}	$P_{loss}^{m-\alpha}$	$P_{loss}^{m+\alpha}$	$P_{loss}^{m-\alpha}$	$P_{loss}^{m+\alpha}$
1 - 2	4.8093	5.7078	4.7499	5.7462	1.23	0.67
1 - 3	2.5636	3.0551	2.5506	3.0565	0.51	0.05
2 - 5	2.6588	3.0715	2.6609	3.0617	0.08	0.32
5 - 7	0.0962	0.0972	0.0959	0.0975	0.36	0.36

Table 5. Voltages (p.u.)

	NRLF		FNRLF	
Node	V^{Pmin}	V^{Pmax}	$V^{m-\alpha}$	$V^{m+\alpha}$
2	1.043	1.043	1.043	1.043
3	1.023	1.021	1.023	1.021
4	1.015	1.012	1.015	1.012
5	1.010	1.010	1.010	1.010
11	1.082	1.082	1.082	1.082
12	1.058	1.055	1.058	1.055
13	1.071	1.071	1.071	1.071
14	1.045	1.039	1.045	1.039
29	1.010	1.002	1.010	1.002
30	0.999	0.990	0.999	0.990

Table 6. Angles (degrees)

	NRLF		FNRLF	
Node	θ^{Pmin}	θ^{Pmax}	$\theta^{m-\alpha}$	$\theta^{m+\alpha}$
2	-5.25	-5.75	-5.25	-5.75
3	-7.63	-8.37	-7.63	-8.36
4	-9.21	-10.1	-9.21	-10.1
5	-13.8	-15.0	-13.8	-14.9
11	-14.1	-15.5	-14.1	-15.5
12	-15.1	-16.6	-15.1	-16.6
13	-15.1	-16.6	-15.1	-16.6
14	-15.9	-17.5	-15.9	-17.5
29	-16.6	-18.3	-16.6	-18.3
30	-17.4	-19.2	-17.4	-19.2

Table 7 presents the active power in the slack node. The error is less than 0.1%.

Table 7. Active power (MW)

	NRLF		FNRLF	
Slack Node	P_g^{min}	P_g^{max}	$P_g^{m-\alpha}$	$P_g^{m+\alpha}$
1	249.45	272.62	249.53	272.47

Table 8 presents the reactive power in the slack and PV nodes.

Tables 9 and 10 present results for the some active and reactive power flows, respectively.

Table 11 presents some active power losses. The largest errors happen in the lines where the losses are very low.

6 Conclusions

The load flow algorithms were scheduled following the same methodology, with differences just in the parts that contemplate the fuzzy calculations. The table 12 presents the respective amounts of flops, obtained with *MatLab*© for the system IEEE30 [2].

Table 8. Reactive power generation (MVAr)

Node	NRLF Q_g^{min}	Q_g^{max}	FNRLF $Q_g^{m-\alpha}$	$Q_g^{m+\alpha}$	|ERROR|(%) $Q_g^{m-\alpha}$	$Q_g^{m+\alpha}$
1	-15.76	-18.75	-14.42	-20.10	8.48	7.17
2	32.47	39.10	31.36	40.16	3.42	2.71
5	14.79	19.00	15.74	18.03	6.41	5.08
8	-3.07	3.67	-3.06	3.59	0.33	2.16
11	14.63	16.45	14.62	16.44	0.06	0.06
13	9.69	12.55	9.68	12.54	0.14	0.12

Table 9. Active power flow (MW)

Line	NRLF P_{km}^{min}	P_{km}^{max}	FNRLF $P_{km}^{m-\alpha}$	$P_{km}^{m+\alpha}$	|ERROR|(%) $P_{km}^{m-\alpha}$	$P_{km}^{m+\alpha}$
1 - 2	170.006	185.785	170.071	185.663	0.04	0.07
1 - 3	79.447	86.837	79.599	86.671	0.19	0.19
2 - 4	43.461	47.609	43.825	47.237	0.84	0.78
2 - 6	59.364	64.849	59.605	64.598	0.40	0.39
4 - 6	68.982	74.460	68.801	74.641	0.26	0.24
5 - 7	-14.213	-14.054	-14.214	-14.059	1.08	1.14
8 - 28	-0.628	-0.464	-0.628	-0.465	0.05	0.15
25 - 27	-4.750	-5.320	-5.067	-5.004	5,35	4.75

Table 10. Reactive power flow (MVAr)

Line	NRLF P_{km}^{min}	P_{km}^{max}	FNRLF $P_{km}^{m-\alpha}$	$P_{km}^{m+\alpha}$	|ERROR|(%) $P_{km}^{m-\alpha}$	$P_{km}^{m+\alpha}$
1 - 2	-20.340	-23.973	-20.355	-23.945	0.07	0.12
1 - 3	4.579	5.218	4.569	5.208	0.22	0.19
2 - 4	2.159	2.711	2.153	2.698	0.27	0.47
2 - 6	-1.092	-1.097	-1.096	-1.103	0.41	0.59
4 - 6	-15.586	-18.498	-16.950	-17.109	8.75	7.51
8 - 28	-3.055	-1.908	-3.002	-1.973	1.71	3.41
25 - 27	-0.297	-0.647	-0.300	-0.644	1.28	0.42

Table 11. Active power losses (MW)

Line	NRLF P_{loss}^{min}	NRLF P_{loss}^{max}	FNRLF $P_{loss}^{m-\alpha}$	FNRLF $P_{loss}^{m+\alpha}$	|ERROR|(%) $P_{loss}^{m-\alpha}$	|ERROR|(%) $P_{loss}^{m+\alpha}$
1 - 2	4.990	5.974	4.994	5.966	0.08	0.13
1 - 3	2.558	3.056	2.559	3.054	0.02	0.07
2 - 4	0.999	1.199	0.998	1.199	0.10	0.01
2 - 6	1.883	2.247	1.882	2.246	0.04	0.01
4 - 6	0.576	0.682	0.577	0.682	0.02	0.04
8 - 28	0.0007	0.0002	0.0003	0.0002	57.1	0.00
25 - 27	0.024	0.030	0.024	0.030	0.42	0.00

Table 12. Flops

NRLF	FNRLF
3,009,189	1,463,066

The amount of flops for NRLF corresponds to NRLF applied to the points $(m - \alpha)$, m and $(m + \alpha)$.

FDCLF showed good results, allowing that analysis techniques for the planning can be implemented with good precision.

FNRLF provides good results too and it solved the problem of the negative losses [9].

With this algorithm it is possible to get results such as *voltage good, low flow, worst loss* to build a base to decision-makers, to get holistic insight for system behaviors. Another applications can integrate symbolic and numeric computing to represent the operators' knowledge.

The representation of the fuzzy numbers through equation (1) provides results with good precision and smaller amount of flops. It allows to enlarge the analysis of the results, once in the interval of minimum and maximum points, the membership function presents values in the interval [0.5; 1], while for the function trapezoidal the membership function is always unitary. The bell shape allows applications that translate the numeric results in linguistic (Fig. 3).

Acknowledgements

This work has been supported by Coordenadoria de Aperfeiçoamento do Ensino Superior (CAPES) and Fundação de Amparo à Pesquisa do Estado de São Paulo (FAPESP), Brazil.

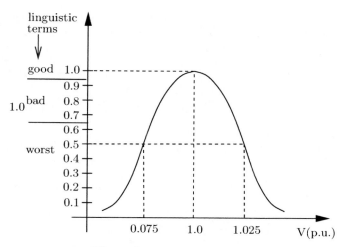

Fig. 3. Interpretation of results

References

1. Zadeh L. A. (1965) Fuzzy sets. Information and Control 8:338-353
2. Freris L. L., Sasson A. M. (1968) Investigation of the load-flow problem. Proceedings of IEE 115:1459-1470
3. Zadeh L. A. (1978) Fuzzy sets as a basis for a theory of possibility. Fuzzy Sets and Systems 1:3-28
4. Dubois D., Prade H. (1978) Operations on fuzzy numbers. International Journal of Systems Science 9:613-626
5. Dubois D., Prade, H. (1980) System of linear fuzzy constraints. Journal of Fuzzy Sets and Systems 13:37-48
6. Goldberg D. E. (1989) Genetic algorithms in search, optimization, and machine learning. Addison-Wesley, New York
7. Baran M. E., Wu F. F. (1989) Network reconfiguration in distribution systems for loss reduction. IEEE Transactions on Power Delivery 4:1401-1407
8. Meliopoulos A. P. S., Cokkinides G. J., Chao X. Y. (1990) A new probabilistic power flow analysis method. IEEE Transactions on Power Systems 5:182-189
9. Miranda V., Matos M. A. A. C., Saraiva, J. T. (1990) Fuzzy load flow - New algorithms incorporating uncertain generation and load representation. Proceedings of the tenth power systems computation conference, Graz, Austria, 621-627
10. Miranda V., Saraiva J. T. (1992) Fuzzy modeling of power system optimal load flow. IEEE Transactions on Power Delivery 7:843-849
11. Kuo Han-Ching, Hsu Yuan-Yih (1993) Distribution system load estimation and service restoration using a fuzzy set approach. IEEE Transactions on Power Delivery 8:1950-1957
12. Kenarangui R., Seifi, A. (1994) Fuzzy power flow analysis. Electric Power Systems Research 29:105-109
13. Pereira M. A., Murari C. A. F., Castro C. A. (1996) A fast on-line three phase power flow for radial distribution systems. Proceedings of IEE Japan Power & Energy 1996, Osaka, Japan, 53-58

14. Pedrycs W., Gomide F. (1998) An introduction to fuzzy sets - Analysis and design, A Bradford Book, The MIT Press, Massachusetts
15. Lin Whei-Min, Chin Hong-Chan (1998) A new approach for distribution feeder reconfiguration for loss reduction and service restoration. IEEE Transactions on Power Delivery 13:870-875
16. Pereira M. A., Murari C. A. F., Castro Jr. C. A. (1999) A fuzzy heuristic algorithm for distribution systems' service restoration. Fuzzy Sets And Systems, Elsevier, 102:125-133
17. Pereira M. A., Murari C. A. F. (1999) Electrical distribution system fuzzy load estimation. Proceedings of ISAP'99, Rio de Janeiro, 370-375

A Fuzzy Adaptive Partitioning Algorithm (FAPA) for Global Optimization: Implementation in Environmental Site Characterization*

Linet Özdamar[1] and Melek Basak Demirhan[2]

[1] Nanyang Technological University, School of Mechanical and Production Engineering, Systems and Engineering Management Division, 50 Nanyang Avenue, Singapore 639798
[2] Yeditepe University, Dept. of Systems Engineering, Kayisdagi, 81120 Istanbul, Turkey.

Abstract. Environmental site investigations are carried out prior to the reclamation of industrially contaminated sites. It is required to take samples and analyze the laboratory reports to find out the distribution of contaminants over the site. In this study it is proposed to modify and implement the global optimization method, Fuzzy Adaptive Partitioning Algorithm (FAPA), in order to identify the topology of polluted areas in a given industrial site. The primary objective is to locate contaminated zones accurately and to reduce the size of the area that is subjected to expensive sampling methods (such as deep drilling). A secondary objective is to decrease the number of samples taken during a quick screening that is carried out as a preliminary investigation. If these objectives are achieved, a considerable cost reduction occurs in the process of site characterization.

1 Introduction

The issue of preserving the natural environment becomes more and more crucial as the process of industrialization gains speed. Brownfields that are ex-industrial sites usually contain health hazardous contaminants in soil and also lead to pollution of groundwater. Brownfields may be re-used for urban purposes and therefore need to be investigated in detail. An accurate characterization of potentially contaminated brownfields is essential before reclaiming them, since they may cause high health risks.

There are two major concerns in site characterization. The first is to determine the contaminated regions in the site as precisely as possible and the second is to reduce the number of samples taken from the site. The first goal minimizes health risks whereas the second reduces site characterization costs.

---

* This research is part of the project PURE-EVK1-1999-00153: Protection of Groundwater Resources at Industrially Contaminated Sites under EU 5th Framework R&D Program.

Conventional techniques for dealing with spatial data are classified as Spatial Interpolation Methods (SIM) [1–5] and they are used in many contexts such as image re-construction, remotely sensed satellite imagery, identification of mining resources, topographical mapping, etc. SIM are interpolation methods for estimating the value of characteristics at unsampled locations. The rationale behind spatial interpolation is the fact that points close together in space are more likely to have similar values than points far apart.

All spatial data analysis algorithms construct surfaces from point observations and therefore, they involve uncertainties. The uncertainties are due to the fact that observed points give an indication about the surface's character at these locations, but not about the actual surface. The surface topology obtained using data points is constructed by generating grids and assigning a unique interpolated value to the whole grid which is not necessarily adequate. Another issue is that interpolation has a smoothing effect that is misleading for identifying the contamination levels of small highly contaminated areas surrounded by clean regions.

2 Environmental Site Characterization: Problem Definition

In environmental site characterization, a number of samples collected from the site are subjected to laboratory tests and the concentration values of one or more contaminants are identified. This information, which consists of the location of the samples and their concentration values, is utilized to determine the spatial distribution of contaminants over the site. The regions that are claimed to have a contamination level above the legal threshold level have to be re-mediated.

Definition 1: Let $\mathbf{w} \in \Re^n$ (for n=1,2 or 3) denote the coordinates of the location of an observation within a given site $\mathbf{S}$. The *concentration value* at $\mathbf{w}$ denoted by $f(\mathbf{w})$ is defined to be the observed contamination value read at $\mathbf{w}$.

Definition 2: A *regular cover*, $\mathbf{C}$, of the site of interest $\mathbf{S} \in \Re^2$ is a rectangle (a regular polygon in general) whose vertices take place at points $\mathbf{v}_0 = (min\ x, min\ y)$, $\mathbf{v}_1 = (max\ x, min\ y)$, $\mathbf{v}_2 = (max\ x, max\ y)$, and $\mathbf{v}_3 = (min\ x, max\ y)$. A regular subset of $\mathbf{C}$ is called a *block* and denoted by α_i.

Definition 3: A site or the sample space, $\mathbf{S}$, is composed of two complementary sets, set of contaminated locations α', having values above threshold (*trsh*) and α having values less than or equal to the threshold.

Definition 4: Any subset, set $\alpha_i \subset \mathbf{S}$ is called a *non-potential block* if it does not contain any $\mathbf{w} \in \alpha'$.

The regions that are classified as contaminated blocks have to be reclaimed. Thus, based on the above definitions, the problem can be stated as follows.

$$\text{Identify all non-potential blocks } \alpha_i \subset \alpha,$$
$$\text{given locations } \mathbf{w}_i \in \alpha_i \text{ and } f(\mathbf{w}_i). \tag{1}$$

In the terminology of Global Maximization, it is required to determine the domain (location boundaries) of near optimum solutions. In other words, exact boundaries of sub-spaces whose contamination levels are above a plateau (*trhs*), should be identified. In Fig. 1, the boundaries of the subsets α' are illustrated, for instance, $\alpha'_1 = [b_1, b_2]$. We note that the distribution of the contaminant over $\mathbf{S}$ is not explicitly given as a mathematical expression, that is, $f(\mathbf{w})$ is unknown. Rather, we have locations as inputs and observations as output. Therefore, the problem expressed by (1) is a black box problem to be solved with only a fixed number of sample data. It is also possible to consider the problem as a classification problem since in a black box problem the input-output relationship is not known.

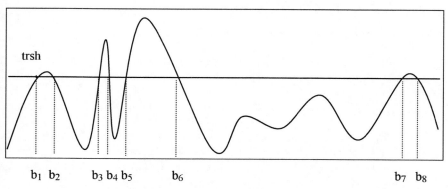

Fig. 1. Illustration of subsets in α' in single dimension

3 Implementation of Fuzzy Adaptive Partitioning Algorithm (FAPA) in Site Characterization

In most spatial data analysis cases it is not adequate to use the conventional statistical approach. The commonly utilized methods in spatial estimation involve geostatistics [1] such as kriging [6,7] and other GIS.

The Fuzzy Tessellation (FT), proposed here, is a modification of Fuzzy Adaptive Partitioning Algorithm and it is a novel approach in site characterization. FAPA is a GO approach that is assumption-free and it does not require any information about the function involved. Consequently, in site characterization, it provides an excellent tool for making spatial inferences given the sample data only.

In FT the spatial concentration values $f(\mathbf{w})$ are treated as function values and plateaus with contamination above threshold are identified using a fuzzy regional assessment. The problem defined by (1) is related to the unconstrained GO problem, in the sense that all existing extremities of $f(\mathbf{w})$ have to be identified.

Definition 5: Suppose we choose m sample locations from the site $\mathbf{S}$, which we denote by $\mathbf{S}_0 = \mathbf{w}_1, \mathbf{w}_2, \mathbf{w}_3, \ldots, \mathbf{w}_m$. Using $\mathbf{S}_0$, FT partitions and classifies $\mathbf{S}$ into clusters A and A' which represent the union of blocks claimed to be contaminated and non-contaminated, respectively. The set A is called the *non-potential homogeneous cluster* and constitutes an inference base for α.

Although sets A and A' represent mutually exclusive circumstances, $A \cap A' = \phi$ does not apply here, because both sets involve spatial uncertainty due to the inference-making process in FT. The uncertainty involved in this case can be represented by fuzzy set theory.

3.1 Modifications in FAPA: Fuzzy Tessellation

Fuzzy Tessellation does not use any statistical means such as global or local estimation as interpolation methods do. Rather it provides non-statistical estimates for the sets α and α'.

The two key features of FT that distinguish it from FAPA are as follows.

- It is an *iterative algorithm* which aims at minimizing the uncertainty related to the boundaries of the fuzzy sets A and A' given *a fixed set of observations.* On the contrary, FAPA continues to take samples while refining their enclosures.
- FT involves an overlapping fuzzy partitioning scheme to maximize the utility of available samples.

Partitioning Scheme

In a real world situation, the solution for the problem stated in (1) has to be achieved using a finite number of observations, i.e., we have to tessellate $\mathbf{S}$ into *clusters* A and A' with a fixed number of data points. Hence, the problem can be re-stated as: find A such that A is an estimate of the cluster α. The approach utilized in FT aims at extracting information more efficiently by enabling data re-utilization which is accomplished by an overlapping partitioning scheme.

Definition 6: In any iteration t, a block $\alpha_i(t)$ is partitioned into k *almost disjoint sub-blocks* $\alpha_j(t+1)$ for $j = 1, 2, \ldots k$ if we have, $\alpha_i(t+1) \cap \alpha_j(t+1) = \phi$ except their boundaries.

$\alpha_i(t+1)$ are called *overlapping sub-blocks* if they are not almost disjoint and the extension of their overlap around their common boundary is equal on both sides (Illustrated in Fig. 2).

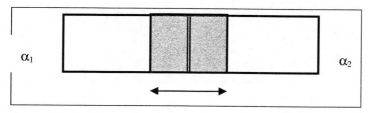

Fig. 2. Overlapped partitioning of a block

In Fig.3 the number of blocks that observations are included in are illustrated for two different overlap rates. The percentage of data that are shared by more than one blocks increases as the number of partitioning iterations increase.

The Algorithm

Step 1. A regular cover of the site, **C**, is determined.

Step 2. **C** is divided into k overlapping blocks as $\alpha_1(t), \alpha_2(t), \ldots, \alpha_k(t)$ (without loss of generality we assume that $k = 2$). Here t represents the iteration index and is equal to 1 for the initial iteration.

Step 3. At any iteration t, $t = 1, 2, \ldots, T$, the potentials of both blocks $\alpha_1(t)$ and $\alpha_2(t)$ are computed and blocks are classified as either in cluster A or not.

If $\alpha_i(t)$ is classified as a member of cluster A, then it is declared as *no trace* and is exempt from being considered in the remaining iterations.

Else, Step 2 is repeated by dividing $\alpha_i(t)$ into two overlapping blocks in the next iteration, $t + 1$.

Step 4. The algorithm stops whenever all blocks are assigned to one of the sets, A, or A', resulting in a final tessellation of the site **S**.

Once the boundaries of $\alpha_i(t)$ are determined in Step 2 of the pseudocode, available sampled data which fall within $\alpha_i(t)$ constitute the sampled set. Thus, unlike the implementation in GO, new samples are not collected in each partitioning iteration, t.

Evaluation Scheme

Definition 7: The membership function $\mu_A(\mathbf{w})$ measures the extent that $\mathbf{w}$ belongs to the fuzzy set A.

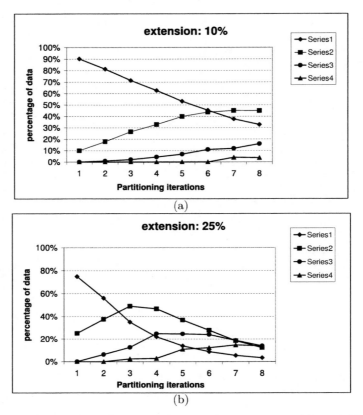

Fig. 3. Data sharing for different overlap rates; 10 percent in (a), and 25 percent in (b). The legend *Serie i*, with $i = \{1, 2, 3, 4\}$, stands for the percentage of data that fall into i blocks.

Similarly, $\mu_{A'}(\mathbf{w})$ is defined to be the membership function that represents the location's degree of belonging to the contaminated homogeneous cluster A'. Hence a location $\mathbf{w}$ may belong to the fuzzy sets A and A' with membership values not necessarily summing up to one. In the context of site characterization, an arbitrarily selected, monotone non-decreasing, linear or non-linear function may serve as a membership function. Linear, Gaussian and Sigmoid membership functions are examples of such functions.

After determining the membership values $\mu_A(\mathbf{w})$ for all observed locations $\mathbf{w}$, a block measure associated with each subset is defined as follows.

Definition 8: The *potential* of a block α_i is defined to be a block measure which aggregates the membership values of observations falling into the boundaries of that block and is also a mapping into the unit interval. The potential of a block can be interpreted as its degree of membership to the fuzzy sets A and A'.

Since the potential of a block is calculated using all observations located in that block, the information provided by a single location is utilized more than once by dividing the site into overlapping blocks. Changing the extent of the overlap makes it possible to obtain blocks of various mobile sizes. Thus, the value of information that an observation provides is increased by utilizing that observation more than once and in a different block each time (Fig. 3).

4 Numerical Results on Hypothetical Sites

FT is tested on 29 hypothetical sites that have been randomly generated as follows. Each test site is constructed on the square $0 \leq x \leq 20$ and $0 \leq y \leq 20$ using the expression given in (2).

$$f(x,y) = \Sigma_i a_{3i} \, exp(-a_{4i}(x-a_{1i})^2) \, exp(-a_{5i}(y-a_{2i})^2) \qquad (2)$$

Here, the summation is taken over a random number i of functions $f_i(x,y)$ where i is uniformly distributed over [3,13]. The coefficients a_{ji} are also uniform random numbers satisfying, $l_j \leq a_{ji} \leq u_j$ for $j = 1, 2, \ldots, 5$.

The parameters a_1 and a_2 determine the location coefficients, a_4 and a_5 are spread parameters, a_3 represents the range of the function. This test function generates sites with many different topologies involving peaks of varying shapes and in-between distances. In Fig. 4, the contours of an exemplary site are illustrated.

In the experiments, data is collected at five density levels: $n_1 = 16$, $n_2 = 41$, $n_3 = 77$, $n_4 = 126$, and $n_5 = 190$, respectively. At each data density level, the data also includes all samples collected previously. Thus, evidence accumulates at each level. The aim of these consecutive experiments is to demonstrate FT's accuracy in identifying contaminated regions using data of varying sparseness.

The experiments are conducted at two threshold values, t_low and t_high. With t_low, 32% of the site area is contaminated on the average. With t_high, this percentage drops to 7%. In Table 1, the average number of "polluted" data with function values above the threshold is also indicated for each data density level. When the threshold value is high, this number is reduced considerably as compared to that of t_low. Consequently, it is more difficult to identify contaminated regions when the threshold value is high and hot-spots of small scale occur in the site.

In Table 1, the numerical results are summarized for two threshold levels and five data density levels. The column under the heading "*Area %*" indicates the (average) percentage of total site area in which FT encloses the contamination. In the column under "*Cover %*", the (average) percentage of the actual contaminated area classified correctly by FT is indicated.

The second column illustrates the accuracy of FT in identifying contaminated regions, α'. It is observed that FT is able to identify contaminated regions with almost 90% accuracy at the first sampling level and it reduces

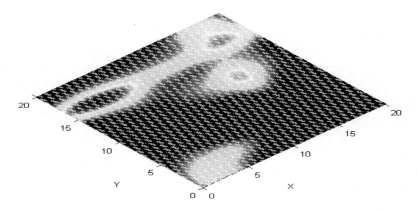

Fig. 4. Contours of an exemplary site generated by equation (2). (Colours other than black indicate contamination above threshold).

the size of the site by more than 30%. Hence, after the third data density level, FT identifies contaminated regions with almost 100% accuracy at both t_low and t_high and reduces the size of the site by half for the next expensive phase of sampling (deep drilling). It is noted that in the context of environmental site characterization, FT is proposed to be coupled with rapid and cheap screening technology such as biosensors with limited intrusive sampling. Consequently, the practical value of FT in terms of the economics of site investigations becomes obvious.

5 Conclusion

The Fuzzy Adaptive Partitioning Algorithm (FAPA) which is originally conceived as a global optimization (GO) approach is modified and adapted to the environmental site characterization context. In the latter context the distribution topology of contaminants over a site is sought for. Since contaminant distribution does not have a mathematical expression, any method used for this purpose works with the available data at hand. FAPA is a GO approach that is fit to be utilized in this context, because it does not require any information about the function involved.

FAPA's partitioning scheme is modified (Fuzzy Tessellation) with the aim of employing the information gained by the limited number of available samples more efficiently. The basic algorithm and the fuzzy assessment scheme remain intact. Tests conducted on hypothetical sites show that FT is able to identify contaminated regions within the site accurately while considerably reducing the area to be investigated and optimizing future site investigation stages.

Table 1. Numerical Results

	t_low average contamined area = 32 %		t_high average contamined area = 7 %	
	Area %	Cover %	Area %	Cover %
	16 data (5 polluted)		16 data (1 polluted)	
Avg	62	89	68	88
Std	21	14	34	19
	41 data (13 polluted)		41 data (2 polluted)	
Avg	60	95	53	92
Std	19	7	36	12
	77 data (24 polluted)		77 data (5 polluted)	
Avg	57	97	43	92
Std	19	3	35	13
	126 data (39 polluted)		126 data (8 polluted)	
Avg	54	99	38	97
Std	19	2	34	7
	190 data (59 polluted)		190 data (13 polluted)	
Avg	51	100	37	98
Std	19	1	34	1

References

1. Isaaks E.H., Srivastava R.M. (1989) An Introduction to Applied Geostatistics. Oxford University Press, New York
2. Cressie N.A. (1993) Statistics for Spatial Data. John Wiley & Sons, New York
3. Englund E.J. (1993) Spatial simulation: Environmental applications. In: Goodchild M.F., Parks B.O., Steyaert S. (eds) Environmental Modelling with GIS. Oxford University Press, New York
4. Heywood I., Cornelius S., Carver S. (1998) An Introduction to Geographical Information Systems. Longman Science and Technology
5. Burrough P.A., McDonnell R. (1998) Principles of GIS (Spatial Information Systems). Oxford University Press, New York
6. Krige D.G. (1951) A statistical approach to some mine valuations and allied problems at witwatersrand. MSc. Thesis, Univ. of Witwatersrand, South Africa
7. Deutsch C.V., Journel A.G. (1998) GSLIB: Geostatistical Software Library and User's Guide. Oxford University Press, New York

Capacitated Vehicle Routing Problem with Fuzzy Demand

Brigitte Werners and Michael Drawe

Faculty of Economics and Business Administration
Ruhr-University Bochum
Universitätsstr. 150
D-44780 Bochum
Germany

Abstract. The capacitated vehicle routing problem with fuzzy demand is considered. Since customers' demand is not precisely known in advance, but is given as uncertain quantities, i.e. as fuzzy numbers, a recommended route may not meet each demand for capacity reasons. Route failure will result in losing the part of demand which could not be satisfied. Consequently, the possibility, or indeed necessity, to satisfy all customers' demand has to be high. Optimization should additionally consider several conflicting objectives, e.g. minimizing total travel costs as well as maximizing sales.

In this article, a fuzzy multi-criteria modeling approach, based on a mixed integer linear mathematical programming model, is presented. In order to solve even larger problems, the established savings heuristic for the classical vehicle routing problem is modified appropriately with regard to fuzzy demand and multi-criteria optimization. A compromise solution is determined interactively by the decision maker, who adjusts the degree of satisfaction with the different goals. The solution method is demonstrated in a 75-customer example.

1 Introduction

The capacitated vehicle routing problem (CVRP) assumes certainty for customers' demand, which has to be satisfied by a homogeneous fleet of delivery vehicles of fixed capacity from a single depot at minimal cost. A wide range of real-world problems has been modeled by the CVRP. For an overview, see Crainic and Laporte [8]. In particular, vehicle routing problems (VRP) with various modifications and different approaches are considered by Laporte [21], Desrosiers et al. [9], and Fisher [14]. Bibliographies in VRP can be found in Laporte and Osman [23] and in Laporte [22]. The VRP with or without side-constraints is an $\mathcal{NP}$-hard problem [24], which makes it difficult to optimally solve larger problems in a reasonable amount of computing time. For problems with more than 50 customers, solution procedures generally utilize heuristics. Stochastic vehicle routing problems (SVRP) are introduced by Tillman [31] in 1969, who suggests a savings approach for the multi-depot VRP with stochastic demand. Not until two decades later, research focuses on stochastic models of SVRP. For an overview, see Powell et al. [26], Bertsimas and Simchi-Levi [4], and Gendreau et al. [16]. In particular, models for vehicle routing problems with stochastic demand (VRPSD) are considered by Dror et al. [12], Bertsimas [2], and Dror [10]. Recourse versions of VRPSD are solved to optimality by Dror et al. [11], who restrict the number of route failures, in 1993 and by Gendreau et al. [15] in 1995. A real world application is presented by Cheung and Powell [5] in 1996. Heuristic procedures to solve the VRPSD are just as rarely considered: See Stewart and Golden [28] and Dror and Trudeau [13] for modified savings algorithms, Bertsimas et al. [3] for a priori heuristics, Gendreau et al. [17] for a tabu search algorithm, Secomandi [27] for neuro-dynamic programming algorithms, and Yang et al. [37] for two-phase methods.

This article deals with the CVRP in a fuzzy environment: The customers' demand is not precisely known in advance, but can be modeled as uncertain quantities, i.e. as fuzzy numbers. Because stochastic models assume stochastic quantities, i.e. random numbers, precise information about the probability functions of demand is needed. To handle imprecise and vague information, an approach which is based on fuzzy sets theory is appropriate. Nevertheless, the capacitated vehicle routing problem with fuzzy demand (CVRPFD) has not yet been published except for Teodorović and Kikuchi [29], who developed a savings algorithm for the CVRP with fuzzy travel times, Teodorović and Pavković [30], who suggest an approximate reasoning algorithm for routing one vehicle with regard to fuzzy demand, and Werners and Kondratenko [36], who present a fuzzy multi-criteria model for tanker routing problems with fuzzy demand.

In the following section the CVRPFD is described in detail, and a fuzzy multi-criteria modeling approach, based on a mixed integer linear mathematical programming model, is presented. An interactive approach is suggested to find a compromise solution, which is suitable for small problems because optimal solutions of the crisp equivalent can be found by using standard mixed integer linear programming software. In Sect. 3, a heuristic, which is adequate for larger problems, is developed and demonstrated in a 75-customer problem. Some conclusions are given in the last section.

2 A Fuzzy Multi-Criteria Modeling Approach

2.1 General Description of the Problem

At a single depot 0, a homogeneous fleet of m delivery vehicles is available: The vehicles are identical and have the same fixed capacity Q. A set of n customers has to be served, who know their respective demand approximately. There are enough vehicles to satisfy any actual total demand. All distances between depot and customers are crisp and given. Each vehicle starts and finishes its route at the depot. We assume late information policy: No customer knows his or her demand precisely until a vehicle has completely served it at the demand location. The demand can only be ordered on the basis of experience and intuition by using such vague terms as "approximately ...", "about ...", "between ... and ...", "not less than ...", "not more than ...", and it is presented using fuzzy sets. We model the uncertain demand using triangular fuzzy numbers $\tilde{d}_j = (\underline{d}_j, \hat{d}_j, \overline{d}_j)$ where $\underline{d}_j < \hat{d}_j < \overline{d}_j$ for each customer j as shown in Fig. 1, see e.g. Zimmermann [38]. The minimal possible, most plausible, and maximal possible demand is denoted by $\underline{d}_j$, $\hat{d}_j$, and $\overline{d}_j$ respectively. We assume full delivery service: Despite demand uncertainty, the actual demand of each customer is equal to or less than the vehicle capacity, i.e. $\overline{d}_j \leq Q$ $(j = 1, \ldots, n)$, so that no split delivery is planned.

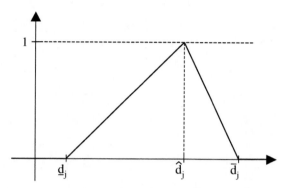

Fig. 1. Triangular fuzzy number $\tilde{d}_j = (\underline{d}_j, \hat{d}_j, \overline{d}_j)$

In general, it is too expensive to require that the maximal possible total demand of every customer has to be served. Consequently, suggested routes must account for the possibility that vehicle capacity may be exceeded by the actual demand at some points on the route. The result is what is called a route failure. It is assumed that it is too time-consuming to return to the depot for restocking and to revisit the customer again. If the demand of a customer cannot be satisfied completely during the planned trip, route failure will result in losing the excess demand because no corrective action is possible. The recourse is lost sales. If returns for restocking are possible, preventive breaks may be planned in anticipation of possible route failures, see e.g. Yang et al. [37].

Modeling the fuzzy total demand $\tilde{D} = (\underline{D}, \hat{D}, \overline{D})$ with $\underline{D} < \hat{D} < \overline{D}$, it is suggested to consider the possibility $\text{Pos}(\tilde{D} \leq Q)$ that the actual total demand is equal to or less than the vehicle capacity to a certain degree $\alpha \in [0, 1]$. The approach to handle these fuzzy constraints is similiar to chance constrained programming in stochastic optimization and is refered to as minimax chance constrained programming model, see Liu [25]. The decision maker has to determine α in advance. Considering a fuzzy number as a method of representing uncertainty in a given quantity by defining a possibility distribution for the quantity is analyzed by Jamison and Lodwick [19]. An even stronger condition is to determine a certain degree of necessity $\beta \in [0,1]$ that the actual total demand can be served. That is $\text{Pos}(\widetilde{D} \leq Q) \geq \alpha$ and $\text{Nec}(\widetilde{D} \leq Q) \geq \beta$. As the fuzzy total demand $\widetilde{D}$ is modeled using triangular fuzzy numbers, crisp equivalent formulae can be developed as follows, cf. e.g. Klir and Yuan [20, p. 199]:

$$\text{Pos}(\widetilde{D} \leq Q) = \sup_{x \leq Q} \mu_{\widetilde{D}}(x) = \begin{cases} 1 & \text{if } \widehat{D} \leq Q, \\ \frac{Q-\underline{D}}{\widehat{D}-\underline{D}} & \text{if } \underline{D} \leq Q < \widehat{D}, \\ 0 & \text{if } Q < \underline{D}, \end{cases} \quad \text{(see Fig. 2)},$$

$$\text{Nec}(\widetilde{D} \leq Q) = 1 - \text{Pos}(\widetilde{D} > Q) = \begin{cases} 1 & \text{if } \overline{D} \leq Q, \\ \frac{Q-\widehat{D}}{\overline{D}-\widehat{D}} & \text{if } \widehat{D} \leq Q < \overline{D}, \\ 0 & \text{if } Q < \widehat{D}, \end{cases} \quad \text{(see Fig. 3)}.$$

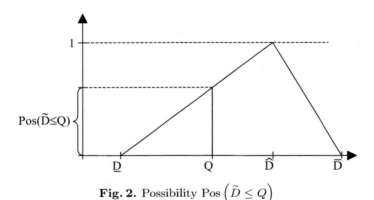

Fig. 2. Possibility $\text{Pos}\left(\widetilde{D} \leq Q\right)$

For each possibility and associated necessity measure and each set $A \subset X$ the following implication is satisfied, see e.g. Klir and Yuan [20, p. 189]:

$$\text{Pos}(A) < 1 \Longrightarrow \text{Nec}(A) = 0.$$

The consequence is that it is more demanding to request the necessity to be greater than 0 than to request the possibility to be equal to or less than 1. In the fuzzy environment considered here, a necessity value of 1 ensures that the entire demand can be satisfied, but it means that many vehicles must travel such long distances that their capacity is high enough to satisfy every possible demand even if the membership degree is very low. Consequently, these membership degrees, too, have to be taken into account when developing a fuzzy multi-criteria approach for the CVRPFD. Optimization takes

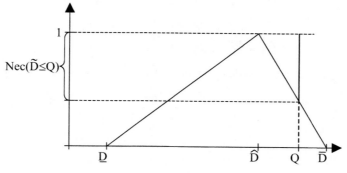

Fig. 3. Necessity $\text{Nec}\left(\tilde{D} \leq Q\right)$

place with respect to the above-stated cost–service trade-off. A compromise solution has to take into account the following objectives:

- minimizing total travel costs in terms of minimal total vehicle travel distance with regard to the number of vehicles used,
- maximizing total sales, i.e. minimizing unsatisfied demand,
- maximizing delivery service in terms of maximal service possibility and necessity that each demand is completely satisfied,

To a certain degree, the second and third goal are complementary to each other whereas the first one is in conflict to them. A *good* compromise solution for the CVRPFD is found in interaction with the decision maker, who can modify the degree of satisfaction with the different goals. Compromise solutions are calculated for small problems by an optimizing algorithm, see Sect. 2.3, and for larger problems by an efficient heuristic, see Sect. 3.

2.2 A Fuzzy Multi-Criteria Model

In accordance with the description of the CVRPFD above, the following three-index vehicle flow model is chosen, see Werners and Kondratenko [36].

Indices and index sets:

$i, j \in I = \{0, 1, \ldots, n\}$ depot and customers

$k \in K = \{1, \ldots, m\}$ vehicles

Parameters:
- Q vehicle capacity
- c_{ij} travel distance from location i to j
- $\tilde{d}_j = (\underline{d}_j, \hat{d}_j, \overline{d}_j)$ fuzzy demand of customer j with $\underline{d}_j < \hat{d}_j < \overline{d}_j$
- $\alpha \in [0,1]$ minimum service possibility for every route
- $\beta \in [0,1]$ minimum service necessity for every route

Variables:
- x_{ijk} binary variable,

    equal to 1 if and only if vehicle k travels from location i to j
- y_{jk} binary variable,

    equal to 1 if and only if customer j is served by vehicle k
- u_{jk} sequence number of customer j on the route of vehicle k,

    not necessarily integer variable

Fuzzy Multi-Criteria Model:

$$\text{minimize } z_1 = \sum_{i \in I} \sum_{j \in I} \sum_{k \in K} c_{ij} x_{ijk} \tag{1}$$

$$\text{maximize } z_2 = \sum_{k \in K} \widetilde{\min} \left\{ Q, \tilde{D}_k \right\} \tag{2}$$

subject to

$$\tilde{D}_k = \sum_{j \in 1}^{n} \tilde{d}_j y_{jk} \ \forall k \in K \tag{3}$$

$$\tilde{D}_k \tilde{\leq} Q \ \forall k \in K \tag{4}$$

$$\sum_{k \in K} y_{jk} = 1 \ \forall j \in I \setminus \{0\} \tag{5}$$

$$\sum_{i \in I} x_{ijk} = y_{jk} \ \forall j \in I, \forall k \in K \tag{6}$$

$$\sum_{j \in I} x_{ijk} = y_{ik} \ \forall i \in I, \forall k \in K \tag{7}$$

$$x_{iik} = 0 \ \forall i \in I, \forall k \in K \tag{8}$$

$$u_{jk} \geq u_{ik} + 1 - (1 - x_{ijk})n \ \forall i, j \in I, i \neq j \neq 0, \forall k \in K \tag{9}$$

$$u_{0k} = 1 \ \forall k \in K \tag{10}$$

$$x_{ijk} \in \{0,1\} \ \forall i \in I, \forall j \in I, \forall k \in K \tag{11}$$

$$y_{jk} \in \{0,1\} \ \forall j \in I, \forall k \in K \tag{12}$$

$$u_{jk} \geq 0 \ \forall j \in I, \forall k \in K \tag{13}$$

In this formulation, the first objective function (1) states that the total vehicle travel distance is to be minimized. The second objective function (2) seeks to maximize sales, which means, in this fuzzy context, to determine and maximize for each vehicle k a fuzzy set sales $\widetilde{S}_k = \widetilde{\min}\left\{Q, \widetilde{D}_k\right\}$. This is the extended minimum of the vehicle capacity Q, considered as a fuzzy set with membership function $\mathbb{1}_Q$, and the fuzzy total customer demand $\widetilde{D}_k$ on vehicle tour k, which is defined by constraint (3) as

$$\widetilde{D}_k = \sum_{j=1}^{n} \widetilde{d}_j y_{jk} = \left(\sum_{j=1}^{n} \underline{d}_j y_{jk}, \sum_{j=1}^{n} \widehat{d}_j y_{jk}, \sum_{j=1}^{n} \overline{d}_j y_{jk} \right) = (\underline{D}_k, \widehat{D}_k, \overline{D}_k).$$

Constraints (4) state that $\widetilde{D}_k$ is fuzzy equal to or less than the vehicle capacity, i.e. to a certain degree. Constraints (5)–(13) are commonly used for modeling CVRP. Constraints (5) state that each customer belongs to exactly one of the tours. Constraints (6) and (7) express the requirement that each customer location must be entered and exited exactly once by the vehicle to which tour it belongs. Constraints (8) state that no customer can follow himself on the tour. Constraints (9) ensure that subtours are forbidden, and constraints (10) that each tour starts at the depot. Constraints (11)–(13) declare the variable types.

As suggested above, constraints (4a)–(4b) are modeled as crisp equivalents of the fuzzy constraints (4):

$$\text{Pos}(\widetilde{D}_k \leq Q) \geq \alpha \ \forall k \in K \tag{4a}$$

$$\text{Nec}(\widetilde{D}_k \leq Q) \geq \beta \ \forall k \in K \tag{4b}$$

For $\alpha > 0$ and $\beta > 0$ the following formulae hold $\forall k \in K$:

$$\text{Pos}(\widetilde{D}_k \leq Q) \geq \alpha$$
$$\Leftrightarrow \alpha \widehat{D}_k + (1-\alpha)\underline{D}_k \leq Q \tag{4A}$$

$$\text{Nec}(\widetilde{D}_k \leq Q) \geq \beta$$
$$\Leftrightarrow \beta \overline{D}_k + (1-\beta)\widehat{D}_k \leq Q \tag{4B}$$

Now, the membership function of the fuzzy set possible sales $\widetilde{S}_k$ for each vehicle k can be calculated. It is shown in Figs. 4 and 5 respectively. A vehicle can satisfy the fuzzy demand as long as the total quantity is less than the vehicle capacity. It is possible to sell an amount equal to the total vehicle capacity. The membership value is determined by the possibility that total demand is equal to or greater than the vehicle capacity. Route failure will result in selling total vehicle capacity and losing the excess demand. Obviously, it is impossible to sell more demand than vehicle capacity:

$$\mu_{\widetilde{S}_k}(x) = \begin{cases} \mu_{\widetilde{D}_k}(x) & \text{if } x < Q, \\ \text{Pos}(\widetilde{D}_k \geq Q) & \text{if } x = Q, \\ 0 & \text{if } x > Q. \end{cases}$$

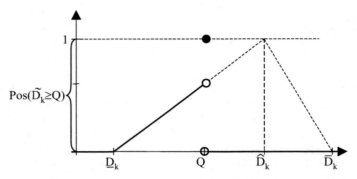

Fig. 4. Membership function of the fuzzy set $\widetilde{S}_k$ if $\underline{D}_k < Q \leq \widehat{D}_k$

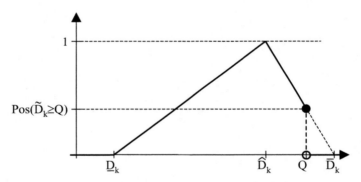

Fig. 5. Membership function of the fuzzy set $\widetilde{S}_k$ if $\widehat{D}_k < Q < \overline{D}_k$

The membership function of the fuzzy set total sales $\bigoplus_{k=1}^{m} \widetilde{S}_k$ can be calculated by extended addition of all fuzzy sales:

$$\mu_{\bigoplus_{k \in K} \widetilde{S}_k}(z) = \sup_{\sum_{k \in K} x_k = z} \min_{k \in K} \left\{ \mu_{\widetilde{S}_k}(x_k) \right\}.$$

For the extension principle and extended operations, see e.g. Zimmermann [38]. A great number of calculations has to be made to determine the fuzzy total sales. An easy to calculate defuzzification method is suggested to determine a crisp approximation $D\widetilde{S}_k$ for the fuzzy sales $\widetilde{S}_k$ on tour k:

$$D\widetilde{S}_k = \frac{1}{3}\min\{Q,\underline{D}_k\} + \frac{1}{3}\min\{Q,\widehat{D}_k\} + \frac{1}{3}\min\{Q,\overline{D}_k\} \quad \forall k \in K.$$

Now, a crisp approximation $D\widetilde{S}$ of the fuzzy total sales can be calculated:

$$D\widetilde{S} = \sum_{k \in K} D\widetilde{S}_k. \tag{14}$$

2.3 An Interactive Approach

The interactive approach by Werners [33], [34], and [35], with some modifications, is used for the solution of the fuzzy multi-criteria model of the CVRPFD presented. Due to the fact that objective function (2) and constraints (4a) and (4b) respectively are not independent of each other, they should be handled differently and adequately. For (4a) and (4b) the crisp equivalent formulae (4A) and (4B) are used.

First, the individual optimum of z_1, i.e. the minimal total vehicle travel distance $\underline{z_1}$, is determined so that at least the minimal demand $\sum_{k=1}^{m}\underline{D}_k$ is served. The solution of the model "minimize (1) subject to $\underline{D}_k \leq Q \; \forall k \in K$ and constraints (5)–(13)" is x_{ijk}^1, y_{jk}^1, and $u_{jk}^1 \; \forall i,j \in I, \forall k \in K$. This solution is used by (14) to calculate the lower bound $\underline{z_2}$ for the second objective z_2.

To determine the individual optimum of z_2, i.e. the maximal total sales $\overline{z_2}$, the crisp approximation (14) is used instead of (2). It is easy to show that for each fuzzy demand $\tilde{d}_j$ of customer j the following inequality holds: $D\widetilde{S}_k \leq \frac{1}{3}\sum_{j=1}^{n}(\underline{d}_j, \widehat{d}_j, \overline{d}_j) \; \forall k \in K$. In case that the number of vehicles is not restricted and each demand is less than the vehicle capacity, a solution can be found which takes this total value. Therefore, this individual optimum can be calculated without solving a mixed integer linear programming model. But in general, there are several solutions, not all of them are fuzzy efficient. A solution is fuzzy efficient if it is not possible to improve one of the values without deteriorating some of the other ones, see Werners [34, p. 137]. To determine a fuzzy efficient solution, a solution with minimal z_1-value has to be found, which is optimal with respect to z_2, i.e., constraints (4B) are satisfied with $\beta = 1$. The solution of the model "minimize (1) subject to $\overline{D}_k \leq Q \; \forall k \in K$ and constraints (5)–(13)" is x_{ijk}^2, y_{jk}^2, and $u_{jk}^2 \; \forall i,j \in I, \forall k \in K$ with the upper bound $\overline{z_1}$, which is used by (14) to calculate the upper bound $\overline{z_2}$.

Now, individual optima and pessimistic solutions can be used to model membership functions for the two goals:

$$\mu_{z_1}(z) = \begin{cases} 1 & \text{if } z \leq \underline{z_1}, \\ \frac{\overline{z_1}-z}{\overline{z_1}-\underline{z_1}} & \text{if } \underline{z_1} < z \leq \overline{z_1}, \\ 0 & \text{if } z > \overline{z_1}, \end{cases} \text{ and } \mu_{z_2}(z) = \begin{cases} 0 & \text{if } z \leq \underline{z_2}, \\ \frac{z-\underline{z_2}}{\overline{z_2}-\underline{z_2}} & \text{if } \underline{z_2} < z \leq \overline{z_2}, \\ 1 & \text{if } z > \overline{z_2}. \end{cases}$$

The first compromise model is suggested:

$$\text{maximize } \lambda$$
$$\text{subject to } (\overline{z_1} - \underline{z_1})\lambda + z_1(x,y,u) \leq \overline{z_1} \quad (15)$$
$$(\underline{z_2} - \overline{z_2})\lambda + \sum_{k=1}^{m} D\widetilde{S}_k \geq \underline{z_2} \quad (16)$$
$$\text{constraints } (5)-(13)$$
$$0 \leq \lambda \leq 1$$

Constraints (4a) and (4b) are omitted because they are to some extend complementary to the nonlinear constraint (16). It can be substituted by the following constraints (17)–(21) to get an equivalent linear model:

$$(\underline{z_2} - \overline{z_2})\lambda + \sum_{k=1}^{m} \frac{1}{3}(\underline{t}_k + \widehat{t}_k + \overline{t}_k) \geq \underline{z_2} \quad (17)$$

$$\underline{t}_k \leq \underline{D}_k \quad \forall k \in K \quad (18)$$
$$\widehat{t}_k \leq \widehat{D}_k \quad \forall k \in K \quad (19)$$
$$\overline{t}_k \leq \overline{D}_k \quad \forall k \in K \quad (20)$$
$$0 \leq \underline{t}_k, \widehat{t}_k, \overline{t}_k \leq Q \quad \forall k \in K \quad (21)$$

One of the main advantages of this model is its generality. In case that the number of customers to be served is rather large, the model cannot be solved optimally and a heuristic has to be chosen. In principle, established heuristic approaches for the CVRP can be used, if they are appropriately modified to face fuzzy demand and multiple criteria and to find a compromise solution very efficiently.

3 A Fuzzy Multi-Criteria Savings Heuristic

3.1 Description

First introduced by Clarke and Wright [7], the savings algorithm has become an established heuristic for the CVRP in the meantime. Certainly, a reason for this is its simplicity and the ease, with which it can be adapted to handle a variety of vehicle routing problems. The savings approach solves the CVRP as follows. First, the algorithm assumes that each customer is assigned to one specific vehicle so that each vehicle makes a round trip from the depot to its customer. The total vehicle travel distance for this route system is $\sum_{i=1}^{n}(c_{0i} + c_{i0})$. An improvement can be realized if two customers are joined on the same route. The savings are calculated for each pair of customers (i,j) as

$$s_{ij} = c_{i0} + c_{0j} - c_{ij} \quad \forall i,j \in I \setminus \{0\}.$$

The algorithm orders these savings in a savings list and proceeds to join customers according to the largest saving available. Joinings are feasible as long as the combined route has a demand, which does not exceed Q, and customers i and j are border customers at the beginning or the end of two different routes. Otherwise, this saving cannot be realized, and the combination with the next savings item is examined. Van Breedam [32] analyzes the influence of the geographical problem structure on the performance of different heuristics for the CVRP. The savings heuristic, as a parallel route-building method, is considered, too. A certain problem independency of the heuristic can be indicated. Another main advantage is its ability to keep routes well-separated and a high capacity utilization. The savings heuristic has, although fast, the drawback that once it has joined two customers, it never considers disconnecting them. This myopia is typical for many heuristics and might be reduced by more sophisticated versions of the algorithm, which relax to some extent the irrevocability of the added stops.

In a first step, the deterministic version of the savings heuristic is modified with regard to fuzzy demand. The fuzzy savings algorithm uses constraints (4a) and (4b) respectively to decide whether two current routes may be feasibly joined. If the degree of a possible route failure is too high, i.e., the possibility and necessity that demand can be satisfied falls below the fixed degrees α and β respectively, the current saving cannot be realized. Otherwise, the operation of the fuzzy savings heuristic is identical to the deterministic version described.

Then, the fuzzy savings heuristic is modified with regard to multiple criteria. To solve the fuzzy multi-criteria model (1)–(13), the fuzzy savings procedure starts to solve the model (1),(3),(4A),(5)–(13) by parametrically increasing α from 0 to 1. Afterwards, the constraints (4A) are replaced by (4B), which are stronger than (4A) for every $\beta > 0$. Analogously, the fuzzy savings algorithm is executed by parametrically increasing β from 0 to 1. To reduce the large number of calculations, although the savings algorithm runs very fast, it might be more efficient to calculate the maximal degree of possibility or necessity each time a solution has been determined by the algorithm. The next solution can then be determined by improving this maximal degree by a small proportion and including this new value into the constraints. After eliminating all dominated solutions, the result is a comprehensive survey of all fuzzy efficient solutions with respect to minimal total vehicle travel distance and maximal possibility or necessity meeting the fuzzy demand respectively.

Now, a compromise solution can be found heuristically by using the interactive approach described above. Each time a compromise model is formulated, the constraints can be reduced to a simple inequality system, for which α and β are parametrically varied. The procedure is demonstrated in 3.2. To our knowledge, the only heuristical approach for multiple objective combinatorial optimization in vehicle routing problems is presented by Hong and Park [18]. They suggest an utility function approach for the bi-objective

VRP with time windows. The conflicting objectives are minimizing both the total vehicle travel time and the total customer wait time.

3.2 Example

Benchmark tests for the CVRPFD are not yet available in literature. Therefore, data sets for vehicle routing problems with stochastic demand are modified for this study. Dror and Trudeau [13] solve a stochastic vehicle routing problem, which is an extended 75-customer VRP of Christofides and Eilon [6, problem 9]. The original demand quantity is used as the mean demand, and a normally distributed demand standard deviation is added. Here, the data are modified for a fuzzy environment as follows. The customers' coordinates remain unchanged, the mean demand is taken as the most plausible demand, and minimal and maximal possible demand are calculated with regard to the given standard deviation in demand. The data of the 75-customer CVRPFD are presented in Table 1. The distances between depot and customers are calculated by using Euclidean metric, which results in a symmetric cost matrix $(c_{ij})_{i,j \in I}$. Consequently, the savings are calculated only for each pair of customers (i, j) with $i < j$ because $s_{ij} = s_{ji} \; \forall i, j \in I \setminus \{0\}, i < j$.

It is assumed that the decision maker requests the possibility to satisfy each demand to be equal to 1, i.e. $\alpha = 1$. Otherwise, delivery service quality is regarded to be too low. The 75-customer CVRPFD is solved by the fuzzy multi-criteria savings heuristic by parametrically increasing the minimal degree β of necessity for complete delivery from 0 to 1. Exemplarily, the designed route system for $\alpha = 1.00$ and $\beta = 0.50$ is presented in detail in Table 2. The number of routes in this solution, for which the total demand $\widetilde{D}_k = (\underline{D}_k, \widehat{D}_k, \overline{D}_k)$ is given, is 10. The necessity to satisfy demand Nec $= \text{Nec}(\widetilde{D}_k \leq Q)$ varies from 0.50 to 1.00. The total vehicle travel distance $z_1 = z_1(x, y, u)$ is 876, and the crisp approximation of the fuzzy total sales $D\widetilde{S}$ is 1340.

The results of all runs are presented in Table 3. They are arranged according to increasing β-values. A backward analysis of the results finds all dominated route systems with respect to four goals: maximizing β / minimizing number of vehicles used / minimizing total vehicle travel distance z_1 / maximizing possible sales. Whenever a route system is compared with its predecessor, there is, of course, a deterioration $(-)$ of β. The number of vehicles remains unchanged $(*)$ in most cases, the only improvements $(+)$ are from $\beta \geq 0.80$ to $\beta < 0.80$ and from $\beta \geq 0.25$ to $\beta < 0.25$. Concerning total vehicle travel distance or possible sales respectively, it depends on the computed results of the heuristic. If the dominance check results in a combination only of $(-)$ and $(*)$, there is no improvement at all. The predecessor is dominated by the present route system and is omitted $(\ldots)$. The remaining route systems are non-dominated, at least not by any known feasible solution.

Table 1. CVRPFD with 75 cust., depot coord.: (40,40), vehicle capacity: 160

No.	x	y	$\underline{d}$	$\hat{d}$	$\overline{d}$	No.	x	y	$\underline{d}$	$\hat{d}$	$\overline{d}$	No.	x	y	$\underline{d}$	$\hat{d}$	$\overline{d}$
1	22	22	15	18	21	26	41	46	15	18	21	51	29	39	8	12	16
2	36	26	18	26	34	27	55	34	16	17	18	52	54	38	18	19	20
3	21	45	8	11	14	28	35	16	26	29	32	53	55	57	16	22	28
4	45	35	27	30	33	29	52	26	10	13	16	54	67	41	15	16	17
5	55	20	18	21	24	30	43	26	15	22	29	55	10	70	5	7	9
6	33	34	18	19	20	31	31	76	23	25	27	56	6	25	20	26	32
7	50	50	14	15	16	32	22	53	24	28	32	57	65	27	10	14	18
8	55	45	14	16	18	33	26	29	19	27	35	58	40	60	18	21	24
9	26	59	21	29	37	34	50	40	13	19	25	59	70	64	22	24	26
10	40	66	23	26	29	35	55	50	8	10	12	60	64	4	12	13	14
11	55	65	35	37	39	36	54	10	10	12	14	61	36	6	13	15	17
12	35	51	12	16	20	37	60	15	10	14	18	62	30	20	17	18	19
13	62	35	10	12	14	38	47	66	17	24	31	63	20	30	10	11	12
14	62	57	30	31	32	39	30	60	13	16	19	64	15	5	22	28	34
15	62	24	7	8	9	40	30	50	31	33	35	65	50	70	7	9	11
16	21	36	15	19	23	41	12	17	11	15	19	66	57	72	31	37	43
17	33	44	19	20	21	42	15	14	9	11	13	67	45	42	28	30	32
18	9	56	9	13	17	43	16	19	14	18	22	68	38	33	9	10	11
19	62	48	10	15	20	44	21	48	14	17	20	69	50	4	7	8	9
20	66	14	17	22	27	45	50	30	18	21	24	70	66	8	8	11	14
21	44	13	19	28	37	46	51	42	25	27	29	71	59	5	2	3	4
22	26	13	11	12	13	47	50	15	14	19	24	72	35	60	0	1	2
23	11	28	5	6	7	48	48	21	16	20	24	73	27	24	5	6	7
24	7	43	19	27	35	49	12	38	4	5	6	74	40	20	10	10	10
25	17	64	10	14	18	50	15	56	18	22	26	75	40	37	19	20	21

Exemplarily, the dominance check $-/*/+/+$ from $\beta \geq 0.70$ to $\beta < 0.70$ is presented. There is a decrease of the β-value, which is a deterioration $(-)$. In both cases the number of vehicles used is 10, which remains unchanged $(*)$. There is a decrease in total vehicle travel distance from 895 to 891, which is an improvement $(+)$, and an increase of possible sales from 1349 to 1351, which is also an improvement $(+)$, and might be confusing. A closer look to the results shows the meaning of β. For $\beta \in [0.65, 0.70)$, there are 10 routes

Table 2. Result of the 75-customer problem for fixed $\alpha = 1.00$ and $\beta = 0.50$

k	$(\underline{D}_k, \widehat{D}_k, \overline{D}_k)$	Pos	Nec	z_1	$D\widetilde{S}_k$	cust. number route sequence
1	(115,145,175)	1	0.50	118	140	5,47,36,69,71,60,70,20,37,15,57
2	(115,142,169)	1	0.67	127	139	61,22,64,42,41,43,56,23,63
3	(111,140,169)	1	0.69	140	137	49,24,50,18,55,25,31,10,72
4	(133,148,163)	1	0.80	97	147	11,65,66,59,14,35
5	(120,146,172)	1	0.54	94	142	58,38,53,19,54,13,27,52
6	(107,136,165)	1	0.83	70	134	12,39,9,32,44,3,16
7	(120,146,172)	1	0.54	81	142	68,74,21,28,62,1,73,33
8	(108,125,142)	1	1.00	44	125	4,34,46,8,7,26
9	(114,141,168)	1	0.70	58	138	6,2,30,48,29,45,75
10	(86,95,104)	1	1.00	46	95	51,40,17,67
			Total	876	1340	

with the following Nec($\widetilde{D}_k \leq 160$)-values, $k = 1, \ldots, 10$: 0.67, 0.69, 0.74, 0.80, three times 0.83, and three times 1.00, whereas for $\beta \in [0.70, 0.75)$, there is another route system for 10 vehicles with the Nec($\widetilde{D}_k \leq 160$)-values: twice 0.73, 0.74, 0.75, 0.80, 0.81, twice 0.83, and twice 1.00. Although for $\beta < 0.70$ there are two routes with necessity values less than 0.70, the effect on sales is compensated by six more satisfying routes, which overall dominate the possible sales for $\beta > 0.70$. Here again, the service–sales trade-off becomes obvious. In a first step, the set of all non-dominated route systems is presented to the decision maker in Table 4.

The interactive approach finds a first compromise model as follows. The constraints (15), (16), (5)–(13) can be simplified by regarding only the fuzzy efficient solutions from Table 4. By this, the objective functions $z_1(x, y, u)$ and $z_2(x, y, u)$ become dependent only on β:

$$843 \leq z_1(\beta) \leq 932 ,$$
$$1293 \leq z_2(\beta) \leq 1364 .$$

The first compromise model is found:

$$\text{maximize} \quad \lambda$$
$$\text{subject to} \quad 89\lambda + z_1(\beta) \leq 932$$
$$-71\lambda + z_2(\beta) \geq 1293$$
$$0 \leq \lambda \leq 1, \, 0 \leq \beta \leq 1$$

Its optimal compromise solution achieves $\lambda = 0.65$ and is printed in bold type in Table 4. The decision maker may prefer compromise solution No. 5

Table 3. Parametric necessity optimization and dominance check

α	β	No. of veh.	$z_1(x,y,u)$	$D\widetilde{S}$	dominance
1	$[0.00, 0.05)$	9	843	1293	$-/*/+/-$
(1	$[0.05, 0.10)$	9	858	1309	$-/*/-/-)$
(1	$[0.10, 0.15)$	9	858	1309	$-/*/-/-)$
(1	$[0.15, 0.20)$	9	854	1311	$-/*/*/*)$
1	$[0.20, 0.25)$	9	854	1311	$-/+/+/-$
(1	$[0.25, 0.30)$	10	873	1318	$-/*/-/-)$
1	$[0.30, 0.35)$	10	870	1330	$-/*/+/-$
(1	$[0.35, 0.40)$	10	878	1335	$-/*/-/-)$
1	$[0.40, 0.45)$	10	874	1340	$-/*/+/+$
(1	$[0.45, 0.50)$	10	876	1339.67	$-/*/*/*)$
1	$[0.50, 0.55)$	10	876	1339.67	$-/*/+/-$
(1	$[0.55, 0.60)$	10	891	1351	$-/*/*/*)$
(1	$[0.60, 0.65)$	10	891	1351	$-/*/*/*)$
1	$[0.65, 0.70)$	10	891	1351	$-/*/+/+$
1	$[0.70, 0.75)$	10	895	1349	$-/*/+/-$
1	$[0.75, 0.80)$	10	903	1355	$-/+/+/-$
(1	$[0.80, 0.85)$	11	931	1358	$-/*/-/-)$
(1	$[0.85, 0.90)$	11	934	1361	$-/*/-/-)$
(1	$[0.90, 0.95)$	11	935	1362	$-/*/-/-)$
1	$[0.95, 1.00)$	11	924	1363	$-/*/+/-$
1	$[1.00, 1.00]$	11	932	1364	

with $\lambda = 0.63$, which is given in detail in Table 2 and shown in Fig. 6, because the minimal service necessity of this route system $\beta < 0.55$ is significantly better while the other goals are achieved comparably.

In case that the decision maker is not satisfied with the compromise solution, he or she can select one of the goals and change its membership function by giving a restricting bound. For example, if the total vehicle travel distance is requested to be equal to or less than 903, the compromise solution changes because the degree of satisfaction with the goal z_1 changes:

$$843 \leq z_1(\beta) \leq 903 .$$

By this interaction, a second compromise model results. The effect is illustrated in Figs. 7 and 8 by sketching the modified $\lambda_{z_1}(\beta)$-function, which

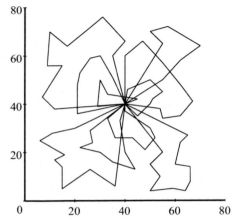

Fig. 6. Route system for $\alpha = 1.00$ and $\beta \in [0.45, 0.55)$

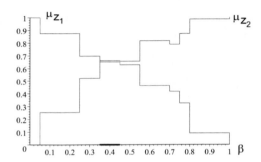

Fig. 7. The degree of satisfaction with z_1 and z_2

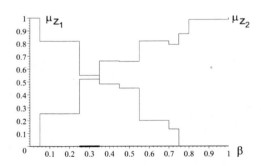

Fig. 8. The degree of satisfaction with "z_1 should be less than 903" and z_2 unchanged

Table 4. Fuzzy efficient solutions

No.	α	β	No. of veh.	$z_1(x,y,u)$	$D\widetilde{S}$	λ
1	1	[0.00, 0.05)	9	843	1293	0.00
2	1	[0.05, 0.25)	9	854	1311	0.25
3	1	[0.25, 0.35)	10	870	1330	0.52
4	1	**[0.35, 0.45)**	**10**	**874**	**1340**	**0.65**
5	1	[0.45, 0.55)	10	876	1339.67	0.63
6	1	[0.55, 0.70)	10	891	1351	0.46
7	1	[0.70, 0.75)	10	895	1349	0.42
8	1	[0.75, 0.80)	10	903	1355	0.33
9	1	[0.80, 1.00)	11	924	1363	0.09
10	1	[1.00, 1.00]	11	932	1364	0.00

is stretched to the baseline because the satisfication preference function of z_2 remains unchanged. The next optimal compromise solution is No. 3 with a membership degree of 0.52. Alternatively, the decision maker can modify the set of admissible solutions by requesting a restricted number of vehicles used or a necessity degree to be met. Depending on the remaining alternatives, it might be useful to calculate modified membership functions for the goals. In the same manner, all interactions of the decision maker can be analyzed and sketched for decision support.

4 Conclusion

In this article, the capacitated vehicle routing problem with fuzzy demand is considered. A heuristic solution method, based on a fuzzy multi-criteria modeling approach, has been developed, which is well-suited even for large problem instances. This method is demonstrated in a 75-customer example. One of the assumptions is that there is not enough time to return to the depot for restocking and then proceeding to serve the customers of a route. An extension of the model formulation and integration of preventive breaks will be interesting. Additionally, alternative consequences of unsatisfied demand can be considered. Presumably, other heuristics can be modified appropriately to solve the CVRPFD. A special focus should be on fuzzy multi-criteria modeling approaches. It is of interest to compare the behavior and the results of different fuzzy heuristics with a set of test examples. Future research will examine dynamic routing algorithms in more detail, and different aspects of uncertainty modeling are given particular attention.

References

1. Ball M.O., Magnanti T.L., Monma C.L., Nemhauser G.L. (Eds.) (1995) Network Routing. Handbooks in Operations Research and Management Science 8, Elsevier, Amsterdam et al.
2. Bertsimas D.J. (1992) A vehicle routing problem with stochastic demand. Operations Research 40:574–585
3. Bertsimas D.J., Chervi P., Peterson M. (1995) Computational approaches to stochastic vehicle routing problems. Transportation Science 29:342–352
4. Bertsimas D.J., Simchi-Levi D. (1996) A new generation of vehicle routing research: Robust algorithms, addressing uncertainty. Operations Research 44:286–304
5. Cheung R.K.-M., Powell W.B. (1996) Models and algorithms for distribution problems with uncertain demands. Transportation Science 30:43–59
6. Christofides N., Eilon S. (1969) An algorithm for the vehicle-dispatching problem. Operational Research Quarterly 20:309–318
7. Clarke G., Wright J.W. (1964) Scheduling of vehicles from a central depot to a number of delivery points. Operations Research 12:568–581
8. Crainic T.G., Laporte G. (1997) Planning models for freight transportation. European Journal of Operational Research 97:409–438
9. Desrosiers J., Dumas Y., Solomon M.M., Soumis F. (1995) Time constrained routing and scheduling. In: [1], 35–139
10. Dror M. (1993) Modeling vehicle routing with uncertain demands as a stochastic program: Properties of the corresponding solution. European Journal of Operational Research 64:432–441
11. Dror M., Laporte G., Louveaux F.V. (1993) Vehicle routing with stochastic demands and restricted failures. ZOR – Methods and Models of Operations Research 37:273–283
12. Dror M., Laporte G., Trudeau P. (1989) Vehicle routing with stochastic demands: Properties and solution frameworks. Transportation Science 23:166–176
13. Dror M., Trudeau P. (1986) Stochastic vehicle routing with modified savings algorithm. European Journal of Operational Research 23:228–235
14. Fisher M. (1995) Vehicle routing. In: [1], 1–33
15. Gendreau M., Laporte G., Séguin R. (1995) An exact algorithm for the vehicle routing problem with stochastic demands and customers. Transportation Science 29:143–155
16. Gendreau M., Laporte G., Séguin R. (1996a) Stochastic vehicle routing. European Journal of Operational Research 88:3–12
17. Gendreau M., Laporte B., Séguin R. (1996b) A tabu search heuristic for the vehicle routing problem with stochastic demands and customers. Operations Research 44:469–477
18. Hong S.-C., Park Y.-B. (1999) A heuristic for bi-objective vehicle routing with time window constraints. International Journal of Production Economics 62:249–258
19. Jamison K.D., Lodwick W.A. (1999) Minimizing unconstraint fuzzy functions. Fuzzy Sets and Systems 103:457–464
20. Klir G.J., Yuan B. (1995) Fuzzy Sets and Fuzzy Logic: Theory and Applications, Prentice Hall, Upper Saddle River, NJ

21. Laporte G. (1992) The vehicle routing problem: An overview of exact and approximate algorithms. European Journal of Operational Research 59:345–358
22. Laporte G. (1997) Vehicle routing. In: Dell'Amico M., Maffioli F., Martello S. (Eds.) Annotated Bibliographies in Combinatorial Optimization. John Wiley & Sons, Chichester et al., 223–240
23. Laporte G., Osman I.H. (1995) Routing problems: A bibliography. Annals of Operations Research 61:227–262
24. Lenstra J.K., Rinnooy Kan A.H.G. (1981) Complexity of vehicle routing and scheduling problems. Networks 11:221–227
25. Liu B. (1998) Minimax chance constrained programming models for fuzzy decision systems. Information Sciences 112:25–38
26. Powell W.B., Jaillet P., Odoni A.R. (1995) Stochastic and dynamic networks and routing. In: [1], 141–295
27. Secomandi N. (2000) Comparing neuro-dynamic programming algorithms for the vehicle routing problem with stochastic demands. Computers & Operations Research 27:1201–1225
28. Stewart W.R., Golden B.L. (1983) Stochastic vehicle routing: A comprehensive approach. European Journal of Operational Research 14:371–385
29. Teodorović D., Kikuchi S. (1991) Application of fuzzy sets theory to the saving based vehicle routing algorithm. Civil Engineering Systems 8:87–93
30. Teodorović D., Pavković G. (1996) The fuzzy set theory approach to the vehicle routing problem when demand at nodes is uncertain. Fuzzy Sets and Systems 82:307–317
31. Tillman F.A. (1969) The multiple terminal delivery problem with probabilistic demands. Transportation Science 3:192–204
32. Van Breedam A. (2002) A parametric analysis of heuristics for the vehicle routing problem with side-constraints. European Journal of Operational Research 137:348–370
33. Werners B. (1984) Interaktive Entscheidungsunterstützung durch ein flexibles mathematisches Programmierungssystem, Minerva Publication, München
34. Werners B. (1987a) An interactive fuzzy programming system. Fuzzy Sets and Systems 23:131–147
35. Werners B. (1987b) Interactive multiple objective programming subject to flexible constraints. European Journal of Operational Research 31:342–349
36. Werners B., Kondratenko Y.P. (2001) Tanker routing problem with fuzzy demand. Working Paper in Operations Research 0104, Faculty of Economics and Business Administration, Ruhr-University Bochum, Bochum
37. Yang W.-H., Mathur K., Ballou R.H. (2000) Stochastic vehicle routing problem with restocking. Transportation Science 34:99–112
38. Zimmermann H.-J. (2001) Fuzzy Set Theory and its Applications, 4rd ed., Kluwer Academic Publishers, Boston et al.

An Adaptive, Intelligent Control System for Slag Foaming

Eric L. Wilson and Charles L. Karr

Aerospace Engineering and Mechanics Department
Box 870280
The University of Alabama
Tuscaloosa, AL 35487-0280

Abstract. The field of computational intelligence (CI) is primarily concerned with the development of computer systems that are capable of adapting to and exploiting information about their environments, much like organisms in natural systems are capable of doing. It is no coincidence therefore, that the field of CI relies heavily on computer techniques patterned after natural systems. Many of these techniques (including neural networks, genetic algorithms, and fuzzy logic) have demonstrated their utility in solving problems independent of other methods. However, as the systems we seek to control, design, and improve become increasingly complex, it is unlikely that any single CI technique will prove to be adequate. This paper describes an architecture that combines the three CI techniques listed above to produce process control systems suitable for effectively manipulating complex engineering systems characterized by relatively slow process dynamics. Implementation of the architecture results in an intelligent adaptive control system. The effectiveness of the controller is demonstrated via application to a slag foaming operation at a steel plant.

1 Introduction

Slag foaming is a steel-making process that has become generally accepted for use in steel plants with electric arc furnaces. In an electric arc furnace, the slag foaming process begins after scrap metal has been sufficiently melted in a steel furnace. Oxygen and carbon are then injected into the molten steel and react both with each other, and with the iron oxide present in the steel to form bubbles of carbon monoxide. These bubbles rise to the surface of the molten steel (called the slag) and cause it to foam. As the slag foams, it prevents the electric arc that is coming into the molten steel from sparking and hitting the lining of the furnace (called the refractory), thus reducing noise levels while also preventing both internal furnace wear and energy loss. Foamed slag retains heat in molten steel by blanketing the steel, thus saving on energy costs by retaining heat. Additionally, by removing iron oxide from the steel, slag foaming aids chemically to improve the efficiency of the steel-making process.

Despite its desirable attributes, slag foaming is a highly dynamic process that is difficult to control. Data collected from steel plants indicates that slag foaming performance can vary greatly within a short span of time. This variance is compounded by the human controllers who monitor and adjust the slag foaming process in steel plants. Because each controller utilizes a different set of control rules, slag foaming performance at each steel plant lacks a certain level of consistency and optimality. However, indications are that an effective, automatic process controller for slag foaming could greatly improve the consistency of slag foaming operations while in turn reducing overall steel plant costs and improving steel quality.

In the mid-1980s BCD Technology researchers developed an approach for producing optimizing, adaptive process control systems [1]. The controllers developed using this approach relied extensively on techniques from the field of *computational intelligence* (CI). In recent years, CI has begun to emerge as a powerful, efficient base from which to design intelligent, adaptive process control systems. CI is generally thought to be a part of the larger field of *artificial intelligence*, and is comprised of computer methods that are based on the mechanics of natural systems. Some of the more popular CI techniques are neural networks [2], genetic algorithms [3], fuzzy systems [4], and immunized computational systems [5]. Each of these approaches has demonstrated its utility in the area of process control, primarily via its application to a wide variety of industrial problems. However, the true potential of CI techniques for producing truly intelligent systems lies in the synergy of the various individual methods.

Recently, numerous applications involving the combination of CI techniques have appeared in the literature. These applications include systems in which

- neural networks have been combined with genetic algorithms [6];
- fuzzy systems have been combined with genetic algorithms [7];
- fuzzy systems have been combined with neural networks [8];
- various other combinations have been implemented [9].

These applications have led to comprehensive approaches by which intelligent systems based on CI techniques can be developed for process control applications. The procedure for designing optimizing, adaptive process control developed by the BCD Technologies researchers and presented here has proven to be extremely robust in that it has been shown to be effective across a spectrum of industries. The first successful implementation of this method occurred on a classic academic problem – a cart-pole balancing problem [10]. This problem is often used to model balancing problems associated with two-legged robots. Next, the method was used to solve a difficult pH titration problem from the field of chemical engineering [11]. In this problem, the method was used to develop an adaptive control system that could compensate for the lack of sensory data from the system being controlled. Next, the method was used to solve another control problem from the field of chemical engineering. Specifically, it was used to control an exothermic reaction used for the production of hexamine [12]. In this application, Karr and his co-workers demonstrated the ability to produce a controller that changed strategies based on current, dynamic economic information. The most recent success of this approach involved the development of an adaptive control system for helicopter flight control which required fast response time [13]. In this problem, Phillips, Karr, and Walker developed a helicopter flight control system that was successfully tested at White Sands Missile Base, NM.

For many years, the industry standard in control systems has been stabilizing control. In this type of control system, various process parameters are driven toward and ultimately maintained at pre-defined setpoints, typically through the implementation of Proportional-Integral-Derivative (PID) controllers [14]. A tremendous amount of research has been done recently to move beyond stabilizing control, with a specific focus on developing control systems capable of generating: (1) self-improving tracking errors, (2) adaptive control parameters, and (3) optimized estimates of the performance error [15]. A control system that includes all three of the capabilities listed above is known as an *intelligent adaptive controller*. The control system described in the current paper is such a control system for manipulating the slag foaming process at an electric arc steel plant.

The intelligent adaptive controller consists of three primary elements: (1) the *control element*, (2) the *analysis element*, and (3) the *learning element*. The control element is responsible for directly manipulating the slag foaming controls within the steel plant to achieve a defined system performance goal. In the architecture described here, the control element is based on a fuzzy logic controller. The analysis element is responsible for recognizing when the response conditions of the plant have changed, and for quantifying these response changes. In the adaptive control system developed here, this unit achieves its goals through the use of a neural network model of the plant. The learning element must make adjustments to the control element; adjustments made necessary as a consequence of the changes to the system identified by the adaptive element. In the current system, the learning element employs a genetic algorithm-based search engine that adjusts a copy of the fuzzy logic control system (taken from the control element) using a neural network model of the current slag foaming environment (taken from the analysis element). When fully engaged, the three-element control system performs optimizing, adaptive control.

This paper describes a computer software architecture for achieving the intelligent control of complex industrial systems [16]. This architecture combines the computational power of fuzzy systems, genetic algorithms, and neural networks into what has been defined as an *intelligent adaptive control system*. A CI-based architecture for achieving intelligent adaptive control is described later. The architecture is implemented on the problem of controlling the slag foaming process at an electric arc steel furnace. To provide a context from which to discuss the issues associated with the control system, the problem environment to be controlled is discussed in the next section.

2 The Problem Environment: Slag Foaming in an Electric Arc Steel Furnace

Figure 1 is a schematic of an electric arc steel furnace. The actual slag foaming process within the electric arc furnace begins with scrap metal being placed inside the furnace. The furnace generally will have some molten steel in place, in order to aid in melting the scrap metal. Once the scrap is in place inside the furnace, an electric arc is discharged from electrodes into the furnace in order to melt the scrap metal. After the scrap has been sufficiently melted, oxygen gas and raw carbon are injected into the molten steel. The placement of the injection lances is particular to individual steel plants, but generally occurs near the surface of the molten steel. The oxygen gas and carbon begin reacting within the iron oxide present in the molten steel to create carbon monoxide. This allows the injected oxygen and carbon to both foam the surface of the molten steel (via the creation of carbon monoxide) and to remove iron oxide from the steel bath. This is just one of the noted benefits of slag foaming – the removal of iron oxide from the steel improves the process efficiency of the steel being produced.

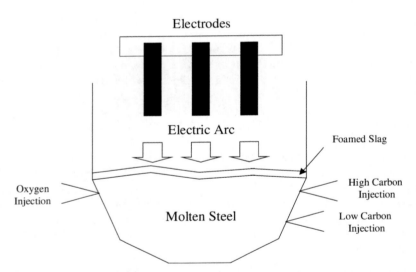

Fig. 1. A schematic of an electric arc furnace. As oxygen and carbon are injected into the molten steel, carbon monoxide bubbles are formed that rise and begin to foam the slag surface of the molten steel.

The overall benefits of slag foaming make it an essential part of the modern steel-making process in electric arc furnaces. Without slag foaming, electricity coming from the electric arc into the furnace would not be transmitted directly into the molten steel, thus resulting in energy loss and increased noise levels. The electric arc coming into the molten steel would also reflect at the surface and hit the lining of the furnace, causing damage and increasing the frequency of regular repairs. Foamed slag also serves as a blanket to insulate the molten steel and retain heat, thus saving on energy costs. Finally, slag foaming serves as a steel composition control mechanism. The chemical reactions in slag foaming aid in the removal of iron oxides from the molten steel, making the steel-making process more efficient. Thus, slag foaming not only serves to reduce the associated costs of making steel, but it also improves the efficiency of the steel-making process.

Slag foaming is a beneficial cost-saving process for the steel industry. However, slag foaming as it is currently practiced at steel plants does not follow a standardized set of rules. Some plants only control the amount of carbon coming into the molten steel, while others add magnesite or dolomitic material into the molten steel as a foaming agent. Other plants may add more scrap metal at later points during the steel-making heat. The control options available to a steel plant and how those controls are implemented may vary widely from plant to plant.

Georgetown Steel (South Carolina) is an atypical steel-making plant that implements a simple slag foaming control process. No magnesia or dolomitic material is used to aid with the foaming process (although direct reduced iron is used). The controller cannot adjust the oxygen input into the molten steel. Additionally, the controller is not able to adjust the placement of the carbon lances – they are considered static. This only leaves the carbon injection switches available to the slag foaming controller. Although these conditions limit what can be controlled with respect to slag foaming, they also scale the control problem down to adjusting a small number of variables for maximum control.

The carbon injection variables are not the only variables measured in the steel plant, however. The mill at Georgetown Steel records data on thirty-five different plant variables every thirty seconds. These variables can be divided into three major subgroups: (1) power/electric variables, (2) physical state variables, and (3) injection variables.

The power/electric variables deal with properties of the electric arc coming into the molten steel. A large percentage of the slag foaming variables fall into this category, mostly because Georgetown Steel records the voltage, amperes, and kilowatts of power that are being spent every thirty seconds. Additionally, Georgetown Steel records the impedance involved with the electric arc, the amount of average current coming from the electric arc, and the accumulated number of megawatt hours being used.

The physical state variables are furnace variables that deal with the physical state of the furnace. Several variables fall into this category, including calculated bath temperature, direct reduced iron (DRI) feed rate, accumulated DRI weight, liquid heel weight, total weight, tap position, and slag height. Calculated bath temperature is a measure in degrees Celsius of how hot the steel furnace is. Direct reduced iron is specially made iron pellets which are added consistently throughout the heat to aid in controlling the foamy slag (although the rate at which the DRI is added is not controlled by the furnace operator). Accumulated DRI weight is the sum of weight due to the addition of DRI pellets. Liquid heel weight notes the total weight due to scrap metal and leftover steel from previous heats of steel-making. Total weight is a measure (in pounds) of the amount of raw material in the steel furnace. Tap position represents a height setting at which the electric arc is set away from the molten steel. Slag height represents the relative height (on a scale of 0 to 100) of the foamed slag, with 0 being no foam and 100 being maximum foam. This particular variable serves a very important role in the adaptive control system. Slag height is the driving factor in slag foaming control at Georgetown Steel – the higher the foamed slag, the better the slag foaming process is considered to be working.

The injection variables deal with the materials that are being lanced into the molten steel over time. These variables include oxygen flow, high carbon injection, and low carbon injection. The oxygen flow is a measure of how much oxygen is being pumped into the steel furnace. The high carbon injection and low carbon injection variables are carbon injection control knobs that are either on (100) or off (0). All three of these variables play a major role in the slag foaming adaptive control system, either as inputs into the control system or as control variables that are implemented within the slag foaming control system.

Although this data is recorded every thirty seconds, it would be very difficult for human controllers to fully utilize this data in making control decisions for slag foaming. The slag foaming controllers at Georgetown Steel are provided with two mechanisms for slag height control – two different carbon injection controls. In a general sense, the adaptive control system developed for controlling foamed slag at Georgetown Steel can be thought of as one that processes thirty-five variables of data that makes control decisions on two variables to control a single variable. It is this system for which an intelligent adaptive controller will be developed.

3 An Architecture for Achieving Intelligent Adaptive Control

The optimizing, adaptive controller described in this paper combines several CI techniques to form a comprehensive approach to adaptive process control. The three specific CI techniques used to produce the adaptive process control systems are: (1) fuzzy logic, (2) genetic algorithms, and (3) neural networks. This section provides an overview of the architecture used to achieve adaptive process control.

3.1 System Architecture

Figure 2 shows a schematic of the fundamental architecture used to achieve, optimizing, adaptive control. The heart of this control system is the loop consisting of the control element and the problem environment. The control element receives information from sensors in the problem environment concerning the status of the *condition variables* (also termed state variables or controlled variables depending on the field of study) which detail the current state of the system. It then computes a desirable state for a set of *action variables* (also termed manipulated variables). These changes in the action variables force the problem environment toward the desired setpoint. This is the basic approach adopted for the design of virtually any closed loop control system, and in and of itself includes no mechanism for adaptive control.

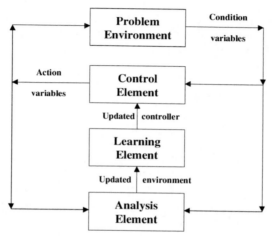

Fig. 2. The architecture used to implement the optimizing, adaptive control system consists of three main elements: (1) control element, (2) analysis element, and (3) learning element.

The adaptive capabilities of the system shown in Figure 2 are due to the analysis and learning elements. In general, the analysis element must recognize when a change in the problem environment has occurred. A "change," as it is used here, consists of a perturbation to the system that causes the problem environment to react differently to control actions. Also, such a change must involve parameters that are not included in the list of condition variables; if they were included in the list of condition variables, then the control element could account for the perturbations. The analysis element uses information concerning the condition and action variables over some finite time period to recognize changes in the environment and to compute the new performance characteristics associated with these changes.

The new environment (the problem environment with altered responses to control actions) can pose many difficulties for the control element, because the control element is no longer manipulating the environment for which it was designed. Therefore, the algorithm that drives the control element must be altered. As shown in the schematic of Figure 2, this task is accomplished by the learning element. The most efficient approach for the learning element to use to alter the control element is to utilize information concerning the past performance of the control system. The strategy used by the control, analysis, and learning elements of the stand-alone, comprehensive adaptive controller used in the current effort is provided in the following sections.

Control Element The control element receives feedback from the problem environment, and prescribes appropriate values of the action variables based on the current state of the environment. Because of the flexibility needed in the control system as a whole, a fuzzy logic controller (FLC) is employed. Like conventional rule-based systems (expert systems), FLCs use a set of production rules that are of the form:

IF {*condition*} THEN {*action*}

to arrive at appropriate control actions. The left-hand-side of the rules (the *condition* side) consists of combinations of the controlled variables; the right-hand-side of the rules (the *action* side) consists of combinations of the manipulated variables. Unlike conventional expert systems, FLCs use rules that utilize fuzzy terms like those appearing in human rules-of-thumb. For example, a valid rule for a FLC used to manipulate one of the control mechanisms of the slag foaming operation is:

IF {DRI Feed is **HIGH** and O_2 Injection is **LOW**}
THEN {Carbon Injection is **0**}.

The fuzzy terms ("HIGH," "LOW") are subjective; they mean different things to different "experts," and can mean different things in various situations. Fuzzy terms are assigned meaning via fuzzy membership functions [17]. These membership functions are used in conjunction with the production rule set to prescribe single, crisp values of the action variables. Unlike conventional expert systems, FLCs allow for the application of more than one rule at any given time. The crisp actions taken by these rules are computed using a weighted averaging technique that incorporates a *min-max* operator [18].

The most effective condition variables to be used in the FLC of the control element were determined via a statistical analysis [19]. The five variables are the power setting, the temperature in the furnace, the total weight in the furnace, the DRI feed rate into the furnace, and the oxygen flow into the furnace. Each of these variables is characterized using three fuzzy sets. Because the FLC actively controls two variables, the FLC consists of $2*3^5 = 486$ production rules. The values of these condition variables at a given time are used by the control element to determine the changes to the high and low carbon injection controls.

Analysis Element The analysis element recognizes changes in parameters associated with the problem environment not taken into account by the rules used in the control element. These changes can, potentially, dramatically alter the way in which the problem environment responds to the control actions, thus forming what is virtually a new problem environment requiring an altered control strategy. One of the main driving forces behind using this approach is to keep the control element as small and as simple as possible. A small control element (using a limited number of rules) will respond faster; a simple control element (one without special rules designed to handle conditions that are not normal) is easier to maintain. By limiting the size and complexity of the control element, some of the parameters that affect the problem environment are necessarily left out of the rule set. This in turn requires that the control element be regularly altered for those system response changes that are not recognized by the control element. However, before the control element can be altered, the adaptive control system must recognize that the problem environment has changed and in turn compute the nature and magnitude of those changes.

The analysis element recognizes changes in the system parameters by comparing the response of the physical system to the response of a model of the problem environment. In general, recognizing changes in the parameters associated with the problem environment requires the control system to store information concerning the past performance of the problem environment. This information is most effectively acquired through either a database or a computer model. However, storing an extensive database can be cumbersome and can require extensive computer memory. Additionally, equations describing complex industrial systems are often not available for use in modeling the system. A neural network model is well suited to serve as a model of the slag foaming furnace due to the fact that the system is simply too complex to model using first principles.

In the approach to adaptive control adopted here, a neural network-based computational model predicts the response of the actual problem environment. This predicted response is compared to the actual response of the problem environment. When the two responses differ by more than a tolerance amount over a finite period of time, the actual problem environment is considered to have been altered. Figure 3 shows an instance in which the neural network simulation is accurately tracking the performance of the slag height within the steel plant furnace. Here, the relationship between the input and output parameters of the network is adequately represented to allow for accurate simulation. In cases such as this, the control element is adequately tuned to allow for efficient manipulation of the slag foaming operation.

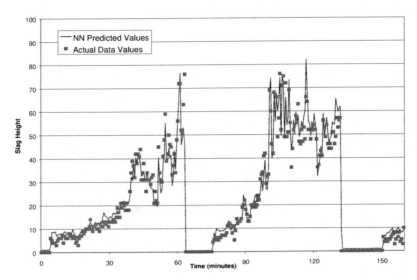

Fig. 3. The neural network-simulated response in the above plot matches the response of the slag foaming system quite well. In the above instance, the unmeasured parameters of the slag foaming system are considered not to have changed.

Figure 3 shows a case when the neural network model is accurately matching the system performance of the slag height over time. However, when it is determined that the neural network model does not match the system performance, parameters associated with the both the model and the control element must be changed. The neural network model is retrained to accurately represent the current system response state. The model is then used by the learning element to adjust the control element for the updated problem state.

Learning Element The learning element alters the control element in response to changes in the problem environment. It does so by altering the production rule outputs of the rule base employed by the FLC of the control element. By altering the production rule outputs, the learning element effectively changes the control strategy of the FLC. These changes reflect the response changes shown in the problem environment, and results in a FLC that produces more effective control decisions.

The current adaptive control system for slag foaming utilizes a genetic algorithm (GA) to adjust the production rule outputs of the FLC for the response changes recognized by the neural network model. The GA attempts to optimize a 486-parameter search problem, where each parameter represents a production rule output value. Thus, each potential solution tested by the GA represented a potential FLC for use within the control element. The effectiveness of each potential solution is tested by implementing the neural network model of the slag foaming environment (taken from the analysis element). Simulations are run using past data in order to analyze the effectiveness control decisions that are made by the potential controller. Once the GA has finished its search, the set of production rule outputs that performed best are sent to the control element for installation and use. By using a GA learning element in conjunction with the neural network analysis element, the FLC control element is adjusted for system response changes in the problem environment.

Summary The computer software architecture for intelligent adaptive control described above consists of three primary elements: (1) a control element, (2) an analysis element, and (3) a learning element. These elements combine the power of fuzzy systems, genetic algorithms, and neural networks into a comprehensive system that is capable of achieving intelligent adaptive control. The next section demonstrates the effectiveness of the system on slag foaming control at a steel plant.

4 Results

The ability of the intelligent adaptive control system to successfully manipulate the slag foaming environment at Georgetown Steel is highly sensitive to the neural network model of the foaming operation. Thus, it is imperative to demonstrative the effectiveness of the computer simulation. A standard backpropagation neural network consisting of thirty-five input nodes, twenty hidden nodes, and one output node (representing slag height) was trained to mimic the slag foaming operation.

Figure 4 shows the effectiveness of the neural network simulation. This figure shows a time-series plot of the slag height, which is a prime indicator of how well the slag foaming operation is performing. The fact that the simulation accurately tracks both the trends and the magnitude of the slag height indicates that the adaptive control system will potentially be effective in this domain.

Once it has been demonstrated that the neural network simulation accurately depicts the operation of the plant, both the learning and control elements can be constructed. An intelligent adaptive control system was developed for implementation at Georgetown Steel. The performance of the adaptive control system was evaluated by running an extended simulation of the slag foaming operation at Georgetown Steel. The adaptive control system was run for a period of seven days, simulating the actual slag foaming

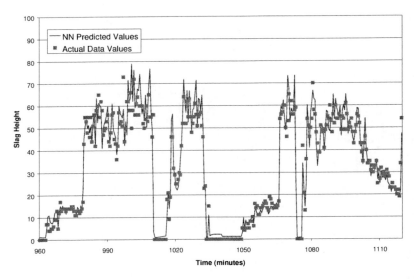

Fig. 4. The neural network simulation of the plant accurately tracks (or predicts) the slag height, which is a prime indicator of how well the plant is performing.

process by using data collected from Georgetown Steel in conjunction with the control decisions made by the control element. Results of this test are shown in Figures 5 and 6.

Figure 5 shows how well the adaptive slag foaming control system performs in comparison to the human controller. In general, the adaptive control system achieves a significantly better slag height than that achieved by the human controller. The adaptive control system shows an increase in slag height of around ten in the middle slag ranges (between thirty and sixty), and achieves a much higher maximum slag (over eighty) than the human controller (under seventy). Even at the lower slag levels, the adaptive control system performs as well if not better than the human slag foaming controller. Although the amount of data shown in Figure 5 is relatively small (two hours out of an entire week of data), the performance of the adaptive control system showed similar results across the entire week of slag foaming data. This fact is supported by Figure 6.

Figure 6 shows a plot of the entire slag foaming data set with respect to the performance of the adaptive control system and the human controller. Each data point in Figure 6 represents the slag height achieved by each controller for that particular time step. Additionally, a line is drawn representing those data points where the adaptive control system would equal the human controller. This line divides the graph into two sections – the section above the line where the adaptive control system outperforms the human controller, and the section below the line where the human controller outperforms the

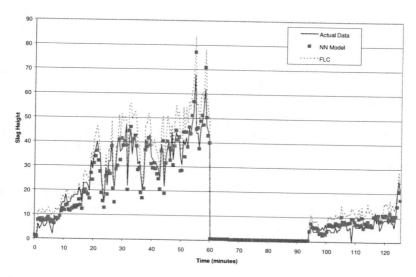

Fig. 5. The adaptive control system outperforms the human controller by achieving a higher slag height on most of the data points.

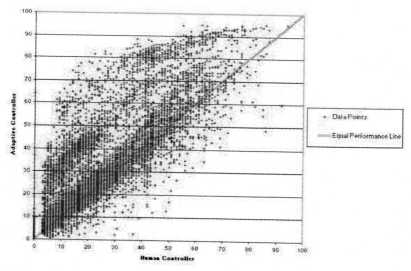

Fig. 6. The above plot shows that the adaptive control system outperforms the human controller over a wide range of possible slag heights.

adaptive control system. Figure 6 shows that the adaptive control system achieves better performance than the human controller across a wide spectrum of slag heights. Thus, the adaptive control system not only outperforms the human controller on a large percentage of data points, but also across a broad range of slag heights. All of these graphs provide confidence in the adaptive control systems ability to effectively control slag foaming in an electric arc steel furnace.

5 Summary

A generic computer software architecture has been described for achieving intelligent adaptive control. This architecture combines the strengths of three techniques from the field of computational intelligence: genetic algorithm, fuzzy logic, and neural networks. The methodology represented by the architecture has been applied to the problem of controlling a highly complex manufacturing system – a slag foaming operation at a steel plant. The intelligent adaptive control system performs in simulation generally at or above the level of a human operator over time.

References

1. Karr, C. L. (1991a). Genetic algorithms for fuzzy controllers. *AI Expert*, **6**(2), 26–33.
2. Fiesler, E., & Beale, R. (Eds.) (1997). *Handbook of neural computation.* New York: Oxford University Press.
3. Back, T., Fogel, D. B., & Michalewicz, Z. (Eds.) (1997). *Handbook of evolutionary computation.* New York: Oxford University Press.
4. Kandel, A., & Langholz, G. (Eds.) (1993). *Fuzzy control systems.* Boca Raton, FL: CRC Press.
5. Dasgupta, D. (Ed.) (1998). *Artificial immune systems and their applications.* Berlin: Springer.
6. van Rooij, A. J. F., Jain, L. C., & Johnson, R. P. (1996). *Neural network training using genetic algorithms.* Singapore: World Scientific.
7. Sanchez, E., Shibata, T., & Zadeh, L. A. (Eds.) (1997). *Genetic algorithms and fuzzy logic systems: Soft computing perspectives.* Singapore: World Scientific.
8. Kosko, B. (1991). *Neural networks and fuzzy systems: A dynamical systems approach to machine intelligence.* Englewood Cliffs, NJ: Prentice Hall.
9. Goonatilake, S., & Khebbal, S. (Eds.) (1995). *Intelligent hybrid systems.* New York: John Wiley & Sons.
10. Karr, C. L. (1991b). Fine tuning a cart pole balancing fuzzy logic controller using a genetic algorithm. *Proceedings of The Applications of Artificial Intelligence VIII Conference*, **1468**, 26–36.
11. Karr, C. L., & Gentry, E. J. (1992). Fuzzy control of pH using genetic algorithms. *IEEE Transactions on Fuzzy Systems*, **1**(1), 46–53.
12. Karr, C. L., Sharma, S. K., Hatcher, W. J., & Harper, T. R. (1993). Fuzzy control of an exothermic chemical reaction using genetic algorithms. *Engineering Applications of Artificial Intelligence*, **6**(6), 575–582.
13. Phillips, C., Karr, C. L., & Walker, G. (1996). Helicopter flight control with fuzzy logic and genetic algorithms. *Engineering Applications of Artificial Intelligence*, **9**(2), 175–184.

14. Medsker, L. R. (1995). *Hybrid intelligent systems*. Boston: Kluwer Academic Publishers.
15. Miller, W. T., Sutton, R. S., & Werbos, P. J. (Eds.) (1991). *Neural networks for control*. Cambridge, MA: The MIT Press.
16. Karr, C. L. (1999). *Practical applications of computational intelligence for adaptive control*. Boca Raton, FL: CRC Press.
17. Zadeh, L.A. (1973). Outline of a new approach to the analysis of complex systems and decision processes. *IEEE Transactions on Systems, Man, and Cybernetics*, **SMC-3**, 28–44.
18. Karr, C. L. (1991b). Fine tuning a cart pole balancing fuzzy logic controller using a genetic algorithm. *Proceedings of The Applications of Artificial Intelligence VIII Conference*, **1468**, 26–36.
19. Wilson, E. (2002). Artificial Intelligence-Based Computer Modeling Tools for Controlling Slag Foaming in Electric Arc Furnaces. Ph.D. Dissertation, University of Alabama.

Druck: Strauss GmbH, Mörlenbach
Verarbeitung: Schäffer, Grünstadt